Reverse Logistics

Springer

Berlin
Heidelberg
New York
Hong Kong
London
Milan
Paris
Tokyo

Rommert Dekker
Moritz Fleischmann
Karl Inderfurth
Luk N. Van Wassenhove
Editors

REVERSE LOGISTICS

Quantitative Models
for Closed-Loop
Supply Chains

With 76 Figures
and 34 Tables

 Springer

Professor Dr. Rommert Dekker

Rotterdam School of Economics
Erasmus University Rotterdam
Burgemeester Oudlaan 50
3062 PA Rotterdam, The Netherlands
rdekker@few.eur.nl

Dr. Moritz Fleischmann

Rotterdam School of Management/Faculteit Bedrijfskunde
Erasmus University Rotterdam
Burgemeester Oudlaan 50
3062 PA Rotterdam, The Netherlands
MFleischmann@fbk.eur.nl

Professor Dr. Karl Inderfurth

Faculty of Economics and Management
Otto-von-Guericke University Magdeburg
Universitätsplatz 2
39106 Magdeburg, Germany
Inderfurth@ww.uni-magdeburg.de

Professor Dr. Luk N. Van Wassenhove

Technology Management Department
INSEAD
Boulevard de Constance
77305 Fontainebleau, France
luk.van-wassenhove@insead.edu

ISBN 3-540-40696-4 Springer-Verlag Berlin Heidelberg New York

Cataloging-in-Publication Data applied for

A catalog record for this book is available from the Library of Congress.

Bibliographic information published by Die Deutsche Bibliothek
Die Deutsche Bibliothek lists this publication in the Deutsche Nationalbibliografie;
detailed bibliographic data available in the internet at *http.//dnb.ddb.de*

Springer-Verlagis a part of Springer Science+Business Media
springeronline.com

© Springer-Verlag Berlin Heidelberg 2004
Printed in Germany

Cover design: Erich Kirchner, Heidelberg

SPIN 10951705   42/3130 – 5 4 3 2 1 0 – Printed on acid-free paper

Preface

Today's supply chains no longer are confined to one-way product flows from producers to consumers, but increasingly also need to deal with flows in the opposite direction. Examples of such 'reverse' streams of products range from end-of-life computer equipment to returned merchandise in online channels, from reusable packaging to defective products requiring rework. Managing these complex interrelated flows confronts companies with novel challenges. At the same time, it calls for a broadening of scientific theory.

In 1997, we initiated a European research network, 'Reverse Logistics and its Effects on Industry' (REVLOG), to address these issues (see also www.fbk.eur.nl/OZ/REVLOG). Our objective was to help establish a sound theoretical basis for reverse logistics. We focused on developing quantitative models that support decision-making. Six universities participated in the network, namely the Aristotle University of Thessaloniki, the University of Piraeus, the Otto-von-Guericke University of Magdeburg, INSEAD, the Technical University of Eindhoven, and the Erasmus University Rotterdam (project coordinator). The European Union generously sponsored this project within the TMR-framework from December 1997 until December 2002.

We structured the field of our study around five clusters, namely production planning and inventory control, distribution, business economics, information and computational aspects, and environmental impact. Our research approach included mathematical modeling, software development, case studies, and literature reviews. This book presents the major results of the REVLOG project from a quantitative modeling perspective. A companion book appearing with the same publisher provides a detailed description of current industrial practice in reverse logistics through a set of case studies.

This book is organized in four parts. The first part, encompassing two chapters, presents a qualitative framework for reverse logistics and locates the topic within the more general field of supply chain management. Chapter 2 also provides a detailed outline of the book. The subsequent parts, each consisting of four or five chapters, address managerial issues and corresponding quantitative models related to distribution management, inventory control and production planning, and broader supply chain management issues. Each chapter opens with an illustrative case, highlights key issues, and then explains in detail the available models and theoretical results. Each chapter is written by a team of authors that have made substantial contributions to the literature on the given topic.

This book is aimed at academics and students in the field of supply chain management. It can be used as a basic text for graduate courses in this field. By bringing together all available knowledge on reverse logistics and adding a substantial body of new theory it is, we believe, unique in its kind.

Even though the book may be primarily aimed at academics and students, it should also be useful to professionals in the field of supply chain management, especially people with a specific responsibility in product recovery. The many decision models and decision support tools discussed in the book should provide them with a solid basis in designing effective systems in practice.

Many individuals have contributed to the REVLOG project and to this book specifically. First of all, we would like to thank all participating researchers for their scientific and personal contributions, enthusiasm, and willingness to collaborate throughout the project. No less important, on behalf of all participants we thank the EU for their generous financial support. Professors Dirk Cattrysse (K.U. Leuven), Paul van Beek (U. Wageningen) and Thomas Spengler (T.U. Braunschweig) did a great job in critically reviewing parts of the book. A special thank you also goes to Michelle Baum for assisting us with copyediting the text.

Rotterdam, July 2003   *Rommert Dekker*
                                                              Moritz Fleischmann
Magdeburg, July 2003   *Karl Inderfurth*
Fontainebleau, July 2003   *Luk N. Van Wassenhove*

Contents

Part IV Supply Chain Management Issues in Reverse Logistics

Appendix

Part I

A Framework for Reverse Logistics

A Framework for Reverse Logistics

Marisa P. de Brito and Rommert Dekker

Rotterdam School of Economics, Erasmus University Rotterdam, P.O. Box 1738, 3000 DR Rotterdam, The Netherlands, debrito@few.eur.nl, rdekker@few.eur.nl

1.1 Introduction

Twenty years ago, supply chains were busy fine-tuning the logistics of products from raw material to the end customer. Products are obviously still streaming in the direction of the end customer but an increasing flow of products is coming back. This is happening for a whole range of industries, covering electronic goods, pharmaceuticals, beverages, and so on. For instance, the automobile industry is busy changing the physical and virtual supply chain to facilitate end-of-life recovery (Boon et al., 2001; Ferguson and Browne, 2001). Besides this, distant sellers like e-tailers have to handle high return rates and many times at no cost for the customer. It is not surprising that the Reverse Logistics Executive Council has announced that US firms have been losing billions of dollars on account of being ill prepared to deal with reverse flows (Rogers and Tibben-Lembke, 1999). The return as a process was recently added to the Supply-Chain Operations Reference (SCOR) model, stressing its importance for supply chain management in the future (Schultz, 2002). Reverse Logistics has been stretching out worldwide, involving all the layers of supply chains in various industry sectors. While some actors in the chain have been forced to take products back, others have proactively done so, attracted by the value in used products. One way or the other, Reverse Logistics has become a key competence in modern supply chains.

In this book, we provide a wide set of quantitative approaches to support reverse logistics decision making, which firms can use to enrich their competence. At the same time, we emphasize the academic contribution to reverse logistics problems, as they pose both modelling and solving challenges. In particular, we focus on the work of the European Network on Reverse Logistics, REVLOG, which during the last five years published more than 100 scientific articles on the topic. Before going into that, we present in this chapter a content analysis of reverse logistics issues. To do so, we propose the following content framework focusing on the following questions with respect to reverse logistics: *why? how? what?* and *who?*, i.e. driving forces and return reasons, what products, how are they being recovered, and who is involved (see Fleischmann et al., 1997).

The remainder of the chapter is organized as follows. First, we go into more detail regarding the scope of Reverse Logistics and we distinguish its domain, relating it with other subjects like green logistics and closed-loop supply chains. Next, we provide a short review of the contributing literature to structure the Reverse Logistics field. After that, we characterize Reverse Logistics by looking into it from the four perspectives of the content framework: why are there reverse flows, i.e. the return reasons and the driving forces; what constitutes these reverse flows, i.e. which are the products and materials characteristics; how can they be recovered, i.e. which are the intricate processes; and who is carrying out these activities, i.e. the supply chain actors. We end this chapter by relating all the issues together.

1.2 Reverse Logistics: Definition and Scope

Definition and a Brief History

In the sweat of your face you shall eat bread
Till you return to the ground,
For out of it you were taken; For dust you are,
And to dust you shall return
Genesis 3:19

Though the conception of Reverse Logistics dates from long time ago (as the aforementioned citation proves), the denomination of the term is difficult to trace with precision. Terms like Reverse Channels or Reverse Flow already appear in the scientific literature of the seventies, but consistently related with recycling (Guiltinan and Nwokoye, 1974; Ginter and Starling, 1978).

During the eighties, the definition was inspired by the movement of flows against the traditional flows in the supply chain, or, as put by Lambert and Stock (1981), "going the wrong way" (see also Murphy, 1986, and Murphy and Poist, 1989).

In the early nineties, a formal definition of Reverse Logistics was put together by the Council of Logistics Management, stressing the recovery aspects of reverse logistics(Stock, 1992):

"...the term often used to refer to the role of logistics in recycling, waste disposal, and management of hazardous materials; a broader perspective includes all relating to logistics activities carried out in source reduction, recycling, substitution, reuse of materials and disposal."

The previous definition is quite general, as is evident from the following excerpt: "the role of logistics in all relating activities." Besides that, it is originated from a waste management standpoint. In the same year, Pohlen and Farris (1992) define Reverse Logistics, guided by marketing principles and by giving it a direction insight, as follows:

"...the movement of goods from a consumer towards a producer in a channel of distribution."

Kopicky et al. (1993) defines Reverse Logistics analogously to Stock (1992) but keeps, as previously introduced by Pohlen and Farris (1992), the sense of direction opposed to traditional distribution flows:

"Reverse Logistics is a broad term referring to the logistics management and disposing of hazardous or non-hazardous waste from packaging and products. It includes reverse distribution...which causes goods and information to flow in the opposite direction of normal logistics activities."

At the end of the nineties, Rogers and Tibben-Lembke (1999) describe Reverse Logistics stressing the goal and the processes (the logistics) involved:

"The process of planning, implementing, and controlling the efficient, cost-effective flow of raw materials, in-process inventory, finished goods, and related information from the point of consumption to the point of origin for the purpose of recapturing value or proper disposal."

The European Working Group on Reverse Logistics, REVLOG (1998-), puts forward the following definition, which we will use in this book:

"The process of planning, implementing and controlling backward flows of raw materials, in process inventory, packaging and finished goods, from a manufacturing, distribution or use point, to a point of recovery or point of proper disposal."

This perspective on Reverse Logistics keeps the essence of the definition as put forward by Rogers and Tibben-Lembke (1999), which is logistics. We do not, however, refer to "point of consumption" nor to "point of origin." In this way, we give margin to return flows that were not consumed first (for instance, stock adjustments due to overstocks or spare parts which were not used) or that may go back to a different point of recovery than the original (e.g. collected computer chips may enter another chain). We employ the expression "point of recovery" to stress the distinction we want to make from pure waste-management activities (see next section). Furthermore, we include the "reverse direction" (backward flows) in the definition to exclude what can be considered forward recovery (for instance, when a consumer gives his/her personal computer to the neighbor).

In summary, the definition of Reverse Logistics has changed over time, starting with a sense of "wrong direction," going through an overemphasis on environmental aspects, coming back to the original pillars of the concept, and coming finally to a widening of its scope. For other discussions on the evolution of the definition of Reverse Logistics, we refer to Rogers and Tibben-Lembke (2001) and to Fernandéz (2003).

Delineation and Scope

Since Reverse Logistics is a relatively new research and empirical area, the reader may encounter in other literature terms like reversed logistics, return logistics, retro logistics, or reverse distribution, sometimes referring roughly to the same. In fact, the diversity of definitions with respect to recovery practices is a well-recognized source of misunderstandings both in research and in practice (Melissen and De Ron, 1999).

We would like to remark that Reverse Logistics is different from waste management as the latter mainly refers to collecting and processing waste (products for which there is no new use) efficiently and effectively. The crux of the matter is the definition of waste. This is a major issue, as the term has severe legal consequences, e.g. it is often forbidden to import waste. Reverse Logistics concentrates on those streams where there is some value to be recovered and the outcome enters a (new) supply chain. Reverse Logistics also differs from green logistics as that considers environmental aspects in all logistics activities and it has been focused specifically on forward logistics, i.e. from producer to customer (see Rodrigue et al., 2001). The prominent environmental issues in logistics are consumption of nonrenewable natural resources, air emissions, congestion and road usage, noise pollution, and both hazardous and non-hazardous waste disposal (see Camm, 2001). Finally, reverse logistics can be seen as part of sustainable development. The latter has been defined by Brundland (1998) in a report to the European Union as "to meet the needs of the present without compromising the ability of future generations to meet their own needs." In fact, one could regard reverse logistics as the implementation of that at the company level by making sure that society uses and reuses both efficiently and effectively all the value which has been put into the products.

The border between forward logistics (from raw materials to end user) and reverse logistics (from end user to recovery or to a new user) is not strictly defined in modern supply chains, as one can wonder about what 'raw materials' are or who the 'end user' is. For instance, used/recovered glass is a substantial input for new production of glass. A holistic view on supply chains combining both forward and reverse logistics is embraced by the closed-loop supply chain concept (Guide and van Wassenhove, 2003). Recovery practices are framed within the supply chain, and the encircling aspect of the process as a whole is therefore stressed: having either 1) a physical (closed-loop) to the original user (see Fleischmann et al., 1997) or 2) a functional (closed-loop) to the original functionality. Thinking in term of closed-loop supply chains emphasizes the importance of coordinating the forward with the reverse streams. Actually, whenever both forward and reverse flows are involved, coordination has to be minded, as will later be discussed in more detail in Chapter 12. This happens, either in closed or open loops (the latter refers to when neither the original user or original functionality are in the reverse logistics process).

Industrial ecology is another field that relates to reverse logistics. As explained in Garner and Keoleian (1995), industrial ecology is primarily dedicated to the study of the interactions between industrial systems and the environment. The underlying aim is to change the linearity of industrial systems (from raw materials to waste) into a cyclical system (with product or materials recovery). In the latter, reverse logistics doubtless plays a major role.

1.3 Reverse Logistics: A Review of Literature on Theoretical Developments

In the literature dealing with reverse logistics one can essentially find quantitative modelling, case studies, and theory building. For a review on the state of affairs of quantitative modelling before the REVLOG project, we refer to Fleischmann et al. (1997). Throughout the whole book one will find many references to articles on quantitative decision making for reverse logistics. For a compilation of reverse logistics case studies, we suggest De Brito et al. (2003) and Flapper et al. (2003) . In addition, for more overviews on reverse logistics, we suggest the following references: Stock, 1992; Kopicky, 1993; Kostecki, 1998; Stock, 1998; Rogers and Tibben-Lembke, 1999; and Guide and van Wassenhove, 2003.

In this review, we focus on contributions that attempt to lay the fundamentals of reverse logistics theory, i.e. a general body of principles to explain reverse logistics issues. Since there are no established references on Reverse Logistics theory, we sometimes depart from literature on related fields that secondarily add to the theoretical growth of Reverse Logistics. In addition, we briefly point out how the content framework presented in this book relates to this literature on the one hand and how it deviates from it on the other.

In the mid-nineties, Thierry et al. (1995) shaped product recovery management by detailedly going over the recovery options, distinguishing 1) direct reuse (and resale), 2) product recovery management (repair, refurbishing, remanufacturing, cannibalization, and recycling), and 3) waste management (incineration and landfilling). The authors characterize the recovery options according to the level of disassembly and the quality required, as well as to the resulting product. Besides this, the paper uses three case studies (BMW, IBM, and an anonymous copy remanufacturer) to illustrate changes in operations when companies become engaged in recovery. In the *How* section of this chapter, we employ a similar recovery option typology, introducing the term redistribution as another form of direct recovery, and using (parts) retrieval instead of the term cannibalization. Besides outlining how products can be recovered, we also put forward the *who* perspective and add two dimensions that help to understand Reverse Logistics: *why* and *what*. Fleischmann et al. (1997) have actually touched upon these perspectives, except for the characterization of the return reasons. The following characteristics are used to show the diversity of reverse logistics systems: 1) motivation (economical and ecological), 2) the type of items (spare parts, packages, and consumer goods, 3) the form of reuse (reused directly, repair, recycling, and remanufacturing), and processes (collection, testing, sorting, transportation, and processing), and 4) the involved actors (members of the forward channel or specialized parties). These issues partially come back in the subsequent sections with further elaboration on their characterization. In addition, we sum up the return reasons in the *why* dimension. The authors concluded that the research up to then was rather narrowed to single questions and accordingly they suggested dedicating more research effort to the impact of return flows on supply chain management. A relevant share of this book is precisely devoted to supply chain management issues (Chapters 12 to 16).

Fuller and Allen (1995) used the recycling of post-consumer goods as the inspiration for the following typology of reverse channels: 1) manufacturer-integrated, 2) waste-hauler, 3) specialized reverse dealer-processor, 4) forward retailer-wholesaler, and 5) temporary/facilitator. In this chapter, we consider a broad range of recovery options, besides recycling, and a broader list of actors besides the ones involved in a pure recycling system. Yet, what we can learn from Fuller and Allen (1995) is that in Reverse Logistics there are likely to appear three types of players: 1) typical forward chain players (manufacturer, wholesaler, retailer); 2) specialized reverse players; and 3) opportunistic players. We come back to this in the *Who* section.

Late nineties, Carter and Ellram (1998) put together a model of stimulating and restraining forces for Reverse Logistics. They identify four environmental forces: 1) from the government (in terms of regulations); 2) from suppliers; 3) from buyers; and 4) from competitors. In the *Why* section we discuss as well driving forces for Reverse Logistics, where the aforementioned items are part of the discussion. One important matter not emphasized by Carter and Ellram (1998) is the direct economic benefit. Another ignored aspect is the company inner-responsibility, denominated as corporate citizenship during our discussion.

Gungor and Gupta (1999) present an extensive review of the literature (more than 300 articles or books) on environmentally conscious manufacturing and product recovery. They subdivide the literature in categories, outlining a framework. This paper looks upon product recovery from the point of view of environmentally conscious manufacturing. Both the regulatory and opinion pressure, respectively by government and customers, are mentioned. We contemplate a tri-fold driving force for Reverse Logistics: corporate citizenship (where the environmental accountability is included), economics, and legislation (see *Why* section). Gungor and Gupta (1999) go over environmental design, stressing the relevance that the constitution of the product has for recovery. They mention, for example, the number of components, the number of materials, and the ease of material separation. We come back to this in the *What* section.

Goggin and Browne (2000) have recently suggested a taxonomy of resource recovery specifically for end-of-life products with a focus on electronic and electrical equipment. Their classification is based on three dimensions, as follows: 1) public vs. private sector, 2) commercial vs. domestic market segments, and 3) large vs. small products. These issues come into the framework presented here, specifically in the *What* and *Who* sections. Essentially, what we retain from them is that a relevant characteristic of a reverse logistics system is the nature of those behind the network, i.e. private vs. public (for more on this, we refer to Chapter 4). Furthermore, other relevant traits are the size of the product and the market segment. With respect to resource recovery, Goggin and Browne (2000) delineate three types: material reclamation, component reclamation, and remanufacturing. The authors feature these groups according to input and output product complexity. In fact, this was earlier brought up by Thierry et al. (1995), who presented a lengthy list of recovery options, which we generally follow (see *How* section).

In contrast with most of the previous contributions, we are not going to go into details for one or another form of reverse logistics. Instead, we bring forward a content framework on reverse logistics as a whole by bringing structure to the fundamental contents of the topic and their interrelations. As mentioned before, we do this by answering four basic questions on reverse logistics: *Why? How? What?* and *Who?* In this way, the reader can generally understand what reverse logistics issues are about and at the same time capture the vast assortment of the matters involved.

1.4 Reverse Logistics: Why? How? What? and Who?

After having briefly introduced the topic of Reverse Logistics, we now go into the fundamentals of Reverse Logistics by analyzing the topic from four essential viewpoints: *why, how, what* and *who*. Former studies have argued that these types of characteristics are relevant to characterize reverse logistics (see e.g. Thierry, 1995; Fleischmann et al., 1997; and Zhiquiang, 2003). In this chapter, we consider the following details.

- *Why* are things returned: we go over the driving forces behind companies and institutions becoming active in Reverse Logistics, *Why-drivers* (receiver), and the reasons for reverse flows (return reasons), i.e. *Why-reasons* (sender).
- *How* Reverse Logistics works in practice: we give a list of processes carried out in reverse logistic systems and focus on how value is recovered in the reverse chain (recovery options).
- *What* is being returned: we describe product characteristics which makes recovery attractive or compulsory and give examples based on real cases (products and materials).

- *Who* is executing reverse logistic activities: we go over the actors and their role in implementing reverse logistics (reverse chain actors).

Why-drivers (Receiver): Driving Forces Behind Reverse Logistics

In the literature of reverse logistics, many authors have pointed out driving factors like economics, environmental laws, and the environmental consciousness of consumers (see e.g. REVLOG, online). Generally, one can say that companies do get involved with Reverse Logistics either 1) because they can profit from it, and/or 2) because they have to, and/or 3) because they "feel" socially motivated to do it. Accordingly, we categorize the driving forces under three main headings:

- Economics (direct and indirect);
- Legislation; and
- Corporate citizenship.

Economics

A reverse logistics program can bring direct gains to companies from dwindling the use of raw materials, from adding value with recovery, or from reducing disposal costs. Independents have also gone into the area because of the financial opportunities offered in the dispersed market of superfluous or discarded goods and materials. Metal scrap brokers have made fortunes by collecting metal scrap and offering it to steel works, which could reduce their production costs by mingling metal scrap with virgin materials in their process. In the electronic industry, many products arrive at the end of useful life in a short period, but still with its components having intrinsic economic value. ReCellular is a US firm that has shown that it has known how to take economic advantage from this from the beginning of the nineties by trading in refurbished cell phones (see Guide and Van Wassenhove, 2003).

Even with no clear or immediate expected profit, an organization can get (more) involved with Reverse Logistics because of marketing, competition, and/or strategic issues, from which are expected indirect gains. For instance, companies may get involved with recovery as a strategic step to get prepared for future legislation (see Louwers et al., 1999) or even to prevent legislation. In face of competition, a company may engage in recovery to prevent other companies from getting their technology, or to prevent them from entering the market. As reported by Dijkhuizen (1997), one of the motives of IBM's getting involved in (parts) recovery was to avoid brokers doing it. Recovery can also be part of an image build-up operation. For instance, Canon has linked the copier recycling and cartridge recycling programs to the 'kyo-sei' philosophy (cooperative growth), proclaiming that Canon is for "living and working together for the common good" (see Meijer, 1998, and www.canon.com). Recovery can also be used to improve customer or supplier relations. One example is a tire producer who also offers customers rethreading options in order to reduce the customer's costs. In sum, the economic driver embraces, among others, the following direct and indirect gains.

- Direct gains:
    - input materials
    - cost reduction
    - value added recovery;
- Indirect gains:
    - anticipating/impeding legislation
    - market protection
    - green image
    - improved customer/supplier relations

Legislation

Legislation refers here to any jurisdiction indicating that a company should recover its products or accept them back. As mentioned before, in many countries home shoppers are legally entitled to return the ordered merchandise (e.g. in the UK, consult the Office for Fair Trading, online). Furthermore, and especially in Europe, there has been an increase of environmental legislation, such as recycling quotas, packaging regulation, and manufacturing take-back responsibility. The automobile industry and industries of electrical and electronic equipment are under special legal pressure (see Chapter 15). Sometimes companies participate 'voluntarily' in covenants, either to deal with or prepare for legislation.

Corporate Citizenship

Corporate citizenship concerns a set of values or principles that in this case impel a company or an organization to become responsibly engaged with reverse logistics. For instance, the concern of Paul Farrow, the founder of Walden Paddlers, Inc., with "the velocity at which consumer products travel through the market to the landfill" pushed him to an innovative project of a 100-percent-recyclable kayak (see Farrow and Jonhson, 2000). Nowadays indeed many firms, like Shell (www.shell.com), have extensive programs on responsible corporate citizenship where both social and environmental issues become the priorities.

Figure 1.1 depicts the previously described driving triangle for Reverse Logistics (compare with Carrol's triple bottom line for corporate social responsibility, Carrol, 1979).

One should notice that these are not mutually exclusive motivations and, in reality, it is sometimes hard to set the boundary. In many countries, customers have the right to return products purchased via a distant seller as mail-order companies or e-tailers. Thus, these companies are legally obliged to give the customer the opportunity to send back merchandise. At the same time, this opportunity is also perceived as a way to attract customers, bringing potential benefits to the company.

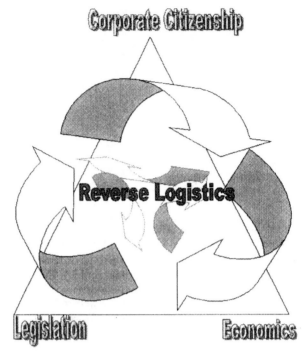

Fig. 1.1. Driving triangle for reverse logistics

Why-reasons (Sender): Return Reasons for Reverse Logistics

Roughly speaking, products are returned or discarded because either they do not function (anymore) properly or because they or their function are no longer needed. Let us next elaborate these return or discard reasons in more detail. We can list them according to the usual supply chain hierarchy: starting with manufacturing and going to distribution until the products reach the customer. Therefore, we differentiate between manufacturing returns, distribution returns, and customer returns.

Manufacturing Returns

We define manufacturing returns as all those cases where components or products have to be recovered in the production phase. This occurs for a variety of reasons. Raw materials may be left over, intermediate or final products may fail quality checks and have to be reworked, products may be left over during production, or by-products may result from production. Raw material surplus and production leftovers represent the 'product not needed' category, while quality-control returns fit in the 'faulty' category. In sum, manufacturing returns include

- raw material surplus;,
- quality-control returns, and
- production leftovers/by-products.

Distribution Returns

Distribution returns refers to all those returns that are initiated during the distribution phase. It refers to product recalls, commercial returns, stock adjustments, and functional returns. Product recalls are products recollected because of safety or health problems with the products, and the manufacturer or a supplier usually initiates them. B2B commercial returns are all those returns where a retailer has a contractual option to return products to the supplier. This can refer to wrong/damaged deliveries, to products with a too-short remaining shelf life, or to unsold products that retailers or distributors return to the wholesaler or manufacturer. The latter include outdated products, such as those products whose shelf life has been too long (e.g. pharmaceuticals and food) and may no longer be sold. Stock adjustments take place when an actor in the chain redistributes stocks, for instance between warehouses or shops, e.g. in the case of seasonal products (see De Koster et al., 2002). Finally, functional returns concern all the products whose inherent function keeps them going backward and forward in the chain. An obvious example is the one of pallets as distribution carriers: their function is to carry other products and they can serve this purpose several times. Other examples are crates, containers, and packaging. Summarizing, distribution returns include

- product recalls,
- B2B commercial returns (e.g. unsold products or wrong/damaged deliveries),
- stock adjustments, and
- functional returns (distribution items/carriers/packaging).

Customer Returns

The third group consists of customer returns, i.e. those returns initiated once the product has at least reached the final customer. Again there are a variety of reasons to return the products:

- B2C commercial returns (reimbursement guarantees),
- warranty returns,
- service returns (repairs, spare parts),
- end-of-use returns, and
- end-of-life returns.

The reasons have been listed more or less according to the life cycle of a product. Reimbursement guarantees give customers the opportunity to

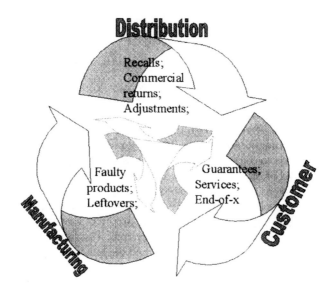

Fig. 1.2. Return reasons for reverse logistics

change their minds about purchasing (commonly shortly after having re-
ceived/acquired the product) when their needs or expectations are not met.
The list of underlying causes is long, for example with respect to clothes, where
dissatisfaction may be due to size, color, fabric properties, and so forth. Independent of the underlying causes, when a customer returns a new product
benefiting from a money-back guarantee or an equivalent, we are in the presence of B2C commercial returns. The next two reasons, warranty and service
returns, refer mostly to an incorrect functioning of the product during use, or
to a service that is associated with the product and from which the customer
can benefit.

Initially, customers benefiting from a warranty can return products that do
not (seem to) meet the promised quality standards. Sometimes these returns
can be repaired or a customer gets a new product or his or her money back,
upon which the returned product needs recovery. After the warranty period
has expired, customers can still benefit from maintenance or repair services,
but they no longer have a right to get a substitute product for free. Products
can be repaired at the customer's site or be sent back for repair. In the former
case, there are many returns in the form of spare parts in the service supply
chain, since in advance it is hard to know precisely which parts will be needed
for the repair.

End-of-use returns refer to those situations where the user has a return
opportunity at a certain life stage of the product. This refers to leasing cases

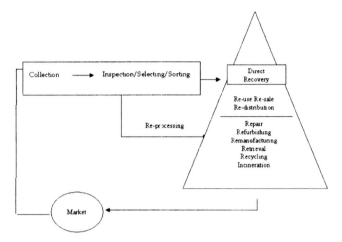

Fig. 1.3. Reverse logistics processes

and returnable containers like bottles, but also to returns to second-hand markets like bibliofind, a division of amazon.com for used books.

Finally, end-of-life returns refer to those returns where the product is at the end of its economic or physical life as it is. They are either returned to the OEM because of legal product take-back obligations or other companies like brokers take them for material and value-added recovery (see *How* section).

Figure 1.2 summarizes the return reasons for Reverse Logistics in the three stages of a supply chain: manufacturing, distribution, and customer.

How: Reverse Logistics Processes

The *how* viewpoint is meant to show how Reverse Logistics works in practice, how value is recovered from products.

Recovery is actually only one of the activities involved in the whole reverse logistics process. First there is collection, next there is the combined inspection/selection/sorting process, thirdly there is recovery (which may be direct or may involve a form of reprocessing), and finally there is redistribution (see Figure 1.3). Collection refers to bringing the products from the customer to a point of recovery. At this point the products are inspected, i.e. their quality is assessed and a decision is made on the type of recovery. Products can then be sorted and routed according to the recovery that follows. If the quality is (close to) as-good-as-new, products can be fed into the market almost immediately through reuse, resale, and redistribution. If not, another type of recovery may be involved that now demands more action, i.e. a form of reprocessing.

Reprocessing can occur at different levels: product level (repair), module level (refurbishing), component level (remanufacturing), selective part level (retrieval), material level (recycling), and energy level (incineration). Note

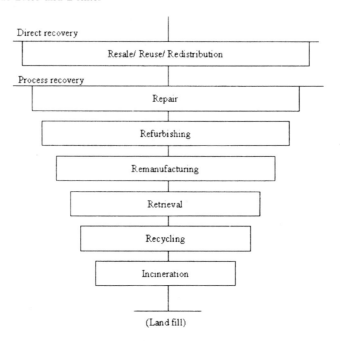

Fig. 1.4. Recovery option inverted pyramid

that remanufacturing is "as-good-as-new" recovery while refurbishing does not have to be so. We refer to Thierry et al. (1995) for complete definitions. For other points of view on recovery/reprocessing levels, see Fleischmann et al. (1997) and Goggin and Browne, (2000). This book does not particularly address repair systems, which is a body of literature on its own. For a review, we refer to Guide and SrivastaveGuide and Srivastava (1997b). At the module level, the product, e.g. a large installation, building, or other civil object, gets upgraded (refurbishment). In the case of component recovery, products are dismantled and used and new parts can be used either in the manufacturing of the same products or of different products (remanufacturing). In material recovery, products are grinded and their materials are sorted out and grouped according to a quality wish, so recycled materials can be put into raw material, such as paper pulp and glass. Finally, in energy recovery products are burned and the released energy is captured (incineration). If none of these recovery processes occur, products are likely to go to landfill.

Figure 1.4 represents recovery options ordered as for the level of reprocess required, in the form of an inverted pyramid. At the top of the pyramid, we have the global levels, such as product and module, and at the bottom more specific levels like materials and energy. Please note that returns in any stage of the supply chain (manufacturing, distribution, and customer) can be recovered according to options both from the top and from the bottom of the pyramid.

One can find overall similarities between the inverted pyramid above and the Lansink's waste hierarchy, introduced in 1979 by this Dutch member of parliament as being prevention, reuse, recycling, and proper disposal (see www.vvav.nl). This principle has been adapted by other EU countries for their Waste Management regulations (see e.g. the German Recycling and Waste Management Act, 1996). At first, one may think that recovery options at the top of the pyramid are of high value and more environmentally friendly than options close to the bottom, which recover less value from the products. We would like to stress that both thoughts are not necessarily true. Originally, the Lansink's hierarchy was put together regarding the environmental friendliness of the recovery option. Yet, the hierarchy is disputable even there. In the case of paper recycling versus landfilling, one may go against recycling by arguing that paper is biodegradable and requires less energy than the de-inking and bleaching processes necessary for recycling. With respect to the economic value of each recovery option, that depends, for instance, on the existence of a matching market. Thus, it is possible that a used product has essentially no market value as such, but is very valuable as a collection of spare parts.

What: Types and Characteristics of Returned Products

A third viewpoint on Reverse Logistics can be obtained by considering what is actually being discarded or returned. Three product characteristics seem to be relevant:

- composition,
- deterioration, and
- use pattern.

Composition

As highlighted by Gungor and Gupta (1999), product composition in terms of the number of components and of materials is one of the many aspects to keep in mind while designing products for recovery. Not only the number, but also how the materials and components are put together, will affect the ease of reprocessing them and therefore the economics of reverse logistics activities (Goggin and Browne, 2000, discuss product complexity). The presence of hazardous materials is also of prime relevance, as it demands special treatment. The material heterogeneity of the product can play a role in recovery, where one tries to obtain separate streams of different materials which are as pure as possible (a problem in the case of plastics). The size of the product has also been pointed out as a significant factor for recovery systems (see Goggin and Browne, 2000). One can see, for example, the impact of this aspect on transportation and handling. Summing up, the intrinsic characteristics of a product are decisive for the recovery, since they effect the economics of the whole process.

Deterioration

Next there are the deterioration characteristics, which eventually cause a non-functioning of the product, but also determine whether there is enough functionality left to make a further use of the product, either as a whole or as parts. This strongly effects the recovery option. Several questions have to be asked in order to evaluate the recovery potential of a product: does the product age during use? (intrinsic deterioration); do all parts age equally or not? (homogeneity of deterioration); does the value of the product decline fast? (economic deterioration). In fact, products may become obsolete because their functionality becomes outdated due to the introduction of new products in the market, as happens with computers. This will restrict the recovery options that are viable. The same can be said for the intrinsic deterioration and whether or not it is homogeneous. If a product is consumed totally during use, such as gasoline, or if it ages fast, like a battery, or if some parts are very sensitive to deterioration, reuse of the product as such is out of the question. If, however, only a part of the product deteriorates, then other recovery options like repair or part replacement or retrieval may be considered.

Use pattern

The product use pattern, with respect to location, intensity, and duration of use, is an important group of characteristics as it affects, for instance, the collection phase (see *How* section). It will make a difference whether the end-user is an individual or an institution (bulk-use), demanding different locations for collection or different degrees of effort from the end-user (e.g. bringing it to a collection point). The use can also be less or more intensive. Let us consider the case of leased medical equipment, which is commonly used for a small time period, and is likely to be leased again (after proper operations like sterilization). Time is not the only component describing intensity of use, but also the degree of consumption during use. Consider for instance the example of reading a book. Quite often one reads a book only once after the purchase and keeps it, but does not do anything with it later on. This has stimulated Amazon to start its successful second-hand trading of books.

The characteristics of the product are related to the type of product in question. Product type in fact gives the first global impression on the potential states of the product when it reaches recovery. The product's type has been used by Zhiquiang (2003) to sketch the planning of reverse logistics activities. Fleischmann et al. (1997) distinguish the following types: spare parts, packages, and consumer goods. A natural addition is the class of industrial goods, which in general are more complex and of a different use pattern than consumer goods. Furthermore, by looking at the United Nations (UN) classification of products (see UN, online), and at the relevant characteristics for reverse logistics (as described previously), we find it important to discriminate a few more classes of products as ores, oils, and chemicals; civil objects; and other, transportable, products. It is common knowledge that civil objects have a long useful life. Besides this, recovery has mostly to be on-site, as objects like bridges and roads are not easily removed and transported as such. Ores, oils, and chemicals are a special category due to their common hazardous composition needing specialized handling during any recovery process. Finally, we attribute a separate category for other materials, such as pulp, glass, and scraps. In sum, the following main product categories are discriminated.

- consumer goods (apparel, furniture, and a vast variety of goods)
- industrial goods (e.g. military and professional equipment)
- spare parts
- packaging and distribution items
- civil objects (buildings, dikes, bridges, roads, etc.)
- ores, oils, and chemicals
- other materials (like pulp, glass, and scraps)

It is a challenging task to describe the current state of recovery for each of these product categories and identify for each the most appropriate reverse logistic chain and recovery action. This is, however, outside the scope of this book, as the focus here is on quantitative modelling. For the state-of-the-art in reverse logistics practice, we refer to Flapper et al. (2003).

Who is Who in Reverse Logistics

There are several viewpoints on actors in reverse logistics. We can make a distinction between (see also Fuller and Allen, 1995)

- forward supply chain actors, (as supplier, manufacturer, wholesaler, and retailer);
- specialized reverse chain players (such as jobbers, recycling specialists, etc.); and
- opportunistic players (such as charity organizations).

Some of the players are responsible for or organize the reverse chain, while other players simple execute tasks in the chain. The final player role we have to add to this is the accommodator role, performed by both the sender/giver and the future client, without whom recovery would not make much sense. Any party can be a sender/giver, including customers. The group of actors involved in reverse logistic activities, such as collection and processing, are independent intermediaries, specific recovery companies (e.g. jobbers), reverse logistic service providers, municipalities taking care of waste collection, and public-private foundations created to take care of recovery. Each actor has different objectives, e.g. a manufacturer may do recycling in order to prevent jobbers reselling his products at a lower price. The various parties may compete with each other.

This viewpoint is made clear in Figure 1.5. The parties at the top of the picture are either responsible or made responsible by legislation. They are from the forward chain, like the OEM. Next we also have parties organizing the reverse chain, which can be the same parties, foundations (in the case of companies working together), or even the state itself. Below these parties we have the two main reverse logistic activities, collecting and processing, which, again, can be done by different parties. After that, products are re-distributed to the market.

1.5 Why, How, What, Who, and Reverse Logistics Issues

In principle, every man-made product or system is returned or discarded at some point in its life. It is the paradigm of product recovery management that some kind of recovery and reverse logistics activities should have been planned for that moment. In many producer responsibility laws, the original manufacturer is made responsible in this respect. In the previous sections, we gave context to Reverse Logistics by presenting brief typologies for the return reasons and driving forces (*why*), for the recovery processes (*how*), for the type of products (*what*), and for the actors involved (*who*). These four basic characteristics are interrelated and their combination determines to a large extent the type of issues arising from the resulting reverse logistics system (see Figure 1.6).

In this context, numerous questions arise for product recovery management which need quantitative support to be appropriately answered. As will be

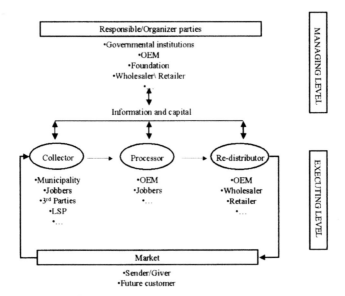

Fig. 1.5. Who is who in reverse logistics

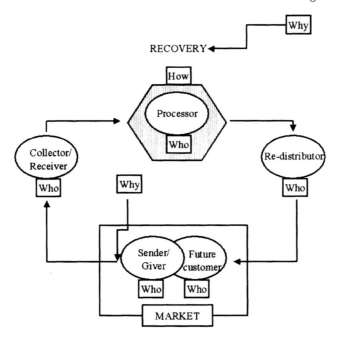

Fig. 1.6. Why, how, what. and who: basic interrelations

further elaborated in Chapter 2, this book gives support for companies to deal with the issues that arise from reverse logistics systems, which in turn are orchestrated by the four main dimensions of the previously introduced framework, i.e. why, how, what, and who. Table 1.1 encapsulates the 12 cases that serve to start the discussion in some of the chapters of this book, the contents of which we highlight next.

Toktay et al. deal with forecasting issues in Chapter 3 and use Kodak's remanufacturing of single-use cameras as an illustration. When the single-use camera is returned, some parts have not deteriorated at all during use (like the circuit boards) and they can feed again the production of remanufactured cameras. Forecasting of the quantity and timing of returns is crucial when Kodak has to decide, for example, 1) the moment to introduce new products into the market (which will make previous circuit boards obsolete), 2) the collection policy, 3) the procurement of new circuit boards, and 4) the inventory management of new circuit boards.

In Chapter 4, Fleischmann et al. discuss network decision making by introducing the IBM case. In many countries, IBM has a take-back program allowing business customers to return used products on top of the take-back responsibility that IBM has for the consumer market. IBM has set a business unit dedicated to the management of recovery with 25 facilities steering repair, remanufacturing, and recycling (also with the involvement of third parties).

Table 1.1. Reverse Logistics (who, why, how, and what): 12 cases

Who	Why-drivers	Why-return reasons	How	What
		The Kodak case (Toktay et al., Chapter 3)		
Kodak Photo-finishing lab	Economics (Corporate Citizenship)	Service	Remanufacturing	Consumer good (single-use camera)
		The IBM case (Fleischmann et al., Chapter 4)		
IBM: Other manufacturers, municipalities (NL), Charities (USA)	Legislation Economics	B2B commercial returns Service End-of-use (End-of-lease) End-of-life	Reuse Repair Remanufacturing Recycling	Consumer and industrial goods Spare parts (electronic goods)
		The case of a specialist recycler in the UK (Beullens et al., Chapter 5)		
Specialist recycler in the UK Car repair centers; return center	Economics (Legislation)	End-of-life	Recycling	Ores, oils, and chemicals (batteries)
		The Wehkamp case (De Brito and De Koster, Chapter 6)		
Wehkamp, a mail-order company in the Netherlands	Economics (also customers relationship); Legislation	Reimbursement	Resale	Consumer goods (clothing and hardware products)
		The Volkswagen case (Van der Laan et al., Chapter 8)		
Volkswagen Car dealers	Economics	Service	Remanufacturing	Spare parts (car parts)
		The Daimler-Chrysler case (Kiesmüller et al., Chapter 9)		
Daimler-Chrysler Car dealers	Economics	Service End-of-life	Remanufacturing	Industrial goods (Mercedes-Benz engines)
		The Xerox case (Inderfurth et al., Chapter 10)		
Xerox	Economics (also Corporate Citizenship)	End-of-use	Remanufacturing	Industrial goods (professional copiers)
		The Schering case (Inderfurth et al., Chapter 10)		
Schering, pharmaceutical manufacturer in Germany	Economics Corporate citizenship	By-product	Retrieval (of valuable substances)	Ores, oils, and chemicals (pharmaceutical)
		The case of a copier producer/remanufacturer (Teunter and van der Laan, Chapter 11)		
Copier producer/remanufacturer	Economics	End-of-use (+ incentives)	Remanufacturing	Industrial goods (professional copiers)
		Lead manufacturer in Greece (Pappis et al., Chapter 14)		
Lead recycler company in Greece; wholesalers; collectors	Economics (Legislation)	End-of-life	Recycling	Ores, oils, and chemicals (batteries)
		The case of a Japanese producer of refrigerators (Bloemhof-Ruwaard et al., Chapter 15)		
Japanese producer of refrigerators (in Europe)	Legislation	End-of-life	Repair, retrieval, remanufacturing, recycling	Consumer goods (refrigerators)
		The case of the pulp and paper industry in Europe (Bloemhof-Ruwaard et al., Chapter 15)		
Paper industry in Europe EU (policy making)	Legislation Corporate responsibility	End-of-life	Recycling	Other materials (pulp and paper)

IBM has to decide on how to organize the collection of product returns, and which parties are involved, depending on the processes required (see *Who* section). Reverse logistics networks have different aspects of forward logistic networks, as is stressed throughout the chapter, and IBM has opted for different solutions depending on the region.

Beullens et al. present several examples of vehicle routing in Chapter 5, among which is the case of a specialist recycler in the UK. This recycler collects products such as car batteries from car repair centers. The collection is organized by sectors. For some specific hazardous content, the collection (and transportation to destination) should not exceed 12 hours. One of the issues confronting this recycler is precisely the delineation of the sectors and the timing of collection. Besides this, different product streams correspond to different containers, complicating the loading and unloading of vehicles and vehicle routing as well.

De Brito and De Koster treat in Chapter 6 warehousing issues in reverse logistics, illustrated by the operations in the warehouse of Wehkamp, a mail-order company in the Netherlands. The assortment consists mainly of fashion products and hardware, which can be returned at no cost for the customer. Wehkamp has to decide whether to combine, or not, forward flows with return flows in the warehouse. This decision can be taken for each operation stage, i.e. 1) transportation to and from the warehouse, 2) receipt at the warehouse, and 3) storage. Furthermore, Wehkamp has to invest in an information system, because returns have to be tracked back so the customer is not billed.

In Chapter 8, Van der Laan et al. consider the joint management of recoverable and serviceable inventories giving the example of Volkswagen's remanufacturing of car parts, which are subsequently sold as spare parts. Demand does not always match the supply of recovered parts, requiring additional production of new parts by Volkswagen. The moments of production and respective quantities need to be decided. Extra complexity is added to the lack of control of the return flow regarding uncertainty of quantity and quality of returned parts. This can lead to disposal due to excess or due to non-remanufacturability (see *What* section).

Kiesmüller et al. in Chapter 9 give attention to dynamic inventory control, with another example in the automobile industry: Daimler-Chrysler. This company remanufactures Mercedes-Benz engines in a facility near Berlin, Germany. The authors go over the dynamic issues resulting from seasonal and life-cycle effects. The engine life cycle is described as having mainly two phases, i.e. a phase of increasing sales followed by one of declining sales. In the first phase, the demand for remanufactured engines is higher than the supply of returned ones (as failures are scarce), and the other way around in the second phase.

As we described in the *How* section, reprocessing may involve several processes, including disassembly. Inderfurth et al. go into the details of the scheduling of operations in Chapter 10, introducing the case of copiers remanufacturing by Xerox. The assembly of new parts and the reassembly of refurbished parts is done in the same assembly line. This involves the careful coordination of production and remanufacturing orders. Besides this, Xerox has to decide to which extent they will start with the disassembly of returned copiers before having an order for refurbished parts. The authors also present another case, Schering, a pharmaceutical company in Germany. During the manufacturing processes, some chemical by-products are retrieved. These by-products contain valuable substances that after being regained can be used again in the production process. Both production and retrieval of substances is processed in batches, which share the same operators and the same multi-purpose containers. Again, coordination here is essential (with respect to production and retrieval operations).

One of the key actors in a reverse logistics system is the sender/giver of returned products, usually a customer (see Figure 6). Yet customers can be more or less active concerning the returns. The following three examples illustrate differences in customer involvement: 1) returning bottles to the supermarket, 2) sending back a toner cartridge via mail (Bartel, 1995), and 3) having a refrigerator collected at home (Nagel and Meyer, 1999). Some companies more or less obvious incentives to get products back, like deposit fees or charity donations (see De Brito et al. 2003). In Chapter 11, Teunter and Van der Laan discuss one implicit incentive, i.e. the valuation of inventories in the case of in-house returns. They illustrate some of the issues with a copier producer/remanufacturer that has a single European plant but several national sales organizations. The latter purchase and return copiers to the plant, which uses internal 'prices' to control the product flows. During the last several years, the plant has tried several price mechanisms in order to optimize profit.

As mentioned in the *What* section, products with hazardous content need special handling during recovery. Not only this, but one may be interested in the best way to handle it so that the environmental impact is minimized. One of the techniques used to assess environmental impact is Life-Cycle Analysis (LCA). Both Chapters 14 and 15 make use of it. Pappis et al. in Chapter 14 measure the environmental performance of the recycling or disposal activities of a Greek lead recycler. In addition, in Chapter 15, Bloemhof-Ruwaard et al. take into account both economic and ecological impact. Two cases are considered in this chapter. The first one is on the composition of the product, more specifically on design decisions for the recovery of refrigerators by a Japanese manufacturer. The second is on the network design for the pulp and paper industry in Europe.

Above we indicated some of the issues arising from reverse logistic systems, by citing twelve of the real-life situations discussed throughout this book. The exact influence of the four dimensions of the framework presented here (*why, how, what,* and *who*) is still an open question that requires more investigation. Goggin and Browne (2000) try to predict the recovery type (in their case, remanufacturing, component recovery, and material recovery) from product characteristics, like input product complexity and output product complexity, together with some other factors. The product complexity is defined using the number of levels and constituents in the Bill of Materials describing the product. Although the paper is an interesting contribution, we feel that more research should be done to validate the hypotheses, especially since more recovery options exist (such as reuse and refurbishing) and more types of products and systems should be considered (the authors restrict themselves to electronic products). De Brito, Dekker, and Flapper (2003) provide an overview of some 60 cases on Reverse Logistics in which they also give an overview of the products handled. Half of their cases deal with metal products, machinery, and equipment. Some 30% of the products processed are transportable goods like wood, paper, and plastic products. Around 20% concern food, beverages, tobaccos, textiles, and apparel. The authors derive several propositions related to the framework put forward here. Furthermore, the individual chapters of the book shed more light on the interrelations of the why, how, what and who.

The next chapter describes in more detail the contribution of this book, which is focused on quantitative models to support reverse logistics decisions. Before that, we conclude with a summary of the matters put forward up to now.

1.6 Summary and Conclusions

In this chapter, we have provided a framework for the basic understanding of Reverse Logistics according to four perspectives: 1) *Why* are things returned? The answer included the forces driving companies to become active in Reverse Logistics, *Why-drivers*, and the reasons for products going back in the supply chain, *Why-reasons*; 2) *How* are returns processed? In this respect, we described the overall activities in a reverse logistics process and gave special attention to recovery options; 3) *What* is being returned? The analysis comprised, on the one hand, crucial product characteristics for reverse logistics and, on the other, a classification of product types; and 4) *Who* are executing reverse logistic activities? Here, the inquiry was on the actors and their role in implementing reverse logistics;.

More specifically, we have differentiated three driving forces for reverse logistics: 1) *Economics*, 2) *Legislation*, and 3) *Corporate citizenship*. In summary, companies become active in reverse logistics 1) because they can profit from it, and/or 2) because they have to, and/or 3) because they "feel" socially motivated to do it. We have also stressed that the economic gains can either be direct or indirect. For instance, a company may directly benefit from having reused materials as raw materials, but it may also be a strategy to improve its relations with customers. The three drivers are also interlinked and boundaries are sometimes blurred, and reverse logistics is often carried out for a mix of motives. The return reasons were organized according to three supply chain stages: 1) *manufacturing returns*, which embrace raw material surplus, quality-control returns, and production leftovers or by-products; 2) *distribution returns*, which include product recalls, returns coming back due to commercial agreements (B2B commercial returns), internal stock adjustments, and functional returns; and 3) *customer returns*, which are comprised of reimbursement guarantees, warranty returns, service returns, and end-of-use and end-of-life returns.

The overall reverse logistics activities were confined to 1) collection, 2) inspection/selection/sorting process; 3) reprocessing, and finally 4) redistribution. The reprocessing can be more or less light. We discriminated two groups: The first is of *direct recovery*, where returned products are in an as-good-as-new condition and so one can directly reuse, resell, and redistribute. The second group, *process recovery*, is where more elaborate reprocessing occurs. When zooming into reprocessing, we distinguished recovery options by occurrence levels (in great concordance with Thierry et al, 1995), that is 1) product level (*repair*), 2) module level (*refurbishing*), and 3) component level (*remanufacturing*), selective part level (*retrieval*), material level (*recycling*), and energy level (*incineration*).

On product characteristics, we qualified three as the main product features affecting reverse logistics: 1) the composition of the product, 2) the deterioration process, and 3) the use pattern. On product types, we were inspired by the work of Fleischmann et al. (1997) and the UN product's classification, and the following types were differentiated: 1) civil objects, 2) consumer goods, 3) industrial goods, 4) ores, oils, and chemicals, 5) packaging and distribution items, 6) spare parts, and finally 7) other materials (like pulp, glass, and scraps).

Regarding actors in reverse logistic systems, we divided them into three groups: 1) *typical forward supply chain actors* (manufacturers, wholesalers, retailers), 2) *specialized reverse chain players* (jobbers, remanufacturers), and 3) *opportunistic players* (entities that have other core 'business' but take an opportunity to benefit from reverse logistic activities, such as charities). With respect to their roles, one finds that 1) there are actors that are actually *responsible* for operations in the reverse logistics chain, 2) others set up or combine operations, having the role of an *organizer*, 3) many parties are busy carrying out the processes, like the collectors, so they play an *executer*, and 4) there are the sender/giver that facilitates the product for recovery and the future client that will acquire recovered products, without which recovery would not make practical sense; so they have an *accommodator* role.

Finally, we have pointed out that the four perspectives of the reverse logistics framework not only give context to reverse logistics systems but their combination determines to a large extent the kind of issues that arise in implementing, monitoring, and managing such a system. The precise influence of the four dimensions of the framework presented here (*why, how, what,* and *who*) is still an open question that requires more investigation. The individual chapters of this book shed more light on their interrelations. Furthermore, we have illustrated with twelve case studies, that appear later in this book, the sort of questions coming from practice, and for which this book gives quantitative support.

Acknowledgements

The first author was financially supported by the Portuguese Foundation for the Development of Science and Technology, i.e. Fundação Para a Ciência e a Tecnologia.

2

Quantitative Models for Reverse Logistics Decision Making

Rommert Dekker[1], Moritz Fleischmann[2], Karl Inderfurth[3], and Luk N. Van Wassenhove[4]

[1] Rotterdam School of Economics, Erasmus University Rotterdam, P.O. Box 1738, 3000 DR Rotterdam, The Netherlands, `rdekker@few.eur.nl`

[2] Rotterdam School of Management / Faculteit Bedrijfskunde, Erasmus University Rotterdam, P.O. Box 1738, 3000 DR Rotterdam, The Netherlands, `MFleischmann@fbk.eur.nl`

[3] Faculty of Economics and Management, Otto-von-Guericke University Magdeburg, P.O.Box 4120, 39016 Magdeburg, Germany, `inderfurth@ww.uni-magdeburg.de`

[4] Henry Ford Chair in Manufacturing, INSEAD, Boulevard de Constance, 77305 Fontainebleau, France, `luk.van-wassenhove@insead.edu`

2.1 Reverse Logistics and Supply Chain Management

After the general structuring of reverse logistics activities outlined in the previous chapter, we now take a quantitative perspective on this field. The purpose of this chapter is to outline the state of affairs of quantitative research in reverse logistics and to explain our contribution. To this end, Section 2.1 locates reverse logistics within the broader context of supply chain management. Section 2.2 characterizes the state of reverse logistics research in comparison to general supply chain management literature and highlights the contribution of the REVLOG network. Finally, Section 2.3 explains the structure of this book and provides an outline of the individual chapters.

As discussed in the previous chapter, reverse logistics reflects a generalization of traditional supply chain concepts. Rather than being a one-way street from producers towards consumers, today's supply chains encompass several types of 'upstream' product flows during the production, distribution, and consumption phases. While these flows have not been considered in traditional supply chain definitions, they do not surpass the concept of supply chain management as such. On the contrary, the call for a holistic perspective that allows for an optimization of the overall value creation by coordinating the individual processes that contribute to it naturally extends to product returns as well. As noted in the previous chapter, this perspective on reverse logistics as an element of supply chain management is emphasized by the recent term 'closed-loop supply chains'. For a discussion of managerial topics in closed-loop supply chains, see Guide and Van Wassenhove (2003).

Recognizing reverse logistics as a part of supply chain management suggests that it should be analyzed, at least initially, by the means that have proven successful in conventional supply chain management. As discussed in Fisher's famous Harvard Business Review article, supply chain management involves a fundamental trade–off between cost and service (Fisher, 1997). A more responsive supply chain, in general, implies higher costs. A given state of technology defines a frontier of efficient solutions of this cost-service trade-off. Which point on this frontier represents the best solution for a specific supply chain depends on its market environment, in particular on the customers' willingness to pay for a higher level of service.

Once the target point on the cost-service efficient frontier is identified, managers have several levers for steering their supply chain in this direction. Chopra and Meindl (2001) group these options into four categories, namely facilities, transportation, inventories, and information. It is easy to see that reverse logistics decisions may relate to any of these levers. For example, inspection and reprocessing of returned products requires corresponding equipment and facilities; transportation decisions relate to the movement of products between the different actors in the reverse chain; inventory buffers provide a means for decoupling consecutive reverse chain processes; and enhanced information on product returns allows for a more responsive recovery strategy. The question therefore becomes how to use these different levers in reverse logistics in such a way as to support the supply chain's overall strategy in an optimal way.

2.2 Quantitative Analysis and Decision Support

In traditional supply chain management, quantitative models have long been recognized as a powerful tool to support decision making. Quantitative models have the potential to contribute to improved decision making in two ways, directly and indirectly. On the one hand, quantitative models may be embedded in software systems to automate and optimize decisions, such as automatic order generation to replenish inventory. Even more important, however, analytic models contribute to a better understanding of the interactions, dynamics, and underlying trade-offs of the various supply chain processes, and thereby enable managers consciously to take these factors into account in their decisions. The goal of our research within the REVLOG project has been to contribute to a sound basis of quantitative models that support reverse logistics decision making in a similar way.

In order to characterize today's state of affairs in this endeavor, we first take a look at the body of quantitative research in conventional supply chain management as a benchmark. We emphasize that this discussion is not meant to provide a detailed literature review, but rather a high-level mapping of

Table 2.1. Quantitative Models for Supply Chain Management – Major Themes

Strategic	Tactical/operational
Distribution:	
Strategic network design	Transportation
Global sourcing	Warehousing operations
Facility layout	
Inventory + Production:	
Capacity planning	Lot sizing
Inventory location	Safety stocks
+ risk pooling	Aggregate production planning
	Production scheduling
Supply Chain Scope:	
Supply chain coordination	Dynamic pricing
+ contracts	+ revenue management
Multi-channel conflicts	Value of information
Auctions	

research areas. For a more detailed discussion of many of these topics we refer to, for example, Tayur et al. (1998) and Simchi-Levi et al. (2002). See also Kopczak and Johnson (2003) for a discussion of major shifts in supply chain thinking from a managerial perspective.

Table 2.1 lists important areas in supply chain management for which a substantial body of quantitative models has been established by the scientific community. We structure these contributions along two dimensions, namely their functional area and their decision horizon. In the first dimension, we distinguish between the areas of distribution management, production planning and inventory control, and research that explicitly focuses on the inter-organizational scope of a supply chain. While it is this last area that has been the focus of recent supply chain management literature in a strict sense, the first two areas have provided important building blocks for these analyses. The second dimension of Table 2.1 refers to the usual distinction between long-term strategic decisions and shorter-term tactical and operational decisions. Let us go briefly through this list of topics.

Logistics network design and facility location are among the classical strategic *distribution* decisions, for which numerous models of different levels of complexity have been developed throughout the decades. In recent years, particular attention has been paid to the global scope of many of today's supply chains. Financial issues such as tax implications and currency risks are becoming increasingly important in this context. Whereas strategic network design relates to external product flows, facility layout addresses similar design decisions on an internal level, considering the flows within a factory or distribution center.

For shorter-term distribution-related models, we refer to transportation issues primarily. A countless variety of models has been developed, aiming to support decisions such as fleet management, route planning, zoning, or real-time vehicle scheduling. Warehousing operations again refer to internal transportation flows.

Production planning and inventory control has been another fruitful area that has seen abundant modeling contributions. On the strategic level, we name capacity planning as the foremost example. This includes decisions on manufacturing and handling equipment and on the workforce size. Other strategic models address the question of where in the supply chain to locate inventory buffers. The trade-off between risk-pooling and response time plays a key role in these approaches.

The literature on production planning and inventory control models for short- and medium-term decisions is even more extensive. On the one hand, we have countless variants of models that address the trade–offs related to different inventory functions, such as cycle stock, safety stock, and seasonal stock. On the other hand, we have an equally broad variety of production planning and scheduling models that consider the timing, sequencing, and batching of manufacturing operations on a given set of equipment.

As noted above, the focus of recent modeling efforts in supply chain management is on capturing the complexities that arise from the *interaction between different organizations* and between different functional areas. In particular, a major stream of research addresses the coordination of multiple decision makers in the supply chain, such as to achieve an overall optimal solution. Interaction between multiple distribution channels, in particular conflicts around disintermediation, also pertains to this theme.

Pricing is another important element of supply chain management that has seen significant modeling contributions during the past few years. Pricing has been recognized as a means for matching supply and demand in an optimal way. Specific topics include the ever-expanding field of revenue management, design of and strategies for auctions, but also the coordination of pricing decisions with operations.

The value of information is another theme within supply chain management that should not go unmentioned here, as it has spurred an important stream of analytic research. As information is becoming ever more easy to collect and to process, the question of which information actually contributes to improving a given set of business processes, and to which extent, is becoming increasingly more relevant.

When we started the REVLOG project in 1997, the equivalent of Table 2.1 for reverse logistics models would have been fairly sparsely populated. We summarized the state of related quantitative literature at that time in a review paper (Fleischmann, 1997). Therein we noted that '...the results published to date are rather isolated. Comprehensive approaches are rare.' While a number of individual contributions had been developed, e.g. facility

Table 2.2. Quantitative Models for Reverse Logistics – State of Affairs and REVLOG Contributions

Strategic	*Tactical/operational*
Distribution:	
Reverse logistics netw. design (4)	**Product return forecasting (3)**
Reusable packaging	**Collection and distr. routing (5)**
	Return handling (6)
Inventory + Production:	
Valuation of recoverable invento-ries (11)	**Lot sizing in PR* operations (7)**
	Safety stocks in PR systems (8)
Product design for reusability	**Dynamic control of PR ops. (9)**
	Production planning for PR (10)
	Remanufacturing operations
	Production planning for bulk recycling
Supply Chain Scope:	
SC coordination + contracts (12)	**Value of information (3, 16)**
Long-term perf. development (13)	**Product acquisition (3, 9)**
Environmental perf. (14, 15)	
Collaborative recycling networks	

* PR = product recovery
boldface = REVLOG contribution
(#) = chapter number

location and inventory control models with product returns, they were not part of a coherent, broader stream of research.

Since then, however, a rapidly increasing number of researchers has brought about significant advances in reverse logistics literature. Therefore, the number of entries in Table 2.2, which outlines the state of affairs to date, is substantial. We structure the topics analogous with the general supply chain management models above. In addition, we highlight in boldface the fields to which the participants of REVLOG have contributed. These are also the topics that we discuss in detail in this book (chapter numbers are listed in braces). We provide a detailed outline of the individual chapters in the next section. Before doing so, we characterize the state of affairs in quantitative reverse logistics research by comparing it to that in general supply chain management. To this end, we go through the entries of Table 2.1 again and comment on corresponding research in a reverse logistics context. In addition, we point out themes that are reverse logistics specific.

We start again with the area of *distribution* management. Network design issues in a reverse logistics context relate to the infrastructure for collecting and reprocessing returned products. A considerable number of corresponding models has been presented in the literature, most of which lean closely towards traditional facility location approaches. We review this area in Chapter 4. Chapter 5 addresses the corresponding transportation operations.

Modeling approaches to the analogous internal logistics issues, such as the impact of product returns on facility layout and internal transportation, are still scarce to date. While internal handling is consistently identified as an important factor in reverse logistics, most available studies on this topic take a qualitative perspective, as we discuss in Chapter 6.

All of the above issues naturally correspond with issues in traditional supply chains. In addition, reverse logistics brings about more particular novel tasks. These include the forecasting of product returns, which provides information that is vital for any reverse logistics operation. The relation between demand and returns is a key element of the corresponding models. Chapter 3 reviews different approaches to this issue.

Another distribution-related issue concerns the management of reusable packaging (see also the discussion under 'functional returns' in the previous chapter). The literature includes models assessing, for example, the number of reuse cycles and an appropriate pool size. While we touch upon these issues on several occasions in this book, we do not dedicate a specific chapter to them.

Product returns also have a significant impact on *inventory control*. Issues relate to the potential discrepancy between supply and demand and to the coordination of returns with other supply sources. It is this area for which the largest number of reverse logistics models has been presented in the literature to date. In particular, many classical inventory control models have been adapted to a reverse logistics setting, including lot-sizing, safety stock, and seasonal stock models. These are reviewed in detail in Chapters 7 through 9.

The impact of product returns on more strategic inventory issues, such as the coordination of inventory and location decisions, remains to be explored. On the other hand, inventory control in a reverse logistics setting gives rise to fundamental issues concerning inventory cost metrics. In particular, it is not always obvious how to assign holding costs to inventories of returned goods. We discuss this issue in more detail in Chapter 11.

The reprocessing operations (see previous chapter) also give rise to particular production planning issues in a reverse logistics context. In this book, we restrict ourselves to generic concepts, such as MRP approaches and disassembly strategies, which we discuss in Chapter 10. More specific production planning aspects that are linked to particular recovery processes, such as control of remanufacturing operations or bulk recycling processes, are beyond the scope of this book. Similarly, we leave the large field of product-design-related decisions outside our discussion.

The bottom part of Table 2.2 lists a number of issues in reverse logistics that exceed the boundaries of a single functional area. Analogous with conventional *supply chains*, the interaction between different organizations with different incentives adds to the complexity of reverse logistics systems. This concerns, for example, transactions between the collector and the reprocessor in the reverse chain and competition between original equipment manufacturers and independent reprocessors (see Chapter 1). Chapter 12 reviews recent approaches in the literature that explicitly model these interactions between multiple decision makers in the reverse chain.

As in conventional supply chains, assessing the value of different types of information is an important issue in reverse logistics. In addition to demand information, as in traditional chains, reverse chains are dependent on information on product returns, which are perceived as a highly uncertain factor in many cases. Chapters 3 and 16 discuss models that provide a starting point for analyzing value of information issues in the reverse chain.

Taking this route one step further, companies are seeking to influence their reverse product flows. In addition to improved information, these organizations are looking for appropriate incentives which allow them to balance supply (in the form of product returns) and demand in an optimal way. For this area, also denoted as 'product acquisition management', a few initial models have recently been proposed which focus on determining appropriate buyback prices to maximize the financial contribution of reverse product streams. These are addressed in Chapters 3 and 9.

Particular supply chain issues arise from the link between reverse logistics programs and environmental management. Given that ecological benefits are among the proclaimed drivers of many reverse chains, environmental issues are particularly relevant here (see Chapter 1). In particular, the relation between economic and environmental performance of reverse logistics systems is a matter of importance. Chapters 13 through 15 present different approaches to this issue.

Comparing Tables 2.1 and 2.2 leads us to the following assessment of the current state of quantitative reverse logistics research. In the past few years, a significant number of reverse logistics models has been developed for several basic functional areas, notably in distribution management and inventory control. In these areas, many of the standard traditional models have been extended in such a way as to match reverse logistics contexts. At the same time, specific mathematical difficulties linked to the incorporation of reverse flows have been identified. The limits of what is tractable are becoming clearer. All in all, reverse logistics modeling is quickly catching up with the level of conventional supply chains in these basic areas. There are a few exceptions of functional areas for which well-established reverse logistics models are still scarce. Notably, these include the area of internal logistics.

As the basic building blocks are available, the next step to extend reverse logistics literature should be in the direction of broader, cross-functional and inter-organizational models, analogous with the concepts that have driven supply chain management for conventional 'forward' chains. To date, such approaches to reverse logistics and closed-loop supply chains remain few and far between. For the issues listed in the rightmost column of Table 2.2, the current state of the literature resembles that of the basic functional areas in

1997. While a number of important contributions have been made, the overall picture is yet to be uncovered.

Research within our REVLOG network has contributed significantly to the advances in reverse logistics literature during the past five years. On the one hand, REVLOG has brought about numerous individual models in all of the areas listed above. Even more important, however, by bringing these models together and relating them to each other and to traditional literature, we are aiming to contribute to a better understanding of reverse logistics in general. In this book we present our view on this field.

2.3 Roadmap

The goal of this book is to contribute to a better understanding of reverse logistics decision making. To this end, we highlight major issues that companies face as they engage in reverse logistics activities. The focus then is on quantitative models that capture these issues and support the decision maker. By analyzing these models, we highlight the underlying key trade-offs. Moreover, by comparing the models with approaches in conventional supply chains, we explain what is particular about reverse logistics. Presenting a state-of-the-art review of quantitative models for reverse logistics decision making, this book is meant to serve as a basis for future research in this field.

We structure the material roughly around Table 2.2. In particular, we group the contributions in three parts of four to five chapters that address distribution issues, production planning and inventory control issues, and broader supply chain issues, respectively. Each chapter is dedicated to a specific reverse logistics topic. The structuring of the material is very similar across all chapters. Each one starts by highlighting the main decision issues, in most cases based on an illustrative case example. The core of the chapter then is dedicated to a presentation and analysis of corresponding quantitative models. The focus is on material developed within REVLOG. However, each chapter refers to other material in the literature to provide a complete state-of-the-art picture.

In what follows, we outline the topic of each chapter. Table 2.3 summarizes the corresponding modeling approaches and links them to related traditional models.

Part II — Collection and Distribution Management Issues
The part of the book following these introductory chapters focuses on the physical flows arising in reverse logistics systems. We open this part in Chapter 3 with a discussion of forecasting methods for estimating product returns. The procedures discussed in this chapter provide the basis for attaining the information on product returns that is used explicitly or implicitly throughout the remainder of the book. Chapter 16 at the end of the book complements this analysis by discussing related information systems. Chapters 4 through

Table 2.3. Modeling Approaches by Chapter

Chapter	Topic	Methodology	Related traditional models
3	product return forecasting	time series analysis	demand forecasting models
4	reverse logistics network design	MILP, stochastic programming	facility location models
		continuous approximation	logistics cost models
5	collection and distribution vehicle routing	MILP, continuous approximation	vehicle routing models
6	return handling and warehousing	—	facility layout models, order-picking models
7	lot sizing in product recovery	continuous optimization, MILP	EOQ, dynamic lot-sizing models
8	stochastic inventory control in product recovery	Markov decision processes	stochastic inventory control models
		stochastic optimization	news vendor model
9	dynamic product recovery	optimal control theory	optimal control models
10	production planning for product recovery	MILP, continuous optimization	MRP, aggregate production planning
11	inventory valuation	net present value analysis	EOQ, stochastic inventory control models
12	coordination in closed-loop supply chains	game theory	supply chain contracts
13	long-term analysis of closed-loop supply chains	system dynamics	industrial dynamics models
14	environmental management	life cycle analysis	life-cycle analysis
15	economic and environmental performance	MILP, LP	network flow / facility location models
16	information technology	—	—

6 address traditional areas of distribution management, namely network design, vehicle routing, and internal logistics handling. Each of them extends approaches from conventional supply chains to a reverse logistics setting. On each level, the potential integration of forward and reverse flows turns out to be a key issue. Chapters 4 and 5 use very similar methodologies. Both consider traditional mixed integer linear programming (MILP) models, which can be readily extended to include reverse product flows. In addition, both chapters discuss continuous approximation methods which facilitate sensitivity analyses and reveal the impact of key parameters on system performance. Chapter 6 focuses on a structuring of the relevant issues, since corresponding quantitative models are not yet available. Specifically, the chapters of Part II proceed as follows.

Chapter 3 addresses the forecasting of product returns. It analyzes time series models that represent returns as a function of past sales, under different levels of information. Based on these models, it compares different forecasting approaches with respect to their performance in terms of robustness to errors in parameter estimates and of propagation of variability. The chapter argues that using past sales information can significantly improve return forecasts. In addition, it appears that more advanced procedures are more sensitive to inaccurate parameter estimates.

Chapter 4 is dedicated to reverse logistics network design. It discusses three alternative modeling approaches that rely on MILP, stochastic programming, and continuous approximation, respectively. Based on these models, it analyzes the robustness of logistics networks with respect to varying return flows. The chapter concludes that reverse logistics networks are, in many cases, compatible with existing forward networks.

Chapter 5 complements the preceding chapter by zooming in on vehicle routing issues. The focus is on the relation between collection and delivery. Several alternative policies are compared that rely on different levels of integration of both types of shipments. To this end, the chapter discusses both traditional MILP models and novel approaches which rely on a combination of MILP with continuous cost approximations. The analysis highlights how a higher degree of freedom in collection policies as compared to deliveries can be exploited to increase transportation efficiency.

Chapter 6 addresses the internal handling of return streams. The chapter reviews the issues that arise in this context, such as inspection and storage decisions. For most of these issues, no quantitative models are available to date that explicitly take reverse flows into account. However, existing traditional models provide a promising starting point for future modeling contributions.

Part III — Inventory Control and Production Planning
The next part of the book focuses on the coordination of product recovery processes. This leads us to inventory control and production planning issues

in reverse logistics. Particular inventory control tasks in this context concern the potential mismatch between supply and demand and the coordination of product recovery with alternative supply sources, such as conventional production. Chapters 7 through 9 provide a detailed analysis of these issues. They all consider essentially the same inventory system, which includes a single stock of end-items fed by product recovery and regular production, and potentially an additional stock of returned items prior to recovery. The chapters differ in their assumptions on the demand process and hence in the role of the resulting inventories. Chapter 7 assumes deterministic demand and considers inventories as cycle stock. Chapter 8 assumes stochastic demand and therefore also includes safety stock considerations. Chapter 9 focuses on seasonal stock by assuming non-stationary, deterministic demand. Each of these cases leads to an extension of traditional inventory control models with an exogenous reverse product flow. Moreover, it is the coordination of product recovery with the regular production source that turns out to be the main challenge throughout. All of these inventory models assume that holding cost parameters are given for the different stock points at hand. Chapter 11 reveals that it is a non-trivial task to choose these parameters such that they reflect the financial consequences of alternative decisions appropriately. The chapter applies a financial net present value approach to address this issue. Chapter 10 complements the preceding material by considering the recovery processes in more detail. Specifically, it applies traditional production planning concepts such as material requirements planning (MRP) and linear programming to the reprocessing operations in the reverse chain. Summing up, Part III is structured as follows.

Chapter 7 analyzes lot-sizing decisions in the reverse chain. It identifies two main factors that distinguish this issue from the setting of traditional supply chains, namely the choice between alternative supply sources and the choice between alternative return dispositions. The chapter illustrates how these factors can be incorporated in traditional lot-sizing models. It reveals that these extensions significantly increase the mathematical complexity since they affect some of the structure that is key to traditional solution approaches. An analysis of the extended model provides insight into the optimal coordination of production and recovery lot sizes.

Chapter 8 reviews stochastic inventory control models in a reverse logistics context. The discussion distinguishes between push- and pull-driven recovery. In the first case, the recovery process is entirely exogenous to the decision maker, whereas in the second case the timing of individual reprocessing steps is a decision variable. This distinction turns out to be a key determinant of the mathematical complexity of the corresponding models. The chapter presents a decomposition approach, which allows the incorporation of autonomous return flows in traditional inventory control in many cases. In contrast, managed return flows in general complicate the model structures to an extent that is only tractable through heuristics. In particular, lead time differences between alternative supply sources are a key complicating factor in this context.

Chapter 9 focuses on dynamic aspects of reverse supply chain flows. It addresses the question of how to manage product recovery processes given the variability of product returns along seasonal and life-cycle patterns. In particular, a key decision is when to stock product returns for future use and when to dispose of them. The chapter presents deterministic optimal control models to address the trade-offs underlying these decisions. The results reveal the value of product returns at different points in time.

Chapter 10 is dedicated to production planning issues in product recovery management. Specifically, it reviews managerial planning tasks related to disassembly operations, coordination of recovery and manufacturing operations, and rework of manufacturing defectives. The chapter argues that conventional production planning tools such as MRP systems can be effective for managing product recovery operations in many cases, provided that they are appropriately adjusted. The high uncertainty inherent in many product recovery environments complicates the matter.

Chapter 11 considers holding cost metrics for recoverable inventories. To this end, the common average cost criterion is compared with a more detailed net present value approach. The analysis reveals that, unlike in many traditional inventory control systems, both approaches are not equivalent, in general, in a product recovery setting. Therefore. the selection of appropriate holding cost rates requires additional thought. The chapter introduces a holding cost framework that is based on opportunity costs rather than on cost price.

Part IV — Supply Chain Management Issues
The last part of the book addresses reverse logistics issues in a broader supply chain setting. Chapter 12 opens this discussion by considering coordination issues between multiple decision makers in a reverse logistics environment. It applies game theoretic supply chain management concepts to analyze these issues. Chapters 13 through 15 all address long-term, strategic supply chain design issues in reverse logistics. Moreover, they all combine economic and environmental perspectives. The chapters differ in their methodologies and in the focus of their contributions. Chapter 13 takes a systems dynamics approach that allows the analysis of the migration of reverse logistics systems during product life cycles and the dynamic response to external and internal policy changes. Chapter 14 addresses the environmental impact of reverse logistics systems in considerable detail through a life-cycle analysis (LCA). Chapter 15 uses rougher, aggregate measures of environmental performance and integrates them into traditional MILP network design models. In this way. it complements the pure economic analysis of Chapter 4. Each of the three above chapters illustrates the complexity of combining economic and ecological aspects in supply chain design decisions. Chapter 16 rounds off the discussion of reverse logistics systems by reviewing corresponding information

technology solutions. Complementary to Chapter 3, it presents tools that help satisfy the information needs of the various examples discussed in this book. Summing up, Part IV encompasses the following contributions.

Chapter 12 addresses coordination issues in closed-loop supply chains. These include, in particular, pricing decisions as a means for managing reverse product flows. Corresponding game theoretic models are presented that relate to the design of the reverse channel and to market segmentation between new and recovered products. Pointing out that research on the coordination of closed-loop supply chains is yet at an early stage, the chapter suggests avenues for further investigations.

Chapter 13 studies the long-term behavior of closed-loop supply chains. Specifically, it considers the dynamic development of product recovery activities over time that results from the complex interaction of numerous external and internal factors. A system dynamics approach is used to model these interrelations. The model provides a tool for evaluating long-term strategies of both companies and environmental regulators.

Chapter 14 addresses reverse logistics from an environmental management perspective. It discusses the application of LCA methods to evaluate the environmental impact of reverse supply chains. An extensive example concerning the recycling of car batteries illustrates this approach.

Chapter 15 considers the relation between economic and environmental performance in closed-loop supply chains. To this end, quantitative models are reviewed that explicitly take both of these factors into account. This allows the highlighting of the synergies and potential trade-offs between different performance criteria. In addition, the models are used in this chapter to analyze the impact of different types of environmental regulation. Examples illustrate that simplistic environmental targets can have counter-productive effects.

Chapter 16 addresses the role of information technology in closed-loop supply chains. The chapter reviews factors of uncertainty, which complicate reverse logistics decision making, and discusses how information technology can help reduce these hurdles. In addition, the chapter provides a detailed review of available information systems that support the different business processes in the reverse supply chain.

Part II

Collection and Distribution Management Issues in Reverse Logistics

3

Managing Product Returns:
The Role of Forecasting

L. Beril Toktay[1], Erwin A. van der Laan[2], and Marisa P. de Brito[3]

[1] Technology Management, INSEAD, Boulevard de Constance,
77305 Fontainebleau Cedex, France, `beril.toktay@insead.edu`
[2] Rotterdam School of Management / Faculteit Bedrijfskunde,
Erasmus University Rotterdam, P.O. Box 1738, 3000 DR Rotterdam,
The Netherlands, `elaan@fbk.eur.nl`
[3] Rotterdam School of Economics, Erasmus University Rotterdam, P.O. Box 1738,
3000 DR Rotterdam, The Netherlands, `debrito@few.eur.nl`

3.1 Introduction

An important challenge arising in reverse logistics supply chains is the effective use of returns so as to maximize the value of this resource. To this end, decisions at strategic, tactical, and operational levels should explicitly incorporate information about return flow characteristics, primarily quantity and quality. In this chapter, we review informational issues concerning product returns. We then investigate the value of accurate forecasting at an operational level by focusing on inventory control.

Most remanufacturable products are sold to the customer and are returned when their useful life is over or when the customer wants to trade in the product for an upgrade. In the former category are products such as single-use cameras, toner cartridges, and tires. In the latter category are durable products such as personal computers, cars and copiers. Predicting the return flow characteristics is important for decisions at all levels, for example, for network design at the strategic level; for procurement decisions, capacity planning, collection policy, and disposal management at the tactical level; and for production planning and inventory control at the operational level. We start by discussing the information needs for these decisions as modelled in the remainder of this book.

Chapter 4 considers network design for a reverse logistics supply chain. The return flow information required by proposed models consists of the aggregate return volume in each period over the planning horizon. These volumes must be estimated before the product is launched and before any sales information is available. They are therefore based on demand forecasts and an estimate of the proportion of sales that will be returned. A robustness analysis in Chapter 4 shows that the solution to the network design problem is robust in the proportion of returns, so a rough estimate based on expert judgment should be sufficient. In countries where legislation stipulates target return rates, this target can be taken as the proportion of products that will be returned.

At a tactical level, Chapter 9 develops a model to determine new product production rates (and therefore the procurement and capacity needs) over the life-cycle of a remanufacturable product, taking the trajectory of demand and return volumes as input. Building warehousing capacity to handle returns, also a mid-term decision, depends on expected return volumes (Chapter 6). Within a particular network design, the relationships with the parties involved can be designed in various ways so as to impact the quantity and timing of returns. We discuss this idea in more depth in Section 3.2. While such tactical decisions will initially be made based on expert judgement, they are typically reversible, so they can be reset iteratively, if necessary, as new information about return volumes becomes available.

At an operational level, one can assume that strategic and tactical-level decisions have been made and are fixed. The information required by models in Chapters 8 and 10 on inventory control and production planning, respectively, are the expectation and variance of returns in future periods. As time goes on, detailed data becomes available regarding sales volumes and return quantities and qualities. In this case, accurate data-based predictions of the quantities to be returned in each period, as well as the quality of these returns, can be made for use in inventory control and production planning. Note that some products are leased to customers (e.g. Xerox copiers to corporate customers) and are collected by the manufacturer at the expiration of the lease. In this case, the timing and the quantity of products to be returned are easier to predict. The major uncertainty is about the condition of the product.

One example of a successful remanufactured product line is the Kodak single-use flash camera (Goldstein, 1994). The reusable parts (the circuit board, plastic body, and lens aperture) of the returned cameras are put back into production after inspection. The circuit board is the primary cost driver for this product. Used boards are valuable to Kodak as long as the product design allows them to be reused, and have minimal salvage value otherwise. Let us consider a number of strategic, tactical, and operational issues that arise in managing this product line.

At a strategic level, the timing of new product introductions should take return flow characteristics into account. For example, the initial design of the product was constrained by the size of the circuit board. Subsequently, Kodak introduced a pocket-size camera that required a smaller circuit board. As a result, a number of larger-size boards would become obsolete by the time they were returned to Kodak. In this setting, forecasting the quantity and timing of returns of the previous generation is an important input in determining the timing of the new product introduction.

At a tactical level, consider the after-use returns. Customers take the used cameras to a photofinishing laboratory, where the film is taken out and processed; only the film and the developed pictures are returned to the customer. Kodak needs to propose a collection policy to the laboratories that is economical and yet encourages them to return the cameras to Kodak. This policy should also take into account differences between laboratories: due to economies of scale in transportation, small labs may wait for a long time before sending a batch back to Kodak or may not send in cameras at all, significantly adding to the return delay and influencing the return percentage, which in turn impacts the profit of the product line.

Again at the tactical level, consider the procurement of new circuit boards. An important decision is whether to source locally or overseas (at a lower cost, but incurring a longer lead time). The proportion of returns will determine the volume of new circuit boards that is required, so estimating this number is important in making the sourcing decision.

Finally, at an operational level, consider the inventory management of new circuit boards. The production facility uses both old and new components to manufacture new cameras. Since circuit boards are costly, recuperating these components from used products is a valuable opportunity. Forecasting returns to a good degree of accuracy and incorporating these forecasts in inventory management decisions improves the value extracted from this resource.

Unlike end-of-life returns that have already been sold for profit and now have the potential of generating additional benefits through value recovery, commercial returns represent a lost margin. In catalog sales, an average return rate of 12% is standard, with return rates varying by product category: 5 – 9% in hard goods, 12 – 18% for casual apparel, 15 – 20% for high-tech products, and up to 35% for high-fashion apparel (Dowling, 1999). Commercial returns impose high costs on retailers and manufacturers alike. For example, at HP Inkjet Imaging Solutions, product returns have been averaging 6.6% of sales dollars and 5.7% of units shipped in North America in 1999 (Van Wassenhove et al., 2002). Like end-of-life returns, an important lever in managing commercial returns is to accurately predict the return quantities for both tactical and operational level decisions.

There are few documented business examples of forecasting specifically for reverse logistics. What has been documented is basic: estimating the return probability by the proportion of cumulative returns to cumulative sales (Goh and Varaprasad, 1986; Toktay et al., 2000). We refer to this method as "naive estimation." It is useful only for tactical decisions such as capacity sizing since it only provides information on the overall proportion of returns but not on their timing. At a more operational level, supply chain planning and inventory software applications that allow for remanufacturing typically do so by providing the capability to do reverse-logistics-specific order processing; to the best of our knowledge, the forecasting of returns, if any, is handled by applying time-series forecasting methods to the historical return stream without exploiting the fact that returns are generated by previous sales (e.g. Genco, ReturnCentral, Xelus).

In this chapter, we discuss ways of actively influencing returns (Section 3.2) and we review data-driven methods for forecasting return flows that exploit the fact that future returns are a function of past sales. In particular, Section3.3 builds the two stages of forecast model building/parameter estimation and return forecasting. We then focus on the value of return forecasting at an operational level, specifically inventory control. Section3.4 uses simulation to evaluate the performance of a periodic-review replenishment policy along two dimensions: the impact of estimation error and order variability. We conclude in Section 3.5 with implications for supply chain management.

3.2 Influencing Returns

An important consideration in extracting value from returns is to actively manage their quantity and timing. Indeed, retailers and manufacturers strive to design reverse logistics systems that increase the visibility and speed of the return process to maximize asset recovery, especially for seasonal or short life-cycle products. Firms vary in how they address this problem. For example, Ingram Micro Logistics, the distribution arm of Ingram Micro, opened the first automated returns facility in the US in early 2001 (Morrell, 2001). Others increasingly rely on third-party reverse logistics providers such as GENCO Distribution System, UPS, USF Processors, and Returns Online (Chapter 16; Gooley, 2001). Various software products that are specifically targeted towards returns processing are now available on the market, provided by such companies as Kirus Inc., Retek.com, ReturnCentral, and The Return Exchange (Gooley, 2001).

This issue has recently started to be addressed in the reverse logistics literature. Hess and Mayhew (1997) develop a regression model for commercial returns that incorporates explanatory variables such as price, product category, and reason for return. This model is then used to predict the cumulative return rate over time. In this case, the analysis is very valuable for identifying the most important reasons for return, for forecasting the profitability of product categories using projected returns, and as input into replenishment orders. Savaşkan et al. (1999) show that decisions about who is responsible for the collection of returns can influence return volumes. Guide and Van Wassenhove (2001) and Klausner and Hendrickson (2000) show that offering differentiated take-back prices to consumers based on the product model and product quality can influence the quality and quantity of returns.

Two different modelling approaches are useful in capturing the dependence of return flow characteristics on system structure and parameters. The first approach is to develop principle-agent models and explore the value of incentive schemes such as return allowances, trade-in offers, and buybacks (as reviewed in more generality in Chapter 12). The second is data-based regression analysis to quantify the impact of several relevant factors in specific settings. We believe that for practical implementation, the data-based approach holds much potential. Developing a good understanding of drivers of return flow characteristics (e.g. product category, life-cycle length, market value of used product, customer segment, ease of return, rebate policy, etc.) would enable better decision making for the following purposes.

- Influencing the return delay. Especially for items that depreciate rapidly, getting the used products back quickly for reprocessing is very valuable. A statistical investigation of which factors are the most relevant can help in determining the most cost-effective levers in reducing the return delay.
- Developing customer- or category-specific return policy. Return policies typically do not distinguish between customer segments or product categories although it is clear that segmentation could result in more effective policies. In the absence of data allowing an evaluation of differentiated service contracts, it is not possible to develop such a policy in an effective manner. This gap can be filled by developing regression models that use product category and customer segment as explanatory variables.
- Trading off customer service level and cost. Typically, it is the marketing department that "develops" the return policy. The focus is on increasing customer service, especially in North America. In addition, competitive pressures drive firms towards offering liberal return policies. However, liberal return policies can be very costly due to direct costs such as transportation, testing, and repackaging, due to opportunity cost and due to legislation that sometimes requires returned products to be labeled as 'used'. In determining return policies, the first step should be to evaluate the cost of alternative return policies. To this end, a model of how return rates will be impacted by the return policy needs to be developed.

3.3 Forecasting Returns

We now turn to the issue of forecasting returns under the assumption that all policies concerning returns have been determined and data collection on returns has started. The first step in any forecasting exercise is to build a forecast model that models the variables to be predicted as a function of the explanatory variables (Box and Jenkins, 1976). For example, the variable to be predicted may be the return quantity in the next period and the explanatory variables can be past sales. This forecast model will have a number of parameters that need to be estimated using historical sales and returns data. Once the validation and estimation phase is complete, we have a fully specified forecast model. As the second step, forecasts of future returns are made using parameter estimates obtained from this forecast model and historical information. The two subsections to follow describe methods for these two steps, respectively. Table 3.1 lists the main notation that is used in this section.

3.3.1 Forecast Models and Parameter Estimation

The key to forecasting returns is to observe that returns in any one period are generated by sales in the preceding periods. A prevalent way of modelling this

Table 3.1. Main notation for Section 3.3

$s(\tau)$	Sales in period τ
$u(\tau)$	Returns in period τ
p	Probability that a sold product will eventually return
r_k	Probability that a sold product returns after k periods, given that it will return eventually
ν_k	Probability that a sold product returns after k periods ($\nu_k = p \cdot r_k$)
$\hat{\nu}_k(\tau)$	The period τ forecast of ν_k
$y_{\tau-i,\tau+j}$	The total returns in period $\tau+j$ originating from sales in period $\tau-i$
$v_{\tau,\tau-i}$	Total returns up to and including period τ originating from sales in period $\tau - i$
$\mathcal{I}(\tau)$	Information set available at the end of period τ to forecast future returns

is to assume that a sale in the current period will generate a return k periods from now with probability ν_k, $k = 1, 2, \ldots$ or will never be returned. We first review methods used in the literature that exploit this structure to postulate a return delay distribution and estimate its parameters; alternative modelling approaches are discussed at the end of Section 3.3.2. A particular characteristic of the return delay data is that it is right-censored: at a given time, if an item has not been returned, it is not known whether it will be returned or not. For accurate estimation, it is important to use an estimation method that takes into account that some items that have not yet been returned will never be returned.

We classify the forecast models used in the literature according to the data that they exploit. We say that *period-level* information is available if only the total sales and return volume in each period are known. For beverage containers, single-use cameras, and toner cartridges, this is typically the only data available. We say that *item-level* information is available if the sale and return dates of each product are known. Electrical motors with electronic data logging technology (Klausner et al., 1998), copiers, and personal computers are typically tracked individually, so this data can easily be obtained for these products. POS (point-of-sale) data technology in retailing also can allow for item-level tracking.

Period-level Information

Let $s(\tau)$ and $u(\tau)$ denote the sales and returns of products in period τ, respectively. Goh and Varaprasad (1986) propose a transfer function model of the form

$$u(\tau) = \frac{\omega_0 - \omega_1 B - \omega_2 B^2 - \ldots - \omega_s B^s}{1 - \delta_1 B - \delta_2 B^2 - \ldots - \delta_r B^r} s(\tau - b) + \epsilon(\tau), \qquad (3.1)$$

where B is the backshift operator, b is the time lag, $\{\omega_i\}$ and $\{\delta_i\}$ are lag parameters, and $\epsilon(\tau)$ is white noise. The determination of the appropriate transfer function model follows the steps of model identification, parameter estimation, and diagnostic checking as described in Box and Jenkins (1976). In this spirit, de Brito and Dekker (2003) use data on commercial returns with individual tracking to test the assumption of exponential delay.

Note that the transfer function model can be rewritten as

$$u(\tau) = (\nu_0 + \nu_1 B + \nu_2 B^2 + \ldots)s(\tau) + \epsilon(\tau). \tag{3.2}$$

Once the parameters $\{\omega_i\}$ and $\{\delta_i\}$ of the transfer function model have been estimated, the parameters $\{\nu_k, k \geq 0\}$ are easily calculated. The statistically significant values of these parameters are used as estimates of the probability of return after k periods, for $k \geq 0$. The probability p that a product is eventually returned is given by $\sum_{k=0}^{\infty} \nu_k$. Goh and Varaprasad (1986) use this method to estimate the return quantities of Coca-Cola bottles.

In practice, the data is augmented in each period as new sales and return information becomes available. The incremental nature of the information received makes Bayesian estimation a natural choice. Toktay et al. (2000) assume that the return process can be modeled by Equation (3.2) where ν_k has the structure $p \cdot r_k$; here p denotes the probability that a product will ever be returned and r_k denotes the probability that the product will be returned after k periods, conditional on ever being returned. In other words, if a product was sold in period τ, the probability that it comes back in period $\tau + k$ is modelled as $p \cdot r_k$.

The type of relation in Equation (3.2) is referred to as a 'distributed lag model' in Bayesian inference (Zellner, 1987). Usually, a specific form of distribution involving one or two parameters is assumed for the lag, which reduces the number of parameters to be estimated. The estimation procedure for a geometrically distributed lag with parameter q (the probability that a sold product is returned in the next period, given that it will be returned; $r_k = q(1-q)^{k-1}$) is illustrated in Toktay et al. (2000). They apply this method to data obtained from Kodak on sales and returns of single-use flash cameras.

Item-level Information

When items are tracked on an individual basis, it is possible to observe the actual return delay of returned items. In a given period τ, define $v_{\tau,\tau-i}$ as the number of items sold in period $\tau - i$ that have been returned up to and including period τ, where $i > 0$. For these items, the return delay is known exactly. For items that have not been returned yet, it is known that the delay is longer than the elapsed time, or possibly infinite (corresponding to a product never being returned). A simple sample average of this sample data would give biased estimates due to the right-censoring of the data. Dempster et al. (1977) introduce the Expectation Maximization (EM) algorithm to compute maximum likelihood estimates given incomplete samples. This algorithm can be effectively used to estimate the return delay distribution using censored delay data. The EM algorithm is illustrated in Toktay et al. (2000) for geometric and Pascal delay distributions.

3.3.2 Forecasting Returns

Given past sales volumes and estimates of the return probability and the return delay distribution, it is possible to forecast future returns based on a number of different information sets. Denote by $\mathcal{I}(\tau)$ the information available at the end of period τ that will be used for forecasting future returns, and by $\hat{\nu}(\tau)$ the period-τ estimate of the vector ν. In particular, Kelle and Silver (1989b) define

- $\mathcal{I}_A(\tau) = \{\hat{p}(\tau)\}$ (estimate of p),
- $\mathcal{I}_B(\tau) = \{\hat{\nu}(\tau), \{s(\tau - i), i = 0, 1, \ldots, \tau\}\}$ (estimate of ν and historical period-level sale information),
- $\mathcal{I}_C(\tau) = \{\hat{\nu}(\tau), \{s(\tau - i), u(\tau - i), i = 0, 1, \ldots, \tau\}\}$ (estimate of ν and historical period-level sale and return information), and
- $\mathcal{I}_D(\tau) = \{\hat{\nu}(\tau), \{s(\tau - i), v_{\tau,\tau-i}, i = 0, 1, \ldots \tau\}\}$ (estimate of ν and historical item-level sale and return information).

Define the random variable $y_{\tau-i,\tau+j}$ as the number of returns in period $\tau + j$ originating from sales in period $\tau - i$. Under the assumption that all period demands are mutually independent and returns from different demand issues are not correlated, Table 3.2 lists the expressions for $E[y_{\tau-i,\tau+j}|\mathcal{I}(\tau)]$ based on one of the information sets $\mathcal{I}_B(\tau)$, $\mathcal{I}_C(\tau)$, or $\mathcal{I}_D(\tau)$. The first two columns of the table forecast future returns based on past and current sales. The third column forecasts future returns based on future sales (and uses the period-τ estimate of ν). In this table, $\hat{c}(i,j)$ is a factor that takes into account the correlation between the observed returns to date and the future returns. An exact expression for $\hat{c}(i,j)$ is not available in general, but Kelle and Silver develop an approximation. The variance of future returns can also be calculated, although these expressions are slightly more complicated (see Kelle and Silver, 1989b). The total number of returns in period $\tau + j$ is simply given as $u_{\tau+j} = \sum_{i=-(j-1)}^{\tau} y_{\tau-i,\tau+j}$, the mean and variance of which can be calculated from those of its mutually independent elements. Note that information set $\mathcal{I}_A(\tau)$ is excluded from Table 3.2. This is because the limited information contained in this information set precludes forecasting at this level of detail. To use $\mathcal{I}_A(\tau)$, we assume that sales over a given time horizon will be returned within the same time horizon. In particular, $E[u_{\tau+j}|\mathcal{I}_A(\tau)] = \hat{p}(\tau)E[s(\tau+j)]$.

There are two recent papers that go beyond using past sales and returns data to forecast future returns. As mentioned in Section 3.2, Hess and Mayhew (1997) develop a regression model for commercial returns that incorporates explanatory variables in predicting the cumulative return rate over time.

Table 3.2. The expectation of $y_{\tau-i,\tau+j}$, the number of returns in period $\tau+j$ from sales in period $\tau-i$, for various information sets

Information set	$\mathrm{E}[y_{\tau-i,\tau+j}\|\mathcal{I}(\tau)]$		
	$i > 0$	$i = 0$	$i < 0$
$\mathcal{I}_B(\tau)$	$\hat{\nu}_{j+i}(\tau)s(\tau-i)$		$\hat{\nu}_{j+i}(\tau)\mathrm{E}(s(\tau-i))$
$\mathcal{I}_C(\tau)$	$\hat{\nu}_{j+i}(\tau)s(\tau-i)+\hat{c}(i,j)$	$\hat{\nu}_j(\tau)s(\tau)$	$\hat{\nu}_{j+i}(\tau)\mathrm{E}(s(\tau-i))$
$\mathcal{I}_D(\tau)$	$\dfrac{\hat{\nu}_{j+i}(\tau)}{1-\sum_{k=1}^{i}\hat{\nu}_k(\tau)}(s(\tau-i)-v_{\tau,\tau-i})$	$\hat{\nu}_j(\tau)s(\tau)$	$\hat{\nu}_{j+i}(\tau)\mathrm{E}(s(\tau-i))$

Marx-Gómez et al. (2002) develop a fuzzy inference system for the forecasting of returns. This approach develops a rule base using expert knowledge and then refines the rule base through training on data.

3.4 Performance Measurement

The previous section introduced some forecast models and related return forecasting methods. In this section, we compare the performance of these forecasting methods in a specific context, that of inventory management.

Although there are many papers on inventory management using returned products (Chapter 8), very few consider the joint forecasting and inventory management of returned products (Kelle and Silver, 1989a, b; Kiesmüller and van der Laan, 2001; and Toktay et al., 2000).

Under fixed ordering costs, Kelle and Silver (1989a) formulate a deterministic dynamic lot-sizing problem taking into account future returns in net demand forecasts. These forecasts are generated by using methods developed in Kelle and Silver (1989b). The impact of future returns is that net demand may be negative. The authors develop a transformation into the non-negative demand case. The Wagner-Whitin deterministic lot-sizing procedure can then be applied to determine procurement quantities in each period.

Kiesmüller and van der Laan (2001) do not focus on forecasting as such, but explicitly model the dependence between the demand and return process by a fixed lag. They conclude that neglecting the dependency structure leads to reduced system performance. Using a simple forecast of expected future returns, depending on the number of items sold in the past and the return probability, considerably improves performance.

Toktay et al. (2000) model a reverse logistics supply chain using a closed queueing network. They develop and implement adaptive estimation and con-

Table 3.3. Supplementary notation for Section 3.4

$ND(\tau)$	Net demand (demand minus returns) in period τ
L	Replenishment lead time
$ND_L(\tau)$	Net demand during lead time (periods $\tau + 1, \tau + 2, \ldots \tau + L$)
$S(\tau)$	Base stock level at the end of period τ
$O(\tau)$	Order quantity at the end of period τ
k	Safety factor
h_s	Holding cost per product per period
c_b	Stockout penalty per occurrence
μ_D	Mean of demand per period
σ_D	Standard deviation of demand per period
$\sigma_{O(\tau)}$	Standard deviation of the order quantity in period τ
$\sigma_{ND(\tau)}$	Standard deviation of the net demand in period τ

trol methods to dynamically determine the procurement quantities of new components. This is the only paper to compare the impact of using different dynamic forecast update mechanisms on supply chain performance.

This section expands on the performance analysis of reverse logistics supply chains. In particular, Section 3.4.1 investigates the robustness of the system cost to parameter estimation error in an inventory management setting. The first part of this discussion is based on de Brito and van der Laan (2002) who investigate the impact of using static but erroneous parameter estimates. The second part extends the discussion to the impact of using dynamically updated parameter estimates. The analysis in Section 3.4.1 is for a single stage. At a supply chain level, the order variability induced in each stage is known to impact the overall performance of the system. This angle has not been studied in a reverse logistics setting. In Section 3.4.2 we analyze the performance of the forecasting methods of Section 3.3.2 with respect to order variability. The additional notation used in this section is listed in Table 3.3.

3.4.1 Robustness of Expected Cost to Parameter Estimation Error

We consider a single-product, periodic-review inventory system with no fixed ordering cost and linear holding and backorder costs. Each individual demand returns according to a distribution ν. The replenishment lead time is a fixed constant L. Demands that cannot be satisfied immediately are fully backordered. In this setting, a simple base-stock policy is optimal when the expectation and variance of the net lead-time demand, $ND_L(\tau) \doteq D_L(\tau) - R_L(\tau)$, are known for each period τ (Kelle and Silver, 1989b). If $ND_L(\tau)$ has a normal distribution, the base-stock level is defined as $S = \mathrm{E}[ND_L(\tau)] + k \cdot \sqrt{\mathrm{Var}[ND_L(\tau)]}$, where $\mathrm{E}[ND_L(\tau)]$ and $\mathrm{Var}[ND_L(\tau)]$ are the expectation and variance of the net demand during the replenishment lead time, respectively (Silver et al., 1998, Ch. 7). The safety factor k is determined according to the desired performance level.

Kelle and Silver (1989b) develop normal approximations for $ND_L(\tau)|\mathcal{I}(\tau)$, the conditional net demand over the replenishment lead time L (periods $\tau+1, \tau+2, ..., \tau+L$), given information up to and including period τ. Kelle and Silver propose using information sets $\mathcal{I}_A(\tau), \mathcal{I}_B(\tau), \mathcal{I}_C(\tau)$ and $\mathcal{I}_D(\tau)$ defined in Section 3.3.2. We refer the reader to Kelle and Silver for the mathematical expressions for $E[ND_L(\tau)|\mathcal{I}(\tau)]$ and $Var[ND_L(\tau)|\mathcal{I}(\tau)]$ under the four information sets. Denote by $S(\tau)$ the base-stock level in period τ. Then

$$S(\tau) = E[ND_L(\tau)|\mathcal{I}(\tau)] + k\sqrt{Var[ND_L(\tau)|\mathcal{I}(\tau)]} \qquad (3.3)$$

and the order quantity in period τ, $O(\tau)$, is given by

$$O(\tau) = \begin{cases} S(\tau) - S(\tau-1) + ND(\tau) \text{ if } & S(\tau-1) - ND(\tau) < S(\tau), \\ 0 & \text{otherwise.} \end{cases} \qquad (3.4)$$

Static Parameter Estimates

De Brito and van der Laan (2002) use simulation to compare the cost performance of the ordering policy given by Equation (3.4) under information sets A-D when parameter estimates are static and erroneous, that is, $\hat{\nu}(\tau) = \hat{\nu} \neq \nu$.

The simulation experiments are conducted in the following manner. In period τ, the cumulative demand $D(\tau)$ is drawn from a normal distribution with mean μ_D and standard deviation σ_D (values are rounded to integers and negative numbers are treated as zero). For each individual item of this cumulative demand, the return probability of the item is drawn from a Bernoulli distribution with parameter p; for those items that will be returned, the time to return is drawn from a geometric distribution with parameter q (expected return time $1/q$). In this case, estimating the vector ν reduces to estimating the parameters p and q since $\nu_k = p \cdot (1-q)^{k-1}q$. In each period, estimates of the expectation and variance of the net demand during the replenishment lead time are computed using these parameters in each of the Methods A–D. These estimates are subsequently used to compute the order size using Equations (3.3) and (3.4). Assuming that the net demand during lead time is normally distributed, the cost optimal value of the safety factor, k^*, satisfies (Silver et al., 1998, Ch. 7)

$$G(k^*) = 1 - \frac{h_s}{c_b}, \qquad (3.5)$$

where $G(.)$ is the standard normal distribution, h_s is the holding cost per item in overstock at the end of each period, and c_b is the stockout penalty per occurrence. At the end of each simulation experiment, the total average cost per period is calculated as the total average holding plus stockout costs per period. More details regarding the simulation setup can be found in de Brito and van der Laan (2002).

In order to investigate the effect of inaccurate parameter information, two types of errors are considered in the parameter estimates. The first is the inaccurate estimation of the overall return probability, p, while the shape of the return distribution is preserved. In other words, estimates $\hat{p}(\tau) = \hat{p} \neq p$ are used in each period in calculating $S(\tau)$. The second is the inaccurate estimation of the expected time-to-return, $1/q$: Estimates $\hat{q}(\tau) = \hat{q} \neq q$ are used in calculating $S(\tau)$. This affects the estimated shape of the time-to-return distribution, but the estimated overall return probability is kept equal to the real return probability.

Inaccurate estimation of the return probability

In de Brito and van der Laan (2002), it is shown that under perfect estimation $(\hat{p} = p, \hat{q} = q)$, Method D is superior to the other methods. This is expected since this method uses all available information correctly. The differences with respect to Methods B and C are rather small unless the return probability is very high. The performance of Method A on the other hand is reported to be very poor. Its naive forecasting procedure gives a rather inaccurate forecast for the mean and variance of the lead time net demand, especially for high return rates and large lead times. As a consequence, Equation (3.5) provides a safety factor that is far from optimal. Due to its poor performance, we will not consider this method in the remainder of this section.

In the case of the inaccurate estimation of the return probability, the authors show that cost increases may occur. For example, an estimation error of 10% in p ($\frac{|\hat{p}-p|}{p} = 0.1$) may lead to a cost increase of 10–30% for Methods B-D; a 20% estimation error may even lead to a cost increase of more than 200%

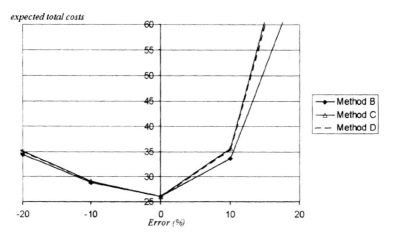

Fig. 3.1. Impact of return probability estimation error
$(\mu_D = 30, \sigma_D = 6. L = 4, p = 0.5, q = 0.6, h_s = 1. c_b = 50)$

relative to the perfect information case (Figure 3.1). Clearly, overestimation of the return probability is far worse than underestimation, since overestimation leads to costly stockouts.

The study furthermore shows that Method B structurally outperforms the other two methods in case of an estimation error of 10% or more, whereas Method C always performs worse. The differences between the methods increase as the return probability increases or the lead time decreases.

Inaccurate estimation of the expected time-to-return

According to de Brito and van der Laan (2002), the inaccurate estimation of the expected time-to-return has little effect if return rates are small. However, as can be observed in Figure 3.2, for $p = 0.8$, Methods B and C may be far more robust with respect to estimation error than the benchmark, Method D. As with the inaccurate estimation of the return probability, the differences in performance with respect to the perfect-information benchmark are reported to increase with the return probability and decrease with the lead time.

Managerial Implications

Since the performance of Method A is very poor in general, this method is not recommended for practical implementation. Method B appears to have a sufficient level of sophistication under perfect information. The cost differences with respect to Methods C and D seem to be too small to justify large investments in data recording and analysis. Furthermore, Method B seems to be much more robust under inaccurate estimation. Method B systematically

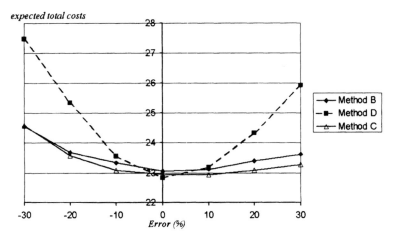

Fig. 3.2. Impact of expected time-to-return estimation error
($\mu_D = 30, \sigma_D = 6, L = 4, p = 0.8, q = 0.6, h_s = 1, c_b = 50$)

outperforms Methods C and D if the return probability estimation error is 10% or more. The cost differences are considerable if the return probability is overestimated. In general, it is better to underestimate the return probability than to overestimate it, since stockouts are usually much more costly than overstocks.

Dynamically Updated Parameter Estimates

De Brito and van der Laan (2002) assume that parameter estimates are erroneous but static. In practice, it may be that return flow parameters are updated over time. In this case, the estimation error would be initially high, but the estimates would eventually converge to the true parameter values. In addition, the speed of convergence of parameter estimates would depend on the nature and volume of the information collected. Consequently, in choosing a particular method, one needs to strike a balance between information requirements and the need for accuracy of the inventory method in question. This section addresses this trade-off.

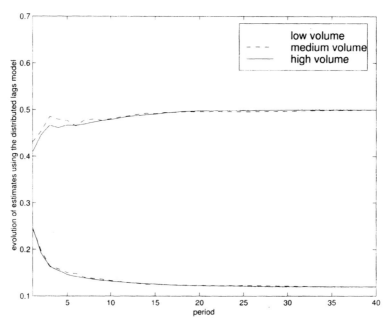

Fig. 3.3. The evolution of return delay parameters using the distributed lags model. The top (bottom) three lines plot the evolution of the estimate of the return probability p ((expected return delay)$^{-1}$ q). The true values of p and q are 0.5 and 0.125, respectively. The number of sales in each period is a Poisson random variable with parameter 200, 2,000, and 20,000, respectively, labelled as low-, medium-, and high-volume scenarios, respectively.

Methods B, C, and D all require an estimate of the return distribution. To generate these estimates, information on past sales and returns per period is required. Therefore, one can eliminate the naive estimation introduced in Section 3.1 from consideration, since it only generates an estimate of the return probability. On the other hand, the distributed lag Bayesian inference model and the EM algorithm both generate the estimates required by these methods, with the former necessitating only period-level information (as do forecasting Methods B and C), and the latter necessitating item-level information (as does forecasting Method D). Toktay (2003) compares the convergence rates in these two models across different demand volumes, as reproduced in Figures 3.3 and 3.4.

With period-level data, the convergence of the estimate depends primarily on the number of periods of data available: in Figure 3.3, the estimate of the return probability converges after eighteen periods of returns for all sales volumes. The sales volume does not impact the point at which the estimate

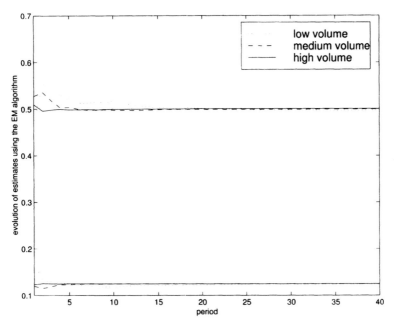

Fig. 3.4. The evolution of the return delay parameters using the Expectation Maximization algorithm. The top (bottom) three lines plot the evolution of the estimate of the return probability p ((expected return delay)$^{-1}$ q). The true values of p and q are 0.5 and 0.125, respectively. The number of sales in each period is a Poisson random variable with parameter 200, 2000, and 20000, respectively, labelled as low-, medium- and high-volume scenarios, respectively.

converges, but it is significant in determining the accuracy of the method in the periods up to that point.

Figure 3.4 shows that the speed of convergence of the EM algorithm does depend on the sales volume per period: in this example, two periods, five periods, and twenty periods, respectively, are needed for the confidence interval of the return probability estimate to include the true value of the parameter in the high-, medium-, and low-volume scenarios, respectively. While it is to be expected that the accuracy of the estimate in the EM algorithm directly depends on the volume of data, it is particularly striking that the algorithm achieves such accuracy after only two periods in the high-volume scenario.

Managerial Implications.

As expected, the EM algorithm clearly outperforms Bayesian inference with a distributed lags model. This is because item-level information is present in the former. However, note that for medium- and high-volume scenarios, the latter method underestimates the return probability, while the former overestimates it. Combining this observation with results of de Brito and van der Laan (2002) that state that overestimation is much more costly than underestimation, we conclude that using the latter method which requires only period-level information may in fact be appropriate for higher-volume products. For low-volume items, investing in collecting item-level information does appear to be beneficial since the EM algorithm is more robust, where a method is said to be more robust if the maximum deviation from the true parameter value over time is lower.

3.4.2 Propagation of Order Variability in Reverse Logistics Supply Chains

Section 3.4.1 considered the impact of erroneous parameter estimates on inventory costs in a single-stage model and discussed the robustness of proposed return forecasting methods to parameter estimation errors. This analysis does not focus on the impact of ordering decisions given by Equation (3.4) on the upstream stages.

One way of measuring the impact of local decisions on upstream stages is to measure the variance of orders placed to the upstream stage. It is well known that order variability at any stage of a forward supply chain tends to be larger than the demand variability; this phenomenon is called the "bullwhip effect." In the literature, several reasons have been suggested to explain the existence of the bullwhip effect in forward supply chains (Lee et al., 1997). In particular, when demand is correlated over time and the base-stock level is dynamically updated, the bullwhip phenomenon is observed. If demand is iid (independently and identically distributed), but its parameters are dynamically updated based on historical data, the bullwhip effect is again observed. However, when demand is iid with known parameters in a forward supply chain, no bullwhip effect is observed. To see this, note that in a forward supply chain, $ND_L(\tau) = D_L(\tau)$ since there are no returns. Therefore, the base-stock level is given by $E[D_L(\tau)] + k\mathrm{Var}[D_L(\tau)]$ and is static when demand is iid with known parameters. In this case, $O(\tau) = D(\tau)$, which is a pure pull system, and the order variance equals the demand variance, in other words, there is no bullwhip effect.

In reverse logistics supply chains, even when demand and return flow parameters are known, and demand is iid, the order-up-to level (3.4) may fluctuate since it is based on estimated lead time *net* demand. As a consequence, the variability of orders will differ from the variability of demands even when demands are iid and its parameters are known.

Table 3.4 compares the ratio of the standard deviation of orders to the standard deviation of demands $(\sigma_{O(\tau)}/\sigma_{ND(\tau)})$ across Methods B, C, and D, and according to the replenishment lead time L, the demand standard deviation σ_D, the return probability p, and the expected time to return $(1/q)$. As with forward supply chains, the magnification in (net) demand variability increases in the replenishment lead time and in the underlying demand variability. We observe that Method C consistently dominates Methods B and D in our simulations. This is because it makes better use of the period-level data than Method B by including return information, and it is more robust than Method D in its calculation of lead-time net demand because it uses aggregate information rather than item-level information. On the other hand, the relative performance of the methods changes depending on the underlying parameters: the largest differences in performance are seen at high return probabilities, at long return delays, and at low demand variances.

Managerial Implications.

Our recommendation concerning minimizing the bullwhip effect in reverse logistics supply chains is to use Method C since it generates the lowest order variability. At the same time, Method B (that requires less information) can be used without generating a much larger order variability when the replenishment lead time is short, the demand variance is high, the return probability is low, and/or the return delay is short.

3.5 Conclusion

In this chapter, we discussed the information needs of a number of strategic, tactical, and operational decisions concerning return flow characteristics and touched on the concept of actively influencing returns. We then reviewed methods for forecasting future returns based on historical sales and returns in two blocks: developing a return delay model and estimating its parameters, and forecasting future returns. Finally, we turned to an operational level decision process: inventory control. The inventory management literature for

Table 3.4. Comparison of the ratio of the standard deviation of orders to the standard deviation of net demand ($\frac{\sigma_{O(\tau)}}{\sigma_{ND(\tau)}}$) with respect to L, σ_D, p, and q when parameters are known and as given in the first column ($\mu_D = 30, h_s = 1, c_b = 50$). All other parameters are known and are as in the base case. The relative performance is calculated with respect to the benchmark case C, and shows how much higher the ratio $\frac{\sigma_{O(\tau)}}{\sigma_{ND(\tau)}}$ is with respect to that of the base case.

	C	B		D	
(L, σ_D, p, q)	$\sigma_{O(\tau)}/\sigma_{ND(\tau)}$	$\sigma_{O(\tau)}/\sigma_{ND(\tau)}$ rel. perf.		$\sigma_{O(\tau)}/\sigma_{ND(\tau)}$ rel. perf.	
$(\mathbf{2}, 6, 0.5, 0.6)$	1.89	1.99	5.63%	2.08	10.56%
$(\mathbf{4}, 6, 0.5, 0.6)$	1.93	2.06	6.88%	2.15	11.62%
$(\mathbf{8}, 6, 0.5, 0.6)$	1.94	2.08	7.11%	2.16	11.62%
$(4, \mathbf{4}, 0.5, 0.6)$	1.66	1.81	8.99%	1.92	15.62%
$(4, \mathbf{6}, 0.5, 0.6)$	1.93	2.06	6.88%	2.15	11.62%
$(4, \mathbf{8}, 0.5, 0.6)$	2.09	2.21	6.00%	2.29	9.87%
$(4, 6, \mathbf{0.5}, 0.6)$	1.93	2.06	6.88%	2.15	11.62%
$(4, 6, \mathbf{0.7}, 0.6)$	2.75	3.16	14.88%	3.36	22.31%
$(4, 6, \mathbf{0.9}, 0.6)$	5.00	6.47	29.33%	6.92	38.24%
$(4, 6, 0.5, \mathbf{0.4})$	1.84	1.95	6.35%	2.21	20.40%
$(4, 6, 0.5, \mathbf{0.6})$	1.93	2.06	6.88%	2.15	11.62%
$(4, 6, 0.5, \mathbf{0.8})$	2.05	2.15	5.25%	2.05	0.35%

remanufacturable products, for the most part, develops the inventory control tools without analyzing the combined performance of the tool and the forecasting methods that drive it. This chapter reports research that attempts to do this and takes it one step further by considering supply-chain performance through an analysis of order variability in reverse logistics.

There are several trade-offs to be made in choosing the parameter estimation model and the forecasting method. Specifically, the data requirements of methods using period-level information are lower than those using item-level information. However, the parameter convergence rate of period-level forecast models is lower, leading to a longer time over the total life-cycle of the product where the parameter estimates are inaccurate. Similarly, comparing forecasting methods requiring differing levels of information (in particular, the four methods proposed by Kelle and Silver. 1999b). we see that these methods differ in their cost performance under inaccurate parameter estimation and in the order variability that they generate. Noting that the estimation method chosen will dictate the extent and duration of the parameter inaccuracy, we conclude that convergence rate and robustness to parameter inaccuracy should be jointly considered.

We first summarize the results of the individual analysis. The single-stage cost performance of Method A that uses only an estimate of the return probability in forecasting and ordering is very poor and is not recommended for practical implementation. Again from a single-stage cost perspective, Methods B, C, and D yield similar cost (with D dominating) under correct parameter information, despite the fact that Method D requires item-level information and Methods B and C require only period-level information. On the other hand, when parameters are unknown and estimated inaccurately, Method D is not the best method. In particular, Method B is more robust under inaccurate estimation of the return probability, with Methods C and D having similar performance. Methods B and C are much more robust under the inaccurate estimation of the expected time to return than Method D. Finally, Methods C, B, and D give rise to increasing levels of order variability, in that order. We also find that for low sales volumes, item-level information clearly dominates period-level information from a parameter convergence perspective, whereas for higher sales volumes, period-level information may be sufficiently accurate.

To conclude, we synthesize these findings in the light of the trade-offs discussed above. In particular, jointly considering these trade-offs leads us to conclude that different methods may be appropriate at different time points and different product volumes. While parameter error is high, it may be worth using Method B due to its superior single-stage cost performance despite the fact that it generates a higher order variance. After parameter estimates converge, switching to Method C is recommended since its cost performance is good and it generates the lowest order variability. For products with low sales volumes, if period-level information is used, parameter estimation error could persist for a longer portion of the life-cycle of the product. Therefore, for such products, we recommend investing in collecting item-level information for parameter estimation, at least initially, and using Methods B and C, in that order, for forecasting. For products with high sales volumes, such an investment is not recommended. Instead, collecting and using only period-level information for both parameter estimation and forecasting is recommended.

Research Directions

The poor behavior of the simplest method, Method A, is mainly due to the fact that the reported optimal safety factor k^* is inappropriate because the net demand during lead time is poorly estimated. Adjusting this value would considerably improve Method A's performance. Another option is to use Method A, but assume that demands and returns are fully uncorrelated instead of fully correlated. This has the advantage that net demand is overestimated rather than underestimated, reducing costly stockouts. Recall that the impact of inaccurate information about the return distribution seems to be rather limited. Based on this observation, one could construct a method whose performance lies between Method A and B by using Method B with a pre-specified return distribution (for instance, the uniform distribution) rather than an estimated one. This considerably reduces the information need, while we expect to maintain a reasonable performance.

An important assumption with respect to the results on the impact of inaccurate estimation is that the system parameters do not change over time. It would be interesting to study the impact of inaccurate estimation in a dynamic environment where the underlying parameters to be estimated change over time, in which parameter forecasts are continuously updated, but not necessarily converging due to changing parameters.

Most contributions in this field have focused on applications in inventory control. Other fields that address operational-level decisions, such as production planning, warehousing, etc., may be affected in a similar way and call for further research on the value of information.

Finally, as discussed in Section 3.2, there is little research on identifying factors that significantly influence return flow characteristics. Developing a good understanding of drivers of return flow characteristics would enable better decision making for influencing return flows.

Acknowledgements

We thank Atalay Atasu for coding the simulations generating Table 3.4.

4

Reverse Logistics Network Design

Moritz Fleischmann[1], Jacqueline M. Bloemhof-Ruwaard[1], Patrick Beullens[2], and Rommert Dekker[3]

[1] Rotterdam School of Management / Faculteit Bedrijfskunde,
   Erasmus University Rotterdam, P.O. Box 1738, 3000 DR Rotterdam,
   The Netherlands, `MFleischmann@fbk.eur.nl`, `JBloemhof@fbk.eur.nl`
[2] K.U. Leuven, Centre for Industrial Management, Celestijnenlaan 300 A,
   3001 Leuven, Heverlee, Belgium, `patrick.beullens@cib.kuleuven.ac.be`
[3] Rotterdam School of Economics, Erasmus University Rotterdam, P.O. Box 1738,
   3000 DR Rotterdam, The Netherlands, `rdekker@few.eur.nl`

4.1 Introduction

The different chapters of this book highlight manifold examples of reverse logistics programs. While these cases vary substantially with respect to products, actors, and underlying motivations, as discussed in Chapter 1, they share a number of fundamental managerial issues. One of these is the need for an appropriate logistics infrastructure.

Analogous with traditional supply chains, the various transformation processes that turn a returned item into a remarketable good need to be embedded in a corresponding logistics network. In conventional supply chains, logistics network design is commonly recognized as a strategic issue of prime importance (see, e.g. Chopra and Meindl, 2001, Chapter 3). The location of production, storage, and cross-dock facilities, and the selection of transportation links between them, are major determinants of supply chain performance. Analogously, logistics network design has a fundamental impact on the profitability of reverse logistics systems. In order to maximize the value recovered from used products, companies need to set up logistics structures that facilitate the arising goods flows in an optimal way. To this end, one needs to decide where to locate the various processes of the reverse supply chain, as introduced in Chapter 1, and how to link them in terms of storage and transportation. In particular, companies need to choose how to collect recoverable products from their former users, where to inspect collected products in order to separate recoverable resources from worthless scrap, where to re-process collected products to render them remarketable, and how to distribute recovered products to future customers.

In this chapter, we take a detailed look at logistics network design in a reverse logistics context. We start in Section 4.2 by highlighting key business issues and contrasting them with logistics network design for traditional 'forward' supply chains. The core part of the chapter then discusses a number of alternative modeling approaches that support the design of reverse logistics networks and allow for a quantitative analysis of the underlying tradeoffs. Section 4.3 addresses integer-programming-based approaches that build upon traditional facility location models. Section 4.4 considers stochastic programming approaches that focus on incorporating the aspect of uncertainty into the network design decisions, and Section 4.5 presents a stream of research based on continuous approximation techniques. Section 4.6 synthesizes the different modeling approaches by exploring them in a common numerical example that highlights the economics of reverse logistics networks. To conclude, Section 4.7 summarizes the key points of this chapter. Before embarking into a systematic analysis of reverse logistics network design, we illustrate some of the main issues in a real-life business example in the remainder of this section.

4.1.1 Illustrative Case: Reverse Logistics Flows at IBM

The electronics industry has been a key sector in the growth of product recovery management. Ever expanding market volumes on the one hand, and shorter product lifecycles on the other, result in huge amounts of used products being disposed of. In this light, it comes as no surprise that electronic waste has been a prime target of environmental regulation, as reflected in enacted or pending take-back obligations in several countries (see also Chapter 15). At the same time, modular product design and a relatively small extent of mechanical 'wear and tear' sustain the reusability of electronic products and components. Together, both developments result in significant value recovery potential.

Business activities of IBM, as one of the major players in this sector, involve several types of 'reverse' product flows, which together cover most of the categories outlined in Chapter 2. From a business perspective, the most important class concerns product returns from expiring lease contracts. To date, leases account for some 35% of IBM's total hardware sales. Furthermore, IBM has implemented take–back programs in several countries in North America, Europe, and East Asia, which allow business customers to return used products for free or for a small fee. For remarketable products, customers may even receive a positive contribution. In the consumer market, IBM is required to take back end–of–life products in several countries in Europe and East Asia, due to environmental regulation. Besides dealing with used products, IBM, as do most companies, faces a 'reverse' stream of new products, which includes retailer overstocks and canceled orders. This flow very much depends on contractual agreements along the supply chain (see also Chapter 12). Finally, it is worth mentioning returns of rotable spare parts as a fairly traditional type of closed-loop flow related to the service business: defective parts replaced in a customer's machine are sent back for repair and may then be stocked as spare parts again (IBM, 2001; Fleischmann, 2001).

Recognizing the growing importance of reverse logistics flows, IBM set up a dedicated business unit in 1998, which is responsible for managing all product returns worldwide. The main goal of this organization, named Global Asset Recovery Services (GARS), is to manage the dispositioning of returned items to maximize the total value recovered. To this end, GARS operates some 25 facilities all over the globe where returns are collected, inspected, and assigned to an appropriate recovery option (see IBM, 2001). Equipment that is deemed remarketable may be refurbished and then put into the market again. For this purpose, IBM operates nine refurbishment centers worldwide, each dedicated to a specific product range. Internet auctions, both on IBM's own website and on public sites, have become an important sales channel for remanufactured equipment. Equipment that does not yield a sufficient value as a whole is sent to a dismantling center in order to recover valuable components, such as hard–disc assemblies, cards, and boards, which can be fed into IBM's spare parts network or sold on the open market (for a detailed description of this channel, see Fleischmann, 2001 and Fleischmann et al., 2003). The remaining returned equipment is broken down into recyclable material fractions, which are sold to external recyclers. In 2000, IBM reports the processing of 51,000 t of used equipment, of which only a residual of 3.2% was landfilled.

The above processes concern equipment from the business market. Given the much lower market value, consumer returns follow a different road. To work around inefficiencies of individual collection, IBM participates in cooperative, industry-wide solutions for this market sector in several countries. In the Netherlands, for example, IBM supports a system organized by the Dutch association of information and communication technology producers, in response to national product take–back legislation. In this case, used machines from different manufacturers are collected by the municipalities and then shipped to recycling subcontractors. Transportation and recycling costs are shared by the member organizations, proportional to their products' volume contribution (see Nederland ICT, 2002). Yet another system has recently been implemented in the USA. Since November 2000, IBM customers have the option to purchase a recycling service together with any new PC. Once the equipment is no longer needed, the customer sends it by UPS to a dedicated recycling center where it is either prepared for donation to charities or broken down into recyclable materials (IBM, 2000).

4.2 Network Design Issues in Reverse Logistics

4.2.1 Delineation of Reverse Logistics Networks

The above example underscores the need for a logistics infrastructure that accumulates used products and conveys them to recovery facilities and eventually to another user. In general terms, such a structure can be viewed as the logistics link between two market interfaces, which provide a supply of used products and demand for reusable products, respectively. Moreover, this link encompasses the reverse channel functions highlighted in Chapter 1, namely collection, testing and sorting, re–processing, and re–distribution. Figure 4.1 depicts a general scheme of this perspective. It is worth pointing out that the two markets involved may coincide, thereby implying a closed–loop network.

From a logistics perspective, one may characterize the structure illustrated in Figure 4.1 as a many–to–many distribution network. Within this layout, one may distinguish a convergent inbound part corresponding to the collection and acquisition function and a divergent outbound part serving a distribution function. The intermediate part of the network hosts the actual transformation processes. Therefore, its structure very much depends on the type of re–processing involved.

One may argue that it is only the inbound part of the network that actually concerns 'reverse' logistics processes, whereas the remainder very much corresponds with a traditional production–distribution network. However, as discussed in Chapter 1, this segregation may hamper a comprehensive analysis since the different product flows are closely interrelated. In fact, in this light one may wish to extend the scope even further and also include the distribution of the original new products (see Figure 4.1). It should be clear that this broad scope does not mean that the entire network is, or should be, managed by a single company. As in a traditional supply chain context, responsibilities may be allocated to multiple players. However, in line with the supply chain management imperative, one should consider the complete picture in order to understand the economics of reverse logistics networks.

Within the above setup, examples of reverse logistics networks are far from identical. Significant differences concern, for example, the players involved and

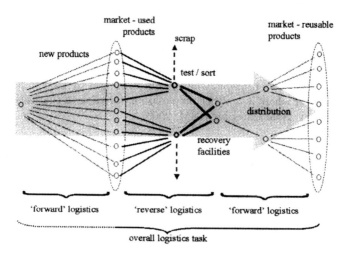

Fig. 4.1. Reverse Logistics Network Structure (adapted from Fleischmann, 2003)

their responsibilities, but also the network structure in terms of centralization and the number of layers. In the literature, several classifications have been proposed for structuring this field.

A first stream of research focuses on the reverse channel structure and the roles of the different players involved. In this vein, Fuller and Allen (1995) distinguish (1) manufacturer–integrated systems, (2) waste–hauler systems, (3) specialized reverse dealer–processor systems, (4) forward retailer–wholesaler systems, and (5) temporary/facilitator systems. The analysis extends earlier suggestions by Guiltinan and Nwokoye (1975) and Pohlen and Farris (1992).

In a different perspective, Bloemhof–Ruwaard and Salomon (1997) and Fleischmann et al. (2000) attribute differences between reverse logistics networks primarily to the form of the recovery process. The authors then distinguish three network types, namely remanufacturing, recycling, and direct reuse networks. Fleischmann (2003) refines this model by including ownership of the recovery process (original equipment manufacturer (OEM) versus third party) and recovery drivers (economic versus legislative) as additional explanatory variables. Based on this analysis, the paper suggests distinguishing the following five network classes: (1) networks for mandated product take–back, (2) OEM networks for value added recovery, (3) dedicated remanufacturing networks, (4) recycling network for material recovery, and (5) networks for refillable containers.

4.2.2 Characteristics of Reverse Logistics Networks

Strategic design decisions related to the type of logistics networks delineated above include the choice of a collection/acquisition method, the location and capacity of the sorting and re–processing operations and corresponding inventory buffers, and the definition of various transportation links in terms of sourcing, modes, and capacities. When comparing these tasks with the design of a conventional production–distribution network, the network structure may seem the most apparent discriminating factor. As pointed out above, reverse logistics implies a many–to–many structure composed of a convergent and a divergent part (see Figure 4.1), whereas production–distribution networks are typically perceived as few–to–many, divergent structures. However, this difference may be a matter of scope rather than an intrinsic element of reverse logistics. Taking a supply chain perspective, and hence taking into account the supplier level in the design of a 'forward' distribution network, one obtains a picture that very much resembles Figure 4.1.

In contrast, the following three factors are specific of reverse logistics networks in a more fundamental sense:

- supply uncertainty,
- degree of centralization of testing and sorting, and
- interrelation between forward and reverse flows.

In traditional supply chains, demand is typically perceived as the main unknown. In a reverse logistics setting, however, it is the supply side that accounts for significant additional uncertainty. Used products are a much less standardized input resource than conventional component supplies or raw materials. As pointed out in Chapter 1, quantity, quality, and timing of product returns are, in general, not known with certainty and may be difficult to influence. Effectively matching demand and supply, therefore, is a major challenge in reverse logistics. Consequently, robustness with respect to variations in flow volumes and composition is an important prerequisite of reverse logistics networks.

The need for testing and sorting operations in reverse logistics is a direct consequence of the above supply uncertainty. The degree of centralization of this stage has a fundamental impact on the transportation needs in a reverse logistics network and is subject to the following tradeoff: testing collected products early in the channel may minimize the total transportation distance since inspected products can be sent directly to the corresponding recovery operation. In particular, this approach helps avoid excessive transportation of worthless scrap. On the other hand, investment costs, for example for advanced test equipment or specially trained labor, may call for centralizing the testing and sorting operations. There appears to be no direct equivalent to this issue in traditional production–distribution networks as product routings are, in principal, known beforehand in this case. Yet, to some extent the underlying trade-off resembles the effect of risk pooling on inventory location decisions.

Another important characteristic of reverse logistics networks concerns potential synergies between different product flows. While traditional distribution networks typically act as one-way streets, closed–loop chains encompass multiple inbound and outbound flows crossing each other's paths. In this setting, it is intuitive to consider integration as a potential means for attaining economies of scale. Opportunities may concern transportation and facilities. For example, integrating the collection of used products with the distribution function may help reduce empty rides. Similarly, integrating operations of the forward and reverse channel in the same facilities possibly reduces overhead costs. At the same time, these opportunities raise a compatibility issue. In many cases, closed-loop supply chains are not designed in a single step but are realized by adding reverse logistics activities to an existing distribution network. It is not clear, however, whether such a sequential approach yields a good solution or whether one should consider an integral redesign of the entire closed–loop network.

In what follows, we review quantitative models that aim at supporting the above network design decisions. Throughout, we pay particular attention to the aforementioned characteristics of reverse logistics networks and discuss how they are captured in the different modeling approaches. In analogy with traditional network design models, we focus on location–allocation decisions and, to a lesser extent, capacity selection. Transportation and the collection strategy are considered on a rather aggregated level here. Chapter 5 zooms in on the transportation operation in more detail. Similarly, Chapter 6 details warehousing and material handling aspects.

4.3 Mixed Integer Location Models for Reverse Logistics Network Design

4.3.1 Literature Review of Reverse Logistics Location Models

The most widespread modeling approach to logistics network design problems in various contexts concerns facility location models based on mixed integer linear programming (MILP). Throughout the decades, an extensive body of literature has been established that ranges from simple uncapacitated plant location models to complex capacitated multi-level, multi-commodity models. At the same time, powerful solution algorithms have been proposed, relying on combinatorial optimization theory. For a detailed overview of models and solution techniques, we refer to Mirchandani and Francis (1989) and Daskin (1995).

Given this extensive body of research, MILP location models appear to be a natural starting point for quantitative approaches to reverse logistics network design. Several authors have followed this route and have presented MILP location models adapted to a reverse logistics context. Table 4.1 provides an overview of the corresponding literature. We distinguish models that encompass the entire network between the two market interfaces sketched in Figure 4.1 and models with a scope restricted to the 'reverse' network part in a strict sense. Moreover, we indicate whether supply of used products is modeled as a push or a pull process, i.e. whether there is a given collection volume that needs to be processed or whether collection primarily responds to demand.

The summary in Table 4.1 indicates that most of the models published to date address the entire network scope and treat supply as a push process. The model of Kroon and Vrijens (1995), which is applied in the context of reusable packaging, essentially is a conventional uncapacitated warehouse

Table 4.1. Reverse Logistics Facility Location Models

	supply push	*supply pull*
integral network	Kroon and Vrijens (1995)	Realff et al. (1999)
	Thierry (1997)	Jayaraman et al. (1999)
	Spengler et al. (1997)	
	Barros et al. (1998)	
	Marín and Pelegrín (1998)	
	Fleischmann et al. (2001)	
reverse network	Berger and Debaillie (1997)	Krikke et al. (1999)

location model with lateral transshipments. Similarly, Marín and Pelegrín (1998) consider a special case of a warehouse location model where each customer's supply equals a fixed fraction of his demand. Jayaraman et al. (1999) analyze a multi–product variant of this model with general supply and demand volumes. Moreover, the supply process is governed by limited core availability rather than by collection obligations.

Thierry (1997) considers a linear programming model that corresponds with the structure outlined in Figure 4.1 with facility locations being fixed. The disposal volume arising at the testing stage is modeled as a fixed fraction of the volume processed. Berger and Debaillie (1997) include location decisions in this model while at the same time limiting its scope to the 'reverse' network part. Moreover, they model the disposal volume as a lower bound rather than a fixed fraction. Krikke et al. (1999) apply a similar model in a case study on copier remanufacturing. In Fleischmann et al. (2001), we analyze a generalization of Thierry's model, including location decisions. This model is discussed in detail in Section 4.3.2 below.

The model presented by Barros et al. (1998) captures a more detailed picture in that it explicitly includes an alternative recovery path rather than an external scrap process for material rejected at the testing stage. Spengler et al. (1997) and Realff et al. (1999) take an even broader perspective by modeling multi–commodity flows in general processing networks. While both cases are motivated by applications in the process industry, they differ in their view of the supply process. The former considers a supply push in a waste recycling context whereas the latter focuses on the recoverable value of the potentially available supply.

The above contributions exhibit much similarity with traditional multi–level location models. From a mathematical perspective, the particular characteristics of reverse logistics identified in the previous section appear to entail only minor modifications. Specific features include additional flow constraints, reflecting supply restrictions. Other variations are due to multiple return flow dispositions and to a possible interaction between forward and reverse channels. As a consequence, most of the models use multi–commodity flow formulations. In the next section, we discuss these aspects in more depth on the basis of a specific model.

4.3.2 A Basic Facility Location Model

To make things specific, let us take a look at a concrete MILP formulation of a reverse logistics network design problem. To this end, we discuss a variant of the model we introduced in Fleischmann et al. (2001). The model picks up the general scheme sketched in Figure 4.1. Specifically, it encompasses the processes between the two market interfaces discussed in Section 4.2. In this setting, the model considers three levels of facilities for a single type of product, namely test centers, factories, and distribution warehouses. Moreover, it

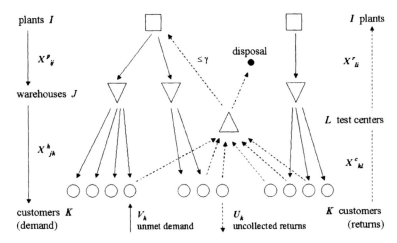

Fig. 4.2. Structure of the Recovery Network Model (adapted from Fleischmann et al., 2001)

includes two generic dispositions for the flow of used products, namely recovery and disposal, where recovery is restricted to a certain maximum yield. Figure 4.2 displays the general structure of this model. The MILP formulation below uses the following notation.

Index sets

$\mathcal{I}$	$=$	set of potential plant locations
$\mathcal{J}$	$=$	set of potential warehouse locations
$\mathcal{K}$	$=$	set of fixed customer locations
$\mathcal{L}$	$=$	set of potential test center locations

Variables

Y_i^p	$=$	indicator opening plant $i \in \mathcal{I}$
Y_j^h	$=$	indicator opening warehouse $j \in \mathcal{J}$
Y_l^r	$=$	indicator opening test center $l \in \mathcal{L}$
X_{ij}^p	$=$	product flow from plant i to warehouse j (in product units)
X_{jk}^h	$=$	product flow from warehouse j to customer k (in product units)
X_{kl}^c	$=$	product flow from customer k to test center l (in product units)
X_{li}^r	$=$	product flow from test center l to plant i (in product units)
V_k	$=$	unsatisfied demand of customer k (in product units)
U_k	$=$	excess supply of used products from customer k (in prod. units)

Costs

f_i^p	$=$	annualized fixed costs for opening plant $i \in \mathcal{I}$
f_j^h	$=$	annualized fixed costs for opening warehouse $j \in \mathcal{J}$

f_l^r = annualized fixed costs for opening test center $l \in \mathcal{L}$

c_{ij}^p = sum of unit production cost at plant i and transportation cost from plant i to customer j

c_{jk}^h = sum of unit handling and storage cost at warehouse j and transportation cost from warehouse j to customer k

c_{kl}^c = sum of unit transportation cost from customer k to test center l and test, inspection, and disposal cost

c_{li}^r = sum of unit transportation cost from test center l to plant i and reprocessing cost minus production cost

c_k^b = unit penalty cost for not serving demand of customer k

c_k^0 = unit penalty cost for not collecting returns of customer k

Parameters

d_k = annual demand of customer $k \in \mathcal{K}$ (in product units)

u_k = annual returns of used products from customer $k \in \mathcal{K}$ (in product units)

γ = average recovery yield

$\bar{p}_i$ = annual capacity of plant $i \in \mathcal{I}$

$\bar{h}_j$ = annual capacity of warehouse $j \in \mathcal{J}$

$\bar{r}_l$ = annual capacity of test center $l \in \mathcal{L}$

We then formulate the general reverse logistics network design (RLND) model as

$$min \,! \quad \sum_{i \in \mathcal{I}} f_i^p\, Y_i^p + \sum_{j \in \mathcal{J}} f_j^h\, Y_j^h + \sum_{l \in \mathcal{L}} f_l^r\, Y_l^r$$

$$+ \sum_{i \in \mathcal{I}} \sum_{j \in \mathcal{J}} c_{ij}^p\, X_{ij}^p + \sum_{k \in \mathcal{K}} \left(c_k^b\, V_k + \sum_{j \in \mathcal{J}} c_{jk}^h\, X_{jk}^h \right)$$

$$+ \sum_{k \in \mathcal{K}} \left(c_k^o\, U_k + \sum_{l \in \mathcal{L}} c_{kl}^c\, X_{kl}^c \right) + \sum_{l \in \mathcal{L}} \sum_{i \in \mathcal{I}} c_{li}^r\, X_{li}^r \qquad (4.1)$$

subject to

$$\sum_{j \in \mathcal{J}} X_{jk}^h + V_k = d_k \qquad \forall k \in \mathcal{K} \qquad (4.2)$$

$$\sum_{l \in \mathcal{L}} X_{kl}^c + U_k = u_k \qquad \forall k \in \mathcal{K} \qquad (4.3)$$

$$\sum_{i \in \mathcal{I}} X_{ij}^p = \sum_{k \in \mathcal{K}} X_{jk}^h \qquad \forall j \in \mathcal{J} \qquad (4.4)$$

$$\sum_{l \in \mathcal{L}} X_{li}^r \leq \sum_{j \in \mathcal{J}} X_{ij}^p \qquad \forall i \in \mathcal{I} \qquad (4.5)$$

$$\sum_{i \in \mathcal{I}} X_{li}^r \leq \gamma \sum_{k \in \mathcal{K}} X_{kl}^c \qquad \forall l \in \mathcal{L} \qquad (4.6)$$

$$\sum_{j \in \mathcal{J}} X_{ij}^p \leq \overline{p}_i Y_i^p \qquad \forall i \in \mathcal{I} \qquad (4.7)$$

$$\sum_{i \in \mathcal{I}} X_{ij}^p \leq \overline{h}_j Y_j^h \qquad \forall j \in \mathcal{J} \qquad (4.8)$$

$$\sum_{k \in \mathcal{K}} X_{kl}^c \leq \overline{r}_l Y_l^r \qquad \forall l \in \mathcal{L} \qquad (4.9)$$

$$Y_i^p, Y_j^h, Y_l^r \in \{0, 1\} \qquad \forall i \in \mathcal{I}, j \in \mathcal{J}, l \in \mathcal{L} \qquad (4.10)$$

$$X_{ij}^p, X_{jk}^h, X_{kl}^c, X_{li}^r, U_k, V_k \geq 0 \qquad \forall i \in \mathcal{I}, j \in \mathcal{J}, k \in \mathcal{K}, l \in \mathcal{L} \qquad (4.11)$$

In this formulation, Equations 4.2 and 4.3 ensure that all customer demand and returns are taken into account. Equations 4.4 through 4.6 represent balance constraints at the warehouse, plant, and test center levels respectively. At the warehouse level, inbound and outbound flows need to be equal. At the plant level, a potential excess outbound volume corresponds with new production. Similarly, the excess inbound volume at the test center level, which is constrained by the recovery yield, corresponds with the disposal volume. Finally, Equations 4.7 through 4.9 are the usual facility opening conditions coupled with capacity constraints.

In order to speed up the solution process, the above formulation can be strengthened by adding the following valid inequalities (compare Bloemhof et al., 1996).

$$X_{jk}^h \leq \min(d_k, \overline{h}_j) \ Y_j^h \qquad \forall j \in \mathcal{J}, k \in \mathcal{K} \qquad (4.12)$$

$$X_{kl}^c \leq \min(u_k, \overline{r}_l) \ Y_l^r \qquad \forall k \in \mathcal{K}, l \in \mathcal{L} \qquad (4.13)$$

It should be noted that this model is rather general and can capture a large variety of reverse logistics situations. For example, closed-loop and open-loop structures both can be represented and are reflected in different settings of the parameters d_k and u_k. Specifically, a closed–loop situation is characterized by $d_k \cdot u_k > 0$ for at least some customer k. Similarly, push and pull drivers for used product collection are reflected in different penalty costs c_k^o. Furthermore, it is worth emphasizing that the 'disposal' route may include any form of recovery that is outsourced to a third party, e.g. material recycling.

Mathematically, the above formulation does not differ much from multi–level facility location models in a more traditional production–distribution context. A particular aspect concerns the two sets of exogenous parameters d_k and u_k, which are linked by the different balance conditions. This reflects the need in reverse logistics for matching market conditions on the supply and the demand side. Another element worth pointing out concerns the additional degree of freedom introduced by the yield condition 4.6. By constraining the disposal volume by a lower bound rather than by a fixed fraction, the recovery strategy and the network design are optimized simultaneously. A relevant question concerns the impact of these features on the performance of specific solution methods. To our knowledge, results on this issue are few to date. Verter et al. (2003) have recently presented a Lagrangean decomposition method that exploits the specific problem structure of a combined forward/reverse logistics network design. Initial numerical results seem promising.

4.3.3 Extensions

The (RLND) model introduced above can be extended in manifold ways. Analogous with traditional facility location models, the formulation can be generalized to a dynamic, capacity selection, multi–product setting. We do not elaborate on these features here since they are well known from other contexts. Instead, we indicate a number of additional elements that appear to be specific to a reverse logistics context. For mathematical details, we refer to Fleischmann et al. (2001).

- *Integrating forward and reverse channel facilities*
  As discussed in Section 4.2, integration versus separation of different processes is an important issue in reverse logistics. For example, co–locating a warehouse and a test center may allow for sharing fixed assets and thereby exploiting economies of scale. This effect can be captured in the model by introducing additional indicator variables for combined facilities.

- *Integrating forward and reverse transportation flows*
  Similar synergies may arise from combining transportation routes for forward and reverse goods movements (see also Chapter 5). In the above setup, this can be modeled by means of additional flow variables representing simultaneous flows in both directions between two locations.

- *Distinguishing demand for new and recovered products*
  The above formulation includes only one class of demand, which may be fulfilled through either new production or recovery. Alternatively, one may wish to distinguish between markets for new and recovered products. In essence, this comes down to explicitly including the leftmost part of the scheme in Figure 4.1. Mathematically, this approach results in a multi–commodity network flow formulation.

- *Multiple recovery options*
  The above formulation uses the most basic representation of a recovery strategy in that it distinguishes two return dispositions, namely internal 'recover' versus external 'disposal'. In order to capture a more refined picture, one may wish to distinguish more recovery options. Mathematically, this extension again results in a multi–commodity formulation.

4.4 Stochastic Location Models for Reverse Logistics Network Design

4.4.1 Stochastic Mixed Integer Modeling Approaches

As discussed in Section 4.2, growing uncertainty on the supply side in particular is frequently named as a major characteristic of reverse logistics networks. In the mixed integer network design approaches presented in the previous section, uncertainty is, in general, addressed by means of scenario analyses. Thus a model is solved repeatedly for a set of scenarios and the solution with the best 'overall performance', according to some multi–criteria measure, is retained. In this section, we review modeling approaches that incorporate the aspect of uncertainty more explicitly.

For a general introduction to stochastic programming, we refer to Birge and Louveaux (1997). A stochastic mixed–integer linear program seeks to minimize the expected costs over a given set of scenarios with associated probabilities, subject to linear and integrality constraints. In the model definition, one needs to specify which decision variables need to be fixed before the realization of a scenario is known and which ones can be adjusted afterwards. Let us denote the vectors of both types of decision variables by Y and X, respectively. Moreover, let $\omega \in \Omega$ denote the set of scenarios. Then a stochastic mixed–integer linear program can be written as

$$\min \ c^T Y + E_\omega[c^*(\omega, Y)] \quad \text{s.t.} \quad Y \geq 0, \ Y_\mathcal{I} \in \{0, 1\} \quad , \qquad (4.14)$$

where c^* is the optimal value of a MILP in decision variables X, which depends on ω and Y, c is a vector of objective coefficients, and $Y_\mathcal{I}$ is some sub–vector of Y. If Ω is finite then (4.14) can be rewritten as an ordinary MILP, though at the expense of an increasing problem size, by introducing scenario-dependent decision variables X_ω. This approach is known as linear programming 'with recourse'.

It is important to note that the optimal solution of (4.14) need not be optimal for any single scenario. In this sense, stochastic programming is more powerful than a simple scenario analysis. This expansion comes at a cost however, since the problem size of the corresponding MILP formulation increases significantly.

In the context of logistics network design, stochastic programming models have been presented to capture the impact of demand uncertainty and price variations (see, e.g. Louveaux, 1986). Typically, these models assume that location decisions are fixed for a longer planning horizon (corresponding to variables Y in our formulation) whereas transportation flows can be adjusted in the short term, according to demand realizations (corresponding to variables X).

Stochastic programming models require a probability to be specified for each scenario. Since in practical applications these probabilities often are hard to define, some authors have argued that other optimality criteria may be more relevant. Instead of expected costs, they suggest considering worst-case criteria, such as minimizing the maximum cost across all scenarios or minimizing the maximum 'regret', i.e. the cost deviation from the corresponding scenario–optimal solution. These approaches do not require any probability specification but seek solutions that provide a good performance guarantee in all cases. For a general introduction to these so–called 'robust' optimization models, we refer to Kouvelis and Yu (1997). It should be noted, however, that despite their name, these approaches may be highly sensitive to the set of scenarios considered since extreme scenarios may strongly dominate the solution.

To our knowledge, two groups of authors have presented robust and/or stochastic extensions to network design models in a reverse logistics context. Realff et al. (2002) report on a case study on the design of a carpet recycling network in the USA. The authors extend a corresponding MILP facility location model to a multi–scenario setting, involving different levels of supply volumes and material prices, and seek to minimize the maximum regret across all scenarios. All binary variables, which represent location choices and capacity levels, are fixed at the beginning of the planning horizon whereas the values of all continuous variables are scenario dependent. In a numerical example, the authors illustrate that the optimal robust solution is not optimal, in general, for any of the individual scenarios considered. Information on the cost deviation between both approaches is not provided, however.

Listes and Dekker (2001) build upon the work of Barros et al. (1998) concerning a case study on the design of a sand recycling network in the Netherlands (see also Section 4.2). The authors extend the original MILP model to a stochastic model that maximizes expected profit under demand and supply uncertainty. In a first approach, they consider uncertain demand locations and volume. Location decisions for cleaning and storage facilities are assumed to be fixed at the beginning of the planning period, whereas all transportation, processing, and storage decisions may be adjusted to the demand realization. In a second approach, supply volumes are also uncertain. Decisions are now taken in three stages as the scenario realization is revealed successively. In a numerical study the authors document that the optimal stochastic solution need not coincide with the solution for any individual scenario. However, the cost deviation between the stochastic solution and the best solution obtained from a scenario analysis is within a few percentages in each of the cases presented.

4.4.2 A Stochastic Location Model for Reverse Logistics

We now apply the above stochastic modeling approaches to the reverse logistics network design model introduced in Section 4.3. To this end, let Ω denote a finite set of scenarios, and for each scenario $\omega \in \Omega$ let π_ω denote its probability. We assume that scenarios differ in terms of demand and return volumes and recovery yields, which we denote by $d_{k\omega}, u_{k\omega}$, and γ_ω, in analogy with Section 4.3. Then a stochastic version of the model in (4.1)-(4.11) can be formulated by adding a scenario index to the continuous variables X^P, X^h, X^c, X^r, V, and U, taking the expected value of the objective function across all scenarios and imposing restrictions (4.2)-(4.9) per scenario.

It is worth noting that the uncertain volume parameters concern the right-hand side of the MILP formulation, whereas the uncertain recovery yield affects the coefficient matrix. In contrast, all cost parameters are assumed to be fixed. Since all continuous variables depend on ω whereas the binary variables do not, location decisions are taken under uncertainty whereas transportation and processing flows can be adjusted to individual scenario realizations, in line with the above motivation. Furthermore, note that setting $V_{k\omega} = d_{k\omega}$ and $U_{k\omega} = u_{k\omega}$ for all k and ω always provides a feasible solution. Hence, each location decision is feasible for all scenarios.

Comparing this formulation with the original deterministic model in Section 4.3, we observe that the number of continuous variables and the number of constraints has increased with a factor of $|\Omega|$. To improve numerical solution procedures, the MILP formulation can be strengthened by means of valid inequalities analogous with Equations (4.12)-(4.13). We illustrate the relation between solutions of the deterministic and the stochastic model in Section 4.6.

4.4.3 Extensions

The above model can be modified in manifold ways, e.g. to allow for different scenario definitions or information evolution. As an illustration, we take a brief look at alternative optimality criteria and multi-stage decision approaches.

If the minimal costs vary largely across scenarios, then the expected cost criterion used in Section 4.4.2 may result in a biased solution in the sense that it is dominated by a few high-cost scenarios. In this case, minimizing the expected 'regret' may be a relevant alternative. To this end, the term $-\sum_{\omega \in \Omega} \pi_\omega c_\omega^*$ should be added to the expected cost function, where c_ω^* denotes the minimal costs for scenario ω in the original deterministic model.

The expected cost criterion may be difficult to apply since estimating the probabilities π_ω may not be straightforward in practical situations. As discussed above, optimizing the worst-case behavior may therefore be a useful alternative. For our model, this so-called 'robust' optimization approach comes down to introducing an additional decision variable Z, which is to be minimized under the additional constraint

$$\sum_{i \in \mathcal{I}} f_i^p Y_i^p + \sum_{j \in \mathcal{J}} f_j^h Y_j^h + \sum_{l \in \mathcal{L}} f_l^r Y_l^r$$
$$+ \sum_{k \in \mathcal{K}} (c_k^b V_{k\omega} + \sum_{j \in \mathcal{J}} c_{jk}^h X_{jk\omega}^h) + \sum_{i \in \mathcal{I}} \sum_{j \in \mathcal{J}} c_{ij}^p X_{ij\omega}^p$$
$$+ \sum_{k \in \mathcal{K}} (c_k^o U_{k\omega} + \sum_{l \in \mathcal{L}} c_{kl}^c X_{kl\omega}^c) + \sum_{l \in \mathcal{L}} \sum_{i \in \mathcal{I}} c_{li}^r X_{li\omega}^r \le Z \quad \forall \omega \in \Omega . \quad (4.15)$$

Analogously, one may choose to minimize the maximum regret by combining both of the above approaches. We illustrate the effect of the different cost criteria in Section 4.6.

Finally, it is worthwhile to take another look at how the scenario is revealed and hence at which information is available for which decision. As explained before, the above formulation implicitly assumes that all location decisions are taken before the actual scenario is known, whereas all other decisions are based on its realization. In this sense, the model captures a two–stage decision process. However, as discussed in Section 4.2, the design of a reverse logistics network may involve more stages, in particular if recovery activities are integrated into an existing 'forward' distribution network. One way to capture such a sequential decision process is to separate the scenario space into two independent sets $\Omega = \Xi \times \Psi$ concerning demand–related information (captured by parameters $d_{k\xi}$) and return–related information (captured by $u_{k\psi}$ and γ_ψ), respectively. The degree of information that is available for the different decisions can then be modeled by indexing the decision variables as $Y_i^p, Y_j^h; \ X_{ij\xi}^p, X_{jk\xi}^h, V_{k\xi}, Y_{l\xi}^r; \ X_{kl\xi\psi}^c, X_{li\xi\psi}^r$, and $U_{k\xi\psi}$ and modifying (4.1) - (4.11) accordingly.

4.5 Continuous Approximation Models for Reverse Logistics Network Design

4.5.1 Approximating Reverse Logistics Costs and Revenues

MILP–based location models as discussed in the preceding sections provide a powerful tool which can be tailored to a variety of different settings. Yet these approaches have some drawbacks when it comes to establishing general insights into the economics of logistics systems. Capabilities for sensitivity analyses in MILP models are limited and, even more importantly, the interrelation between various parameters is not made explicit. Therefore, conclusions on the behavior of a given real-life system often rely on extensive numerical experiments rather than on analytic arguments.

In view of this shortcoming, several authors have considered continuous cost expressions as a basis for alternative approaches to investigating logistics costs and optimizing the design of logistics systems. In particular, Daganzo has promoted this route in what has become known as the 'continuous approximation methodology' (Daganzo, 1999). The key element of this approach is the representation of demand by a continuous density function, as opposed to the discrete demand representations in traditional MILP approaches. If the demand density and other system parameters vary sufficiently slowly across the given service region (which may have spatial and temporal dimensions), logistics costs can be reasonably approximated by appropriately chosen averages, which can be expressed as fairly simple functions in a limited number of parameters. In this way, the cost impact of critical system parameters can be revealed and guidelines for the design of logistics structures can be derived.

In this section, we follow our reasoning in Fleischmann (2003) in applying the 'continuous approximation' approach to the analysis of reverse logistics networks. We consider a setting analogous to the one in Section 4.3. However, for the time being we restrict the modeling scope to the 'reverse' network part in a strict sense, i.e. the logistics structure conveying used products from collection points via inspection and sorting centers to a given recovery facility (compare Figure 4.1). An extension of the model to the entire network, including the redistribution stage, is discussed at the end of this section. Our goal is to approximate the total reverse logistics costs for serving a given area, and eventually to minimize these costs by choosing an appropriate reverse logistics network design. To this end, assume that the return rate of used products per time per unit surface is given by a location–dependent continuous density function, which varies slowly within the service area. The subsequent development is facilitated by considering costs on a per product returned basis. (Note that this criterion differs from total costs just by a scaling factor.) The core idea of the 'continuous approximation' approach then is to express these costs in 'local' problem parameters only and to approximate the overall costs by integrating over the service area.

To assess the unit reverse logistics costs, we distinguish two cases, depending on whether the testing and sorting is carried out at the recovery facility or at a separate location. In what follows, we refer to these cases as 'central' and 'local' testing, respectively. For both cases, one may decompose the total reverse logistics costs into a number of components, namely inbound transportation costs to the test and sort process, outbound transportation costs after sorting, variable sorting and handling costs, and fixed installation costs for the test facility. In what follows, we go through all of these components and discuss the parameters on which they depend. In addition to the symbols introduced earlier, we use the following notation.

A	$=$	overall service area
$\rho(x)$	$=$	return rate of used products per time per unit surface at location $x \in A$
$C_R(\ell, \rho)$	$=$	reverse logistics costs per returned product for a service area with constant return rate ρ at a distance ℓ from the corresponding recovery facility
$C_{RL}(\ell, \rho)$	$=$	— in the case of local testing
$C_{RC}(\ell, \rho)$	$=$	— in the case of central testing
c_t	$=$	low volume vehicle transportation cost per distance
$\tilde{c}_t$	$=$	high volume vehicle transportation cost per distance
v	$=$	low volume vehicle capacity
$\tilde{v}$	$=$	high volume vehicle capacity
c_w	$=$	disposal costs per product
A_R	$=$	size of a test facility's service area

A_R^* =   optimal size of a test facility's service area

ℓ^* =   opt. distance for switching from central to local testing

The first cost component concerns inbound transportation costs to the test and sort process. We assume that used products are collected in milk–runs. The length of a tour can be assessed through a probabilistic analysis of the standard vehicle routing problem. Specifically, it can be approximated by a line-haul distance from and to the test and sort installation plus the sum of the expected distances between two consecutive collection stops (see e.g. Daganzo 1999). Assuming full vehicle loads, we get in the case of central testing

$$\text{unit inbound transp. cost (central)} \approx 2\frac{c_t}{v}\ell + 0.57\, c_t\, \rho^{-1/2}. \quad (4.16)$$

In the case of local testing and sorting, the line-haul distance depends on the size A_R of the area covered by the test facility. Assuming this area to have a circle-like shape with the test facility located at its center, the average line-haul distance approximately equals $2\sqrt{A_R}/3\sqrt{\pi}$ and one obtains

$$\text{unit inbound transp. cost (local)} \approx \frac{4}{3\sqrt{\pi}}\frac{c_t}{v}\sqrt{A_R} + 0.57\, c_t\, \rho^{-1/2}.$$
$$(4.17)$$

The next chapter presents more detailed versions of these expressions, which also take into account inventory accumulation (see Section 5.4).

The second cost component concerns outbound costs from the test location. In the case of central testing, this term encompasses merely the disposal costs for rejected products, which equal $c_w(1-\gamma)$ per unit. For local testing, one also needs to consider the flow of accepted products to the recovery facility. Recognizing the consolidation function of the test centers, we assume those shipments to be line-hauls rather than multi–stop tours. In the same vein, we assume a larger vehicle capacity than for the collection tours, and a corresponding mileage cost. For full vehicle loads, the outbound costs can then be expressed as

$$\text{unit outbound transp. and disposal cost (local)} \approx 2\frac{\tilde{c}_t}{\tilde{v}}\ell\gamma + c_w\,(1-\gamma).$$
$$(4.18)$$

As a third cost term we consider the annualized fixed costs for a local test and sort installation. These can be approximated on a per product basis by

$$\text{unit fixed installation cost (local)} \approx \frac{f_r}{\rho A_R}. \quad (4.19)$$

Finally, any variable handling and processing costs may be aggregated into a term c_h. Summing up the four cost components yields an expression for the reverse logistics cost per collected product. Specifically, in the case of central testing and sorting, one obtains

$$C_{RC}(\ell,\rho) = 2\frac{c_t}{v}\ell + 0.57\, c_t\, \rho^{-1/2} + c_w\,(1-\gamma) + c_h. \quad (4.20)$$

For the local testing case, the corresponding expression still depends on the size of the collection area A_R. Equations (4.17) and (4.19) characterize the optimal size A_R^* of this area. First order conditions imply

$$A_R^* = \left(\frac{3\sqrt{\pi}\, f_r\, v}{2\, c_t\, \rho} \right)^{2/3} \approx 1.92 \left(\frac{f_r\, v}{c_t\, \rho} \right)^{2/3} . \tag{4.21}$$

Inserting this expression for A_R and summing up leads to the following cost function:

$$C_{RL}(\ell, \rho) = 2\, \frac{\tilde{c}_t}{\tilde{v}}\, \ell\, \gamma + 0.57\, c_t\, \rho^{-1/2} + c_w\, (1 - \gamma) + c_h + 1.56 \left(\frac{c_t^2\, f_r}{v^2\, \rho} \right)^{1/3} . \tag{4.22}$$

Comparing $C_{RC}(.)$ and $C_{RL}(.)$ yields an appropriate service area for the central test and sort operation. Specifically, (4.20) and (4.22) define a critical distance ℓ^* from the recovery facility up to which central testing is preferable over local testing. Equating the cost functions yields

$$\ell^* = 0.78 \left(\frac{f_r\, v}{c_t\, \rho} \right)^{1/3} \left(1 - \frac{\tilde{c}_t\, v}{\tilde{v}\, c_t} \gamma \right)^{-1} . \tag{4.23}$$

Putting the above results together, one finally obtains the overall reverse logistics unit cost function $C_R(.)$ as $C_R(\ell, \rho) = \min\{C_{RC}(\ell, \rho), C_{RL}(\ell, \rho)\}$. As discussed above, total reverse logistics costs are then approximated by integrating over the service area $\int_A \rho(x)\, C_R(\ell(x), \rho(x)) dx$.

In Section 4.6 we compare the above cost expressions with the results of the previously discussed discrete models and interpret them in the light of the reverse logistics issues identified in Section 4.2. Before doing so, we discuss a number of extensions and refinements to the above approach.

4.5.2 Extensions

The above formulas reflect a very basic cost model which can be extended in manifold ways. In particular, they do not include any inventory considerations and assume all vehicles to operate at full capacity. These assumptions can be relaxed by including decisions on lot sizes and dispatching frequencies. We refer to Chapter 5 for a detailed discussion of this issue. Furthermore, the formulas can be extended to a multi–product setting. However, since these refinements do not appear to exhibit any particular reverse logistics elements, and since they do not change the core of our argumentation, we content ourselves by referring to Daganzo (1999) for a more in-depth discussion of the 'continuous approximation' technique.

In a similar fashion, we can also derive cost expressions for the 'forward' parts of the network (see Figure 4.1). To this end, assume that products are shipped from the factory to the customers via distribution centers. Following the above analysis, one obtains the same formulas, where ρ is replaced by an appropriate demand density δ and γ equals one. In fact, this is the original model discussed by Daganzo (1999). In what follows, we denote these 'forward' logistics costs by $C_F(.)$.

By putting together $C_F(.)$ and $C_R(.)$ one may address the overall network structure. In particular, by considering $C_R(.)$ as inbound and $C_F(.)$ as outbound costs and including investments one may assess the size of a factory's service area. If $\delta(x)$ and $\rho(x)$ are roughly proportional, one can derive expressions similar to (4.21) with ρ replaced by $\delta + \rho\gamma$. However, a critical look seems advisable. On the one hand, the distance approximations may be less accurate since the number of distribution centers and test centers is much smaller than the number of customer locations in the original model. On the other hand, Equation (4.21) assumes the facility to be located close to the center of its service area. While this seems reasonable for a distribution center, it may not be evident for the location of a factory, which depends on additional factors such as tax rates and labor costs.

Finally, note that we have assumed return and disposal rates to be given and therefore have not included any revenues in the analysis. However, the above cost expressions can also be used to assess profitability of a recovery operation. In particular, the tradeoff between reverse logistics costs and production cost savings or additional revenues can be made explicit. To this end, denote by $C_{RN}(.)$ the unit cost for any used product that is not recovered (which may include, for example, lost revenues and/or fees for local recycling). The unit reverse logistics cost function $C_R(.)$ is then obtained by selecting the cheapest among the three options C_{RC}, C_{RL}, and C_{RN} for each value of ℓ and ρ.

4.6 Quantitative Analysis of Reverse Logistics Network Design Issues

Having reviewed alternative modeling approaches for supporting reverse logistics network design decisions, we now return to the issues highlighted in Section 4.2. In what follows, we exploit the above quantitative tools to analyze these issues and highlight the impact of key parameters on the economics of reverse logistics networks.

We illustrate the analysis in a numerical example adapted from Fleischmann et al. (2001). All computational results are based on an installation of the CPLEX 7.0 standard MILP solver on a Pentium 4, 1495 MHz PC. Consider the situation of an electronic equipment manufacturer operating in the European market (recall the case of IBM from Section 4.1; see also the copier business in Chapter 11 and in Thierry et al., 1995). Assume that used equipment, stemming, for example, from expiring lease contracts, is collected from the customers, remanufactured, and resold. To allow for remanufacturing, used equipment must meet specified quality standards. To this end, all collected equipment is inspected and tested. Rejected equipment is disposed

Table 4.2. Parameter Settings of Network Design Example

Description	Value	Model Parameter	
		discrete	continuous
Fixed cost per factory	5,000,000	f^p	f^p
Fixed cost per warehouse	1,500,000	f^h	f^h
Fixed cost per test center	500,000	f^r	f^r
Transportation costs per km per product			
factory—warehouse	0.0045	c^p	–
warehouse—customer	0.0100	c^h	–
customer—test center	0.0050	c^c	$2\,c^t/v$
test center—plant	0.0030	c^r	$2\,\tilde{c}^t/\tilde{v}$
Penalty cost unsatisfied demand	1,000	c^b	–
Penalty cost uncollected returns	1,000	c^o	–
Capacity factory	500,000	$\bar{p}$	–
Capacity warehouse	150,000	$\bar{h}$	–
Capacity test center	150,000	$\bar{r}$	–
Low volume vehicle capacity	20	–	v
Demand per 1,000 inhabitants	10	$d_k/\#$inh.	$\delta \times$pop.density
Return ratio	[0,0.9]	λ	λ
Recovery yield	0.5	γ	γ
Distance from factory	1,000	–	ℓ

of locally, while the remainder is shipped to the remanufacturing operation, which is co–located with an original manufacturing site.

To implement this example as a MILP model, we assume that customers are located in 50 major European cities (capitals plus cities larger than 500,000 inhabitants) and that demand is proportional to the population size. Moreover, we restrict the potential (re-)manufacturing locations to 20 main metropolitan areas, whereas distribution warehouses and test operations may be located in any of the 50 cities. Table 4.2 summarizes the parameter settings for this example.

We assume that all equipment that passes the test operation has a sufficient contribution margin to be remanufactured rather than disposed of. However, to avoid the cost figures being distorted by large blocks of sunk costs, we do not include variable (re-)manufacturing, handling, and disposal costs. To assess the overall profitability of the remanufacturing operation, these costs as well as sales revenues should be added to the results below.

As a starting point, we compute the optimal 'forward' distribution network for the above example, ignoring any reverse logistics activities. To this end, we solve the conventional two–level facility location model obtained by setting $u_k = 0$ for all k in the MILP model in Section 4.3. The solid lines in Figure

Table 4.3. Results of Network Design Example

Scenario	λ	Test Centers	Min. Cost	Regret in Case of Design		
				Scen. 9	Scen. 3	Robust
			$\text{€} \cdot 10^3$	$\text{€} \cdot 10^3$	$\text{€} \cdot 10^3$	$\text{€} \cdot 10^3$
0	0.0	—	0	4,000	1,500	2,000
1	0.1	D	2,600	2,580	603	951
2	0.2	D	4.700	1,660	206	402
3	0.3	GB,D,E	6,610	933	0	44
4	0.4	GB,D,E,I,HU	8.140	592	182	74
5	0.5	GB,D,E,I,HU	9,550	365	477	218
6	0.6	GB,D,E,I,HU	11,000	139	773	361
7	0.7	S,GB,D,F,E,I,HU	12,200	54	1,210	647
8	0.8	S,GB,D_1,D_2,E,I,HU,BG	13,500	0	1,680	964
9	0.9	S,GB,D_1,D_2,E,I,HU,BG	14,600	0	2,200	1,330
stochastic		GB,D,E,I,HU	ø 8,850			
robust		B,D,E,HU				≤ 2,000

4.3(a) illustrate the resulting network structure, which includes one central manufacturing site in Frankfurt and seven regional warehouses. For the sake of clarity, flows from warehouses to customers are omitted. The corresponding annual costs equal € 44.8m.

4.6.1 Impact of Supply Uncertainty

As discussed in Section 4.2, reverse logistics network design typically faces significant uncertainty concerning the supply of recoverable resources. In the above MILP model, the supply side is characterized by the parameters u_k and γ. In what follows, we analyze their impact on the optimal solution.

In the model formulation (4.1)-(4.11), the volume parameters u_k occur only on the righthand side. Therefore, standard MILP theory implies that the cost function depends on them piecewise linearly (see e.g. Jenkins, 1982). Moreover, for fixed binary variables, i.e. fixed facility locations, the cost function is convex in u_k for each k. A parametric analysis can be carried out by means of Jenkins's heuristic (Jenkins, 1982).

Let us now assume that $u_k = \lambda d_k$ for all k, i.e. the return ratio λ is identical across locations. Table 4.3 summarizes the results for different values of λ. More specifically, we vary the return ratio in steps of 0.1 in the interval [0, 0.9] and compute for each scenario the optimal reverse logistics network while keeping the forward network fixed to the above layout. The solution time for each scenario is in the order of a few seconds. The dashed lines in Figure 4.3(a) illustrate the solution for $\lambda = 0.4$, which encompasses five regional test centers.

Not surprisingly, the optimal number of test centers and the relevant reverse logistics costs increase with the return volume. However, as discussed

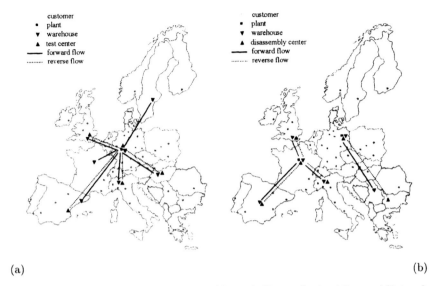

(a)   (b)

Fig. 4.3. Optimal Forward and Reverse Network Versus Optimal Integral Network

before, the actual return volume is not known, in general, when the location decision is to be taken. In Section 4.4, we have discussed modeling approaches that explicitly take this uncertainty into account. The next-to-last row of Table 4.3 characterizes the network design which minimizes the expected costs for the case of a uniform probability distribution across the above scenarios. The solution turns out to be identical to the optimal design for $\lambda \in [0.4, 0.6]$. As an alternative to this stochastic approach, we also compute an optimal 'robust' solution, which minimizes the maximum cost deviation from the scenario–optimal solution across all scenarios (see Section 4.4). Note that this solution is not optimal for any single scenario.

For conventional facility location models, it is well known that the cost function is fairly 'flat' around its minimum, in the sense that a deviation from the optimal network design entails a rather small cost penalty (see, e.g. Daganzo, 1999). An analysis of the continuous cost model developed in Section 4.5 supports a similar conclusion in a reverse logistics context.

To this end, Figure 4.4 illustrates the relation between the discrete and the continuous model for the above example by depicting the corresponding unit reverse logistics costs per product as a function of the return rate. For the discrete model, this cost curve is obtained by dividing the results in Table 4.3 by the return volume. For the continuous model, the curves display the functions C_{RC} and C_{RL} defined in (4.20) and (4.22). The parameter settings are listed in Table 4.2. Note that to make both modeling approaches compatible one needs to adjust c_t and $\tilde{c}_t$ to account for vehicle capacities and line–haul return trips. The values of ℓ and δ approximate the overall averages. Note

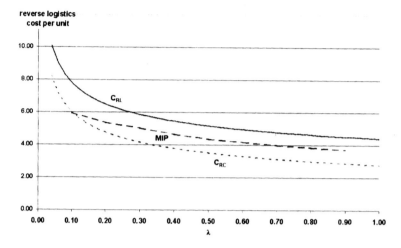

Fig. 4.4. Unit Reverse Logistics Costs

that the eventual unit reverse logistics cost function C_R in the continuous model is a mixture of C_{RC} and C_{RL}. Since the discrete cost function also lies in this interval, Figure 4.4 suggests the results of the discrete model and the continuous model to be compatible.

To quantify the impact of supply uncertainty on the network design, let $C_{RL}(A_R)$ denote the unit reverse logistics costs as a function of the test service area A_R in the case of local testing. From Equations (4.19) and (4.17) one gets that $C_{RL}(A_R)$ can be written as $a + b\sqrt{A_R} + c/A_R$, with some positive constants a, b, and c. Similar to the well-known case of the EOQ formula this function is very flat around its minimum. Specifically, for $\varepsilon > 0$ one gets

$$[C_{RL}((1 + \varepsilon)A_R^*) - C_{RL}(A_R^*)]/C_{RL}(A_R^*) \quad \leq \quad \varepsilon^2/3(1 + \varepsilon). \qquad (4.24)$$

Furthermore, Equation (4.21) implies, for example, that a relative error of ε in ρ causes a relative error of at most $0.67\,\varepsilon$ in A_R^* and therefore by (4.24) a relative cost penalty of at most $0.22\varepsilon^2/(1.5+\varepsilon)$. This implies that a forecasting error of 50% in the return rate results in an eventual cost penalty of less than 3% in the network design decision.

For the design parameter ℓ^*, which characterizes the domain of central testing, one observes a similar robust behavior. Equation (4.23) shows that the impact on ℓ^* of an error in ρ is limited. Moreover, moving to a critical distance ℓ' different from ℓ^* only affects the costs for customers located at a distance between ℓ' and ℓ^*, which again has a dampening effect on the overall cost deviation.

So far, we have restricted our attention to variations in the return *volume*. To round off our analysis, let us take a brief look at the impact of the return quality, characterized by the yield parameter γ. An exact sensitivity analysis in the MILP model is more cumbersome in this case, since γ occurs in the coefficient matrix and the relation with the optimal cost value may therefore be nonlinear. However, the continuous model suggests the network structure to be fairly robust again: the local test area A_R^* turns out even to be independent of γ, whereas for the cost impact through a deviation from ℓ^*, the same argument holds as for ρ above.

The above robustness property is good news from a practical perspective in that it documents that supply uncertainty, which is characteristic of many reverse logistics environments, does not really hamper logistics network design. At the same time, one should note that variations in supply volume and quality do affect total and unit reverse logistics costs, as illustrated in Table 4.3, Figure 4.4, and in Expressions (4.20) and (4.22). Therefore, supply uncertainty is certainly a relevant factor when it comes to estimating the profitability of a reverse logistics operation and taking an overall go/no-go decision. However, its impact is largely independent of the specific network design.

This observation also relates to the proficiency of the different modeling approaches. Specifically, it explains why in this context performance differences tend to be small between the solution of a scenario analysis and those of theoretically more powerful yet computationally more demanding methods, such as stochastic or robust models (see also Table 4.3). It is worth underlining the impact of the scenario selection, however. A scenario analysis may require a much finer gradation of scenarios than a stochastic or a robust model. For an illustration, consider the last three columns of Table 4.3. For a given network structure, the 'maximum regret' will, in general, be assumed for one of the extreme scenarios $\lambda = 0$ or $\lambda = 0.9$. For the robust model, a scenario space consisting of these two cases is therefore sufficient. For a conventional scenario analysis, however, this is not true. Choosing only between the optimal design for $\lambda = 0$ and $\lambda = 0.9$ respectively yields a 100% increase in the maximum regret. However, there does exist an intermediate scenario ($\lambda = 0.33$), whose corresponding optimal solution performs equally well across all scenarios as the optimal robust solution.

4.6.2 Compliance with Forward Networks

The robustness property analyzed in the previous subsection also plays an important role when it comes to the compliance of reverse logistics networks with 'forward' logistics infrastructure already in place. As discussed in Section 4.2, this is an important issue since companies, in many cases, do not set up reverse logistics networks from scratch but on top of an existing 'forward' network.

In this vein, the forward and reverse network parts have been optimized sequentially in the above examples. To assess the consequences of such a two-stage approach, let us compare its outcome with an integral design, which optimizes both network parts simultaneously. Figure 4.3(b) illustrates the optimal solution in this case for $\lambda = 0.4$. All parameters are kept equal to the values in Table 4.2. Comparing Figures 4.3(a) and (b), one observes that an integral design approach indeed leads to a significantly different network structure. However, the costs of both solutions are almost identical, namely € 52.9m in the sequential approach versus € 52.7m in the integral approach. This result generalizes to other values of λ. Specifically, the cost penalty for adding the reverse logistics network on top of a previously designed forward network rather than optimizing both parts together increases from 0% for $\lambda = 0$ to not more than 1.6% for $\lambda = 0.9$. We have observed similar results for many other parameter settings, including the case that demand and return volumes are not proportional (see Fleischmann et al., 2001).

One can explain this observation as follows. First, forward flows outweigh reverse goods flows, in general, in terms of volumes and costs. Therefore, the overall optimal solution can be expected to be 'close' to the optimal forward network. A deviation from this structure must allow for substantial savings per unit in the reverse channel in order to set off against the resulting increase in distribution costs. Second, the flat cost structure highlighted in the previous subsection results in a very limited cost penalty for deviating from the optimal reverse network structure due to constraints imposed by existing infrastructure. This is the more true if demand and returns have a similar geographical distribution.

This observation is again good news from a business perspective since it suggests that setting up an efficient reverse logistics network in many cases does not require a fundamental redesign of a company's existing logistics networks. In addition to limiting investment costs, this conclusion simplifies the

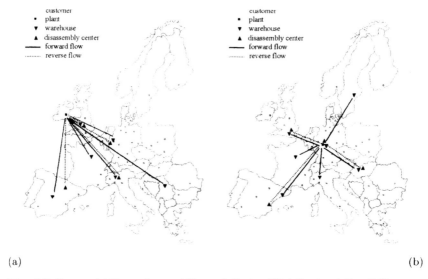

(a)   (b)

Fig. 4.5. Sequential Versus Integral Network Design With Regional Cost Differences

organizational implementation of reverse logistics initiatives. From a modeling perspective, this observation results in a significant reduction of complexity by optimizing forward and reverse network structures separately.

In Fleischmann et al. (2001), we have indicated the limits of the above observation by means of an example of a paper recycling network. In that case, the recycling operation does have a fundamental impact on the entire logistics network by reducing the impact of virgin raw material sources. While the structure of the original forward network is strongly dominated by pulp wood production close to the Scandinavian forests, recycling 'pulls' the business activities closer to the main markets in Western Europe. We illustrate the underlying economics in the context of our previous example.

To this end, consider the potential differences in labor and investment costs across Europe. To make things specific, assume that salaries and tax effects result in a manufacturing cost advantage of €2.- per unit in Ireland and of €1.50 in Eastern Europe, compared to the remaining countries. Moreover, assume that these cost differences are less prominent in the recovery channel due to a lower level of labor skills. For the sake of argument, let us assume that effective differences in remanufacturing costs across countries are negligible. In what follows, we set $c^p = €7.0\text{m}$ and $\gamma = 0.8$, while all other parameters remain unchanged with respect to the previous examples.

Figures 4.5 (a) and (b) depict the optimal network structures according to a sequential and an integral design, respectively $(\lambda = 0.4)$. Moreover, Figure 4.6 shows the corresponding cost functions. Apparently, the cost advantage of an integral design approach is much more significant in this case than in the previous examples. This can be explained as follows. The structure of the

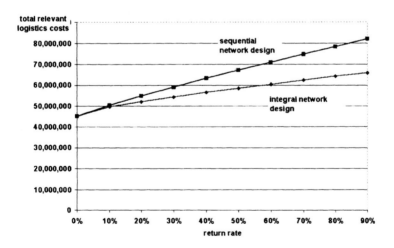

Fig. 4.6. Cost Penalty of Sequential Design With Regional Cost Differences

original forward network is dominated by the labor cost differences, which cause the production facility to be located in Ireland. However, when moving towards remanufacturing this factor is losing importance and the network structure is determined by transportation distances primarily. In other words, it is the *substitution* of virgin products by recovered ones that have structurally different cost drivers, which is key to the above result. It is under these conditions that one can expect reverse logistics to have a more fundamental impact on a company's overall logistics network.

4.7 Conclusions and Outlook

In this chapter, we have considered the design of appropriate infrastructure for companies engaged in reverse logistics programs. In Sections 4.1 and 4.2, we have argued that logistics network design is a key determinant of the overall profitability of closed–loop supply chains. By comparing this setting with more traditional production–distribution networks, we have distilled three important issues that appear to be characteristic of reverse logistics networks. First, the supply side is subject to significant uncertainty. Second, the need for testing and sorting used products before assigning them to an appropriate recovery option leads to a particular centralization–decentralization tradeoff. Third, reverse logistics requires the coordination and integration of different inbound and outbound flows.

The core part of this chapter, encompassing Sections 4.3 through 4.5, reviews quantitative models for supporting reverse logistics network design. Analogous with traditional network design problems, many of the approaches rely on MILP formulations. We pointed out that these models, in general, very much resemble conventional multi–level facility location models. Specific features that can be attributed to reverse logistics are few. They include a set of balance equations that link exogenous conditions on the supply and the demand side and an additional degree of freedom in optimizing the logistics network structure and the recovery policy simultaneously. A few models explicitly incorporate the aspect of uncertainty through a stochastic programming approach. The richer solution space of these models comes at a cost in terms of significantly larger problem sizes. We have pointed out that a network design that maximizes average or worst–case performance may not be optimal for any of the underlying scenarios. However, the eventual cost benefit compared to a scenario–based approach seems small in many cases. Finally, we have discussed how approximate, continuous cost models may be applied in a reverse logistics context. This approach is particularly helpful in making explicit the impact of various context parameters. From a mathematical perspective, the resulting reverse logistics model again turned out to be very similar with its corresponding 'forward' counterpart.

In Section 4.6, we have compared the different modeling approaches on the basis of an extended numerical example. The key observation of this analysis concerns the robustness of reverse logistics costs with respect to moderate changes in the network structure. This result, which concurs with what is known about conventional production–distribution networks, has important practical implications. On the one hand, it limits the impact of the aforementioned supply uncertainty when it comes to choosing an appropriate network structure. On the other hand, it indicates that in many cases reverse logistics networks are flexible enough to comply with existing network structures.

Given the short history of reverse logistics research, it goes without saying that many issues are yet to be explored. We conclude this chapter by highlighting a number of issues that we believe to be important for furthering the understanding of reverse logistics networks.

From a methodological perspective, the research focus to date has been on model formulations and on output analyses. In contrast, attention to algorithmic aspects has been limited so far. As consensus about the modeling foundations is growing, looking for efficient solution algorithms is gaining relevance. In particular, the question arises whether solution methods from traditional location theory are also adequate in a reverse logistics context, and which features may require modified approaches. Although the results in this chapter hint at a close similarity between reverse logistics network design and traditional location models, a thorough understanding of algorithmic implications is still lacking. Recent work by Verter et al. (2003) provides a promising starting point for future research in this direction.

Regarding the modeling choices, an important aspect that has, to date, been left aside in reverse logistics network design is the role of inventories. Note that none of the models reviewed throughout this chapter accounts for inventory effects, except possibly as part of some constant unit handling cost term. Yet, inventory is well known to be an important parameter in distribution network design. Risk-pooling and postponement are major factors in the centralization/decentralization trade-off. We see a clear need for corresponding analyses concerning the effects of inventory considerations on reverse logistics networks.

Another aspect that deserves extra attention in future research concerns the multi-agent character of reverse logistics networks. All of the models presented in this chapter take the perspective of a central decision-maker. Given the lessons learned from supply chain coordination studies (see also Chapter 12), it seems advisable to take a closer look at the incentives and the channel power of the different players involved, such as collectors, logistics service providers, processors, and OEMs. In particular, such an approach would help underpin the characterization of different types of reverse logistics networks, as sketched in Section 4.2.

Finally, we mention globalization as another issue that has not yet received the attention in reverse logistics research that it seems to deserve. Global sourcing has become a key factor in many supply chains. Intuitively, one may doubt whether globalization is equally beneficial for reverse chains. Some of the immediate obstacles include cross-border waste transportation and tax requirements. However, there may be even more fundamental arguments against global reverse logistics flows, which relate to the contribution of each player within the chain. A thorough analysis of these issues, which seems instrumental for a good understanding of the differences between forward and reverse logistics networks, again calls for a broadening of the modeling approaches available to date.

Collection and Vehicle Routing Issues in Reverse Logistics

Patrick Beullens[1], Dirk Van Oudheusden[1], and Luk N. Van Wassenhove[2]

[1] K.U. Leuven, Centre for Industrial Management, Celestijnenlaan 300 A,
3001 Leuven, Heverlee, Belgium, patrick.beullens@cib.kuleuven.ac.be,
Dirk.VanOudheusden@cib.kuleuven.ac.be
[2] Henry Ford Chair in Manufacturing, INSEAD, Boulevard de Constance,
77305 Fontainebleau, France, luk.van-wassenhove@insead.edu

5.1 Introduction

It is not yet known to what extent reverse logistics might increase the total amount of transportation in supply chains — partially since it will also reduce activities related to the use of new, extractive resources. It is clear, however, that the extra transportation will diminish the environmental benefits of closing the loop. Likewise, inefficient or ineffective transport activities limit the economic success of reprocessing products.

The previous chapter dealt with the network design of the complete reverse logistics chain. This chapter focuses on organizing the collection and transport systems that link the chain with the markets. Each facility with either a collection or distribution function is assigned to a geographical district. Depending on the facility's function, one (or more) of the following transport activities has to be organized: the collection of used products from the disposer market, the delivery of recovered products to the reuse market, or the delivery of new products to the user market.

First and foremost, there is the question of novelty. How does the collection of old products differ from the distribution of new products? In some contexts, there aren't any significant new issues. Return flows of low value which may be consolidated to form large batches can be collected by firms deploying container transporters and/or transport by rail, waterways, or sea. Other than the issue of perhaps transporting invaluable goods, there is no significant difference with the transport of bulk materials. Valuable items, such as repairable airplane parts or refillable copier cartridges, can be transported by common carrier, parcel post, or express services. These transport companies do not make a distinction for their planning between 'normal' goods and products that are part of a reverse logistics flow (Romijn, 1999). Such cases hardly need any further discussion, yet it would be unjust to generalize.

Another (false) argument for claiming that there are no significant issues is that of the model's insensitivity to flow direction. Consider, for example, the Capacitated Vehicle Routing Problem (CVRP). In the CVRP (Dantzig and Ramser, 1959), an undirected, complete graph $G = (V, E)$ is given, with vertex set V and edge set E. Every edge has a non-negative cost and every vertex a non-negative demand for service. Given a vehicle capacity, the CVRP consists of finding a set of vehicle routes of minimum cost such that every vertex is serviced by exactly one vehicle, each route starts and ends at a pre-specified vertex (the depot), and the total demand serviced by a route does not exceed vehicle capacity. In the CVRP, it does not matter whether the goods are new or used or whether they need to be delivered or collected. The question, however, is not whether the model is insensitive to flow direction, but whether it is an accurate model for the specific problem faced in reality.

A number of vehicle routing models will be looked at. The models can be roughly classified as being of the normative or the descriptive type. Normative models seek to find the 'optimal' solution for a given problem instance. These models often lead to NP-hard, combinatorial optimization problems. Solution approaches include greedy heuristics, various local improvement heuristics, and the machinery of mathematical programming. Descriptive models study the general behavior of complex systems. Detailed instance data are replaced by concise summaries, and numerical methods by analytic models or simulation runs. These models allow one to identify broad properties of near-optimal solutions, including cost estimates, to investigate the impact of design decisions, and to formulate guidelines for the design of implementable solutions.

This chapter has three themes and is organized as follows. The first theme describes the features of reverse logistics that give rise to new issues and model features. This includes a description of organization options, degrees of freedom, and typical vehicle routing characteristics (Section 5.2). The second theme (Section 5.3) is the presentation of gaps between these new issues and the literature on the normative models. In the trade-off among clarity, brevity, and completeness, we made the decision to reference only the most relevant models, rather than attempt to provide a complete overview of the related vehicle routing literature or to delve into algorithmic aspects. The third theme (Section 5.4) presents descriptive continuous approximation models that capture some essential reverse logistics characteristics. In addition to showing the impact of system design on the collection and vehicle routing cost, these models also confirm that new research is needed to fill the gaps in the routing literature in order to effectively deal with reverse logistics. The chapter ends with conclusions and some directions for future research (Section 5.5).

5.2 Collection and Transport Approaches for Reverse Logistics

This section introduces the issues further investigated in this chapter. The content of this section is based on our own experience, having studied numerous reports on collection schemes and routing models, and having participated in some of these studies. Summaries of some case examples first introduce the key questions. A structured presentation is then given of the options for the design of such systems. A summary of the typical routing model features for reverse logistics concludes this part.

5.2.1 Case Studies

Collecting from industrial firms. A specialist recycler in the U.K. (James, 2001) collects products with a hazardous content from car repair centers, such as oil filters, fluorescent tubes, oil rags, and car batteries. The goods are collected every month or every three months. There are three depots in the U.K., and the country is divided into a set of 30 sectors. Every week, collection vehicles visit some of these sectors. One of the issues the considered company faces is *designing the sectors* so as to meet the desired periodic collection schedules at the lowest cost. There are two cost factors: the number of vehicles needed per week and the variable routing cost. The company wishes to minimize the hiring of extra vehicles and tries to work with a fixed set of company-owned vehicles. Therefore, balancing the workload by sector design is considered an important objective. Sector design models will be reviewed in Sections 5.4.2 and 5.4.3.

Collecting from households and small businesses. The cost of collection per unit of mass varies from city to city and region to region. Empirical findings about the efficiency of refuse collection schemes are reported by, for example, Stedge et al. (1993), Slovin (1994), Kimrey (1996), and McConaghy and Mills (1997). While a part of these differences are the result of differences in demographics and demand, another part can be attributed to the variety in the organization. For example, the costs differ, ceteris paribus, between a system that uses curbside collection and one where the collection occurs from drop-off sites, whether high or low collection frequencies are in place, whether regular or irregular (continually re-optimized) collection schedules are being used, etc. Some of these studies report the observed changes in collection cost and quantities due to the introduction of a *source separation* program. Instead of collecting a heterogenous stream of products at one time, the stream is now divided into distinct classes at the collection point. The primary purpose is to facilitate the separate reprocessing of each class in specific recycling units. One of the issues is the new *collection frequency* of each stream and deciding when to visit each collection point. Decisions should optimize the operational efficiency within the boundaries of the social and political constraints. A related issue is whether to collect those separate streams in a single *combined* visit with one truck, or *separately*, with two stops made by different vehicles. How does source separation, visit frequency, and collection method affect the transport cost? Section 5.4.3 provides some tools that give insight into these relationships.

Integrating deliveries and collections. Bettac et al. (1999) studied several take-back schemes, including printer cartridge recycling in Great Britain, power tools recycling in Germany, worldwide take-back and reuse of single-use cameras, and worldwide collection and refurbishing of IT equipment. In these cases, transportation of old products is combined with delivery of new products. While distribution schedules are usually fixed, there is more freedom in specifying the moment the old products are collected. Returns can be collected during the next delivery, but the collection may also be *postponed* until a larger quantity is available. In some retail chains (De Koster, 2001), deliveries occur daily to every store, and returns are only collected every two or three days. How to determine the optimal postponement policy? In addition, vehicle routing may need to take into account strict time windows for delivery, especially regarding food stores, with possibly a separate compartment in the vehicle reserved for cold storage. Collecting returning goods may get in the way. The issue of integration is studied in Section 5.4.4.

A common observation in these cases is the presence of two decision levels. On the highest level, the decisions concern the design of the system, such as the customer service policy. The challenge is to design the system so as to fully exploit the features specific to reverse logistics, in particular the available degrees of freedom. On the lower level, vehicle routing problems are solved taking into account the specific design and relevant additional constraints.

5.2.2 System Design Options

The design of the system should meet the general requirements of effectiveness and efficiency. Product recovery starts with the effective acquisition from the generators (former users). The aim, clearly, should be to avoid the targeted products ending up in (unwanted) waste streams. The generators must be willing to participate consistently. This behavior can be stimulated by assuring that the collection program delivers good service. The service must be convenient and consistent in time. At best, it will offer the lowest cost alternative. The second requirement for effectiveness is to design the collection and transport system in view of the targeted reprocessing application. Reuse naturally requires that products are returned in the best possible condition and are shielded from any sort of (weather) damage. Material recovery, however, can often bear less careful handling. The products also need to be transported in a cost-efficient way to the facilities of the reverse logistics network. Efficiency may call for the temporary storage and accumulation of products before being shipped, volume compaction, source separation, special vehicle characteristics, etc.

To facilitate further discussion on system design options, a distinction is made between four aspects: the collection infrastructure, the collection policy, the combination level of the collection, and the characteristics of the collection vehicles.

Collection Infrastructure

The collection infrastructure relates to the points at which generators hand over the used products to the network. The three prominent types are as follows.

- **On-site collection.** Used products are collected on the premises of the generator. This type of collection is often applied to commercial firms. Another well-known application is curbside collection of refuse and recyclables from households. The family places the containers/bags/bins at the curbside on the collection day. The containers are emptied on site. Also, large and heavy products, such as washing machines, are commonly collected on site.
- **Unmanned drop-off sites.** The generators bring the products to large storage containers at a designated location in the neighborhood. This is often used as an alternative to curbside collection for refuse or separated recyclables such as glass bottles, paper, or textiles.
- **Staffed and smart drop-off sites.** Staff supervision allows for a more selective acquisition and careful separation. Municipal collection depots, second-hand shops, and even regular shops and retailers can all fulfill this function. Smart drop-locations are 'intelligent' unmanned drop-off sites with a similar purpose. They are relatively new and are currently used for recyclables or reusable packaging (tin cans, bottles). A smart glass collection machine, for example, automatically sorts bottles according to color and does not accept other products. In addition, the machine shreds the bottles to maximize the use of its capacity. Equipped with monitoring devices and telematics, it signals information on the remaining capacity or any malfunction to the collector.

Introducing more systems in parallel usually increases the capture rate; for example, installing drop-off sites both at the retailers and at the municipal refuse collection depots. The integration with existing programs, in particular with the local refuse collection system, can be appropriate and has the advantage of familiarity. The collection might also be organized infrequently and ad hoc. For example, generators can be invited through the local media to bring the products to the municipal depot during a specific 'collection week'. Or, vehicles can make short stops near convenient locations such as schools and churches, providing in this sense 'mobile' drop-off sites. An overview of collection infrastructures and monitoring systems is given by McMillen and Skumatz (2001).

Collection Policy

The collection policy specifies the moment(s) at which a collection point is serviced and the volume collected per visit. The foremost ways to determine the moment of collection are as follows.

- **Periodic schedules.** A periodic schedule is a subset of periods (days) chosen from a base set of consecutive periods, which is repeatedly used. Typically, the collector can initially determine the actual visit periods to be included in the schedule. Once specified, however, visits routinely have to occur during these periods.
- **By monitoring demand.** Smart drop-off sites monitor the generation rate, and the information is used to insert visits just-in-time, to prevent overspill, in a dynamic route-planning model.
- **Call services.** For collections from staffed drop-off sites, visits may be be triggered by a call from the collection point. The collector may specify that a minimum quantity of products should be available before the call is made. In a call service with periodic collection, the collector will plan the visit in the next time period that is part of a pre-arranged periodic schedule for the geographic area to which the collection point belongs. In the call service with a timely collection guarantee, the collector assigns the visit to a period at will, but ensures collection before a predefined time has elapsed from the moment of the call.
- **Triggered by a distribution schedule.** If the integration of collection and delivery is allowed (see also the next section on combination level), the moment of collection could be triggered by a distribution schedule for the customer at the collection point itself or for a customer at a location in its neighborhood.

As for the volume collected per visit, it is standard practice that all goods are collected. The exception occurs when the vehicle has reached its capacity. In that case, the remaining products are either collected (by another vehicle) shortly after or during the next scheduled visit.

Combination Level

Services related to different classes of flows of goods can be combined in various ways.

- **Separate routing of independent resources.** A dedicated fleet of single compartment vehicles collects one (possibly heterogeneous) flow of goods.
- **Separate routing of shared resources.** Two or more classes of flows are collected either with the same crew or by the same set of vehicles, or both, but in any case different classes are never in the same vehicle at any point in time. The vehicle is dispatched to collect one class and it can only

be assigned to a different class after unloading at the depot (and possible cleaning and setup of the vehicle).

- **Co-collecting source-separated flows of goods.** Two or more classes are collected simultaneously, hence two products of different classes are allowed in the vehicle. Multi-compartment vehicles are often needed for practical purposes (see next section on vehicle type).
- **Integrating collection and delivery tasks.** The most common ways are mixing, backhauling, and partial mixing. *Mixing* is the situation in which deliveries and collections can be made at random as long as the vehicle's capacity constraint is respected. A vehicle therefore can collect before all deliveries have been made, resulting in a mixed load. Customers with both a delivery and collection request may require only a single visit. Mixing is not always applicable. First, laws may prohibit mixing certain new and end-of-life products. Second, delivering new products can take priority over collecting old ones. It may be unacceptable to postpone a delivery as a consequence of (unexpected) delays caused by previously made collections. Finally, vehicles are often rear loaded, hence the on-board load rearrangement required by mixing might be difficult, or even impossible, to carry out at customer locations. In those cases, backhauling is an alternative. *Backhauling* means that the vehicle makes all its deliveries before any collection can be made. *Partial mixing* is the approach where mixing is only possible if there is enough shuffling space in the vehicle. In general, a feasible route looks as follows. The (full) vehicle starts with deliveries until the volume occupied by loaded cargo is reduced to a certain level. From this point onward, mixing is allowed, provided that this level is not exceeded. When all deliveries are made, collected goods can fill the vehicle up to its normal capacity.

Vehicle Type

Finally, the characteristics of the collection vehicles have to match the collection infrastructure, policy, and combination level. Different truck designs and assorted collection equipment are described by Graham (2001). There are two remarks of importance for this chapter.

- Traditional collection vehicles have a single compartment and may be equipped with a compaction mechanism. Recent co-collection vehicles are flexible in number and relative size of compartments, or in the compaction rate of different compartments. Dual-compartment vehicles for example, can co-collect two classes, e.g. refuse and recyclables, refuse and yard trimmings, or two streams of recyclables.
- Vehicles focusing on the integration of collections with deliveries are made more accessible than the traditional rear loaders by means of tailgates at the sides, possibly in combination with a conveyor belt covering the full length of the truck's floor.

5.2.3 Features of Vehicle Routing Models for Reverse Logistics

This section summarizes the typical features that may be needed or encountered in (normative) vehicle routing models for reverse logistics.

Single-period Models

The distribution policy in a 'traditional' supply chain context is typically customizable. The benefits of personalized delivery outweigh the additional logistics cost. But an old returned product does not represent a capital cost to the generators or retailers, and there is no such thing as the threat of a 'lost sale' for the retailer if the collector chooses, within some limits, the collection moment. As a result, retailers and generators can more easily agree on a standard policy if this keeps the bill low. In particular, the absence of customer-specific time constraints, visit schedules, and/or bin-packing effects (see 'allowing split collection' below), results in routes that are less likely to overlap. Each vehicle will collect in its own geographical zone. The following modeling features may be encountered.

- **Demand on nodes and arcs.** In distribution settings, typically, demand can be modelled as being located on the nodes of a network, i.e. the extreme points of the arcs. Each arc between two demand nodes represents the shortest or cheapest set of roads connecting these two delivery locations. In some reverse logistics settings, e.g. curbside collection of recyclables, demand may be better modelled as being continuously distributed along (some of) the arcs of a (road) network, given the many stops per vehicle tour.
- **Allowing split collection.** As a rule, split delivery is not allowed in the 'traditional' distribution models, i.e. the total demand of a customer must be delivered in one single visit. Vehicle routing models hence need to cope with bin packing constraints which may interfere with the minimum distance objective, especially when there are large demands relative to the vehicle capacity. In reverse logistics, split collection is more often allowed on the condition that the accumulation capacity of the collection infrastructure is taken into account.
- **Multiple vehicle types.** Professional collectors deploy a mixture of several vehicle types, different in number of compartments, capacity, multimode capabilities, compaction mechanisms, and/or (un)loading mechanisms. The vehicle routing may involve deciding which vehicle to use for which collection tasks (vehicle fleet mix problems).
- **Combining multiple inbound and outbound flows.** The acquisition of low-value flows of goods from a large number of sources requires the use of a low-cost system and, obviously, increasing the combination level can help achieve this objective. The vehicle routing models thus need to be able to cope with the various combination approaches described in Section 5.2.2.

- **Supply uncertainty.** As often mentioned in this book, the availability in timing and volume of used products is, in general, more difficult to predict than in a distribution context. Stochastic vehicle routing techniques apply. In perhaps the simplest way, however, robustness can be obtained by artificially reducing the collection capacity of the vehicle for planning. In refuse collection, the compaction mechanism can be 'abused' to collect a little more than foreseen. Split collection may also be a means to avoid extra vehicle trips.

Multi-period Models

In distribution settings, customer-specific time windows or narrow transport time windows arise frequently. In a typical reverse logistics context, the time windows are not that extremely small, and for low-value products the time span covers typically more than a few (work)days. To exploit the option of postponement of collection to later workdays (periods), *multi-period* vehicle routing models have to be used. The features mentioned for the single-period models may be encountered here as well. In addition to these are the following.

- **Collection policy.** Multi-period vehicle routing models are needed for the various collection policies specified in Section 5.2.2.
- **Minimizing the fixed cost.** In some reverse logistics settings, sizable investments in specialized vehicles or manpower require the consideration of fixed costs as well as variable (routing) costs. In particular, for the design of the collection policy and for solving the multi-period routing problem, the maximum number of vehicles simultaneously deployed over the planning horizon is to be minimized. While fixed costs obviously are also of importance in distribution settings, it is generally easier and cheaper to rent common vehicles or hire common carriers in peak periods.

5.3 Vehicle Routing Models

The objective here is to detect some gaps in the literature on normative vehicle routing models. Since the literature is abundant, we will only list some of the most relevant recent articles. Also, we focus on deterministic models defined on undirected graphs.

5.3.1 Single-period Models

Demand on nodes and arcs. Three different classes can be distinguished based on the location of demand: node, arc, and general routing. The basic formulation for node routing is the Capacitated Vehicle Routing Problem (CVRP) (see the introduction), a well-known NP-hard problem (Garey and Johnson, 1979). It has been widely applied to model collection problems (e.g.

Gelders and Cattrysse, 1991). The literature on CVRP, extensions, algorithms, and applications is abundant. Surveys can be found in, for example, Laporte (1992) and Toth and Vigo (2002).

The arc routing version of the CVRP is the Capacitated Arc Routing Problem (CARP), which is also NP-hard (Golden and Wong, 1981). The CARP captures the essence of curbside collection and collection from densely distributed drop-off sites. Surveys on arc routing algorithms and the various applications can be found in Assad and Golden (1995), Eiselt et al. (1995) and Dror (2000). The solution approaches have only recently raised to a level of node routing algorithmic performances. One of the key features explaining the breakthrough is the more natural and economical representation of a route. While the 'classic' approach is to include the required as well as the deadheading edges (i.e. edges connecting consecutive required edges or vertices in a tour) in the solution vector, the new solution representation takes into account only the sequence of the required edges and the direction in which a required edge is traversed. This brings about a more efficient and effective implementation of local search improvement procedures, described in Muyldermans et al. (2001) and applied for the CARP in a guided local search algorithm in Beullens et al. (2003). Good lower bounds for CARP can be obtained by cutting planes (Belenguer and Benavent, 2000; Benavent et al., 2000).

Certain problems involve (small) collection requests which are dense on some streets and sparse on other streets. It seems appropriate to model the dense demands as one demand along an arc, and the sparse demands as demand on nodes. The uncapacitated (single-vehicle) version of this problem is the General Routing Problem (GRP) (Orloff, 1974), which is NP-hard in general (Lenstra and Rinnooy Kan, 1976), and solvable in polynomial time only if all arcs of the network have a positive demand (reducing the GRP to the Undirected Chinese Postman Problem, Edmonds and Johnson, 1973). The currently best approach for the GRP appears to be the 2-opt and 3-opt variants in combination with guided local search (Muyldermans et al., 2002). Good lower bounds can be obtained by a cutting plane approach (Ghiani and Laporte, 2000; Corberán et al., 2001). It appears that studies on extensions of the GRP, such as the capacitated GRP, are not available. Note that while only demand nodes not incident to demand arcs are relevant in the GRP, this is not so for the capacitated version.

Allowing split collection. Some insights on the effect of split demands for the CVRP are given by Dror and Trudeau (1990). The problem is NP-hard. If the distance matrix satisfies the triangular inequality, it proves that no two routes in the optimal solution can have more than one split demand point in common. Furthermore, it proves that in the optimal solution the following cannot occur: consider any set of k routes, each route $i < k$ including customers $\{p_i, p_{i+1}, ...\}$, and route k including $\{p_k, p_1, ...\}$. In experiments with a heuristic solution approach, savings of one vehicle or more were not uncommon, and average distance reduced by 1.5%. The routes constructed also 'tend to cover cohesive geographical zones'. As can be expected, savings from split delivery routing appeared to be larger for larger demand (as a proportion to vehicle capacity). No comparable studies, however, seem to be available for arc- or general routing, or any of the routing problems further discussed in this section.

Multiple vehicle types. The selection of the vehicle type has seldom received any attention in normative models. In the vehicle fleet mix problem, the (potential) fleet consists of vehicles of different sizes. Salhi and Sari (1997) present a literature review and propose a multi-level construction heuristic. While larger vehicles will lead, in general, to more efficient routing, there are clearly practical boundaries to vehicle capacity by road conditions and/or maximum driving times, and also the trade-off with the fixed costs (maintenance, drivers' salaries, insurance, road tax, etc.) cannot be neglected. The customer's site and how it is implanted in the neighborhood can add so-called 'site-dependent' constraints to the problem, i.e. not all vehicles can visit all sites. If there is only a limited number of different vehicle types required or available, the site-dependent CVRP can also be solved as a Periodic Vehicle Routing Problem by transformation (Cordeau et al., 1997). The arc routing variant is studied in Sniezek (2002). No studies have been found that deal with the issue in a context of multiple flows and a fleet that consists of single- and multi-compartment vehicles, or in problems involving the integration of inbound and outbound flows.

Combining multiple inbound and outbound flows. Multi-compartment vehicles also receive limited attention in the literature. An issue of multi-compartment collection is the effect of randomness on routing efficiency. As pointed out by Sodhi (1999), randomness in the pickup loads on the one hand and rigid, smaller compartments on the other may cause one compartment to fill more rapidly than the others. This effect will therefore decrease the collection efficiency compared to mixed collection in one large compartment. Many vehicle types do not allow for an adjustment of relative compartment volume once the vehicle starts its collection tour. Sodhi and Reimer (1999) model the problem as a newsstand problem, where a capacity limitation is placed on several different items in inventory. It is still unclear when multi-compartment vehicles should be used for the co-collection of source-separated classes (see Section 5.2.1). The issue is further investigated in Section 5.4.

Several generalizations of the CVRP can be associated with the integration strategies presented in Section 5.2.2. These problems are defined on $G = (V, E)$, where $V_d \subseteq V$ represents a subset of vertices each receiving an amount of goods from the depot, and $V_c \subseteq V$ the subset of vertices requiring a pickup of goods to be brought back to the depot. Problem formulations differ in the way V_c and V_d are constructed and in the way the vehicle can sequence the deliveries and collections on its route.

- **Backhauling**. In the Vehicle Routing Problem with Backhauls (VRPB) (Deif and Bodin, 1984), $V_c = V \setminus V_d$, and a vehicle cannot perform a pickup before its last delivery has been made.
- **Mixing**. In the Vehicle Routing Problem with Mixed deliveries and collections (VRPM) (Golden et al., 1985), $V_c = V \setminus V_d$, and collections can be made before all deliveries are made by the vehicle, resulting in a 'mixed' vehicle load. In the Vehicle Routing Problem with Simultaneous Delivery and Pickup (VRPSDP) (Min, 1989), $V_d = V_c = V$, and each node needs to be serviced in a single stop.
- **Partial mixing**. In a slightly relaxed version of the VRPB, the general idea is to allow collections before all deliveries are made if this keeps enough free shuffling space in the vehicle. In the Vehicle Routing Problem with Partially Mixed deliveries and collections (VRPPM) (Casco et al., 1988), $V_c = V \setminus V_d$, and collections may start as soon as the remaining load of deliveries in the vehicle drops below a certain percentage x of the vehicle capacity. Furthermore, as long as there are delivery loads in the vehicle, a collection load is only accepted if the free capacity after the collection would have been made is above a certain percentage y of the vehicle capacity $(x < y)$.

The VRPB, VRPM, and VRPPM formulations can handle the situation of vertices with both a delivery and collection request, henceforth called *exchange vertices*, simply by dividing these vertices into fictitious 'delivery only' and 'pickup only' vertices. The original exchange vertices may, however, when necessary, be serviced twice: one stop for the delivery and another stop for the collection. In the VRPSDP, all nodes are exchange vertices requiring a single visit. The VRPM is a special case of the VRPSDP, where either the delivery or the collection demand of each vertex equals zero (Dethloff, 2002). The VRPSDP can also handle situations where only some of the exchange vertices require a single visit by dividing the other exchange vertices into fictitious 'delivery only' and 'pickup only' vertices, and taking the collection and delivery loads of these vertices to be zero, respectively. All these route planning problems are clearly NP-hard, since the CVRP is a special case (take $V_c = \phi$ or $V_d = \phi$).

Both the VRPB and VRPM have received considerable attention in the literature. An overview is given in Beullens (2001). Yet only a few construction heuristics, largely based on cheapest insertion, are available for VRPPM (Casco et al., 1988; Salhi and Nagy, 1999) and VRPSDP (Dethloff, 2001). Recently, Salhi and Wade (2002) experimented with an ant system metaheuristic for VRP(P)M. A generalization of VRPM and VRPSDP is the Pickup and Delivery Problem (PDP)(see e.g. Dumas et al., 1990). In the PDP, a set of transportation requests is to be carried out, each request having an origin vertex and a destination vertex that can be, and typically are, different from the depot vertex. As the PDP is a somewhat 'extensive' generalization, specialized solution approaches for VRPM and VRPSDP are likely to perform better.

5.3.2 Multi-period Models

These models use a planning horizon of several periods (days) and solution approaches deal with two decisions: the allocation of visits to a period of the planning horizon, and the determination of the individual vehicle routes per period.

Such problems are sometimes referred to as allocation/routing problems, and in other cases as inventory routing problems. In general, in allocation/routing models, the quantities to deliver/collect are either fixed amounts per visit or deterministic values based on the time since the last visit. Periodic vehicle routing models are a special type in which the solutions need to involve a periodic schedule per customer. Inventory routing models typically use (updated) information on the (stochastic) demand or production rate of the customers in order to determine the visit moment and the amounts to deliver/collect per visit as to avoid stockouts. Reviews of models and solution approaches can be found in Ball (1988) and Campbell et al. (1998,2002).

Collection policy. The most studied multi-period problem seems to be the Periodic Vehicle Routing Problem (PVRP). In the PVRP, as defined by Cordeau et al. (1997), an undirected, complete graph $G = (V, E)$ is given, with vertex set V and edge set E. Consider a planning horizon of t periods, and each vertex $v \in V$ specifies a service frequency f_v and a set C_v of allowable combinations of service days. For example, if $f_v = 2$ and $C_v = \{(1,3), (2,4), (3,5)\}$, then v must be serviced twice, either in period 1 and 3, in period 2 and 4, or in period 3 and 5. Every edge has a non-negative cost and every vertex a non-negative demand for service. The PVRP then consists of simultaneously selecting a schedule for each vertex and establishing vehicle routes for each period of t, according to the rules of the CVRP, so that total costs are minimized. The problem is clearly NP-hard since the CVRP is a special case where $t = 1$. The best algorithm so far may well be the tabu search heuristic of Cordeau et al. (1997).

Many authors list the PVRP as the basic model for solving the collection from commercial sites or unmanned drop-off sites. Periodic schedules are particularly suited for steady flows of goods and when it is necessary to make the collection infrastructure accessible to the collector on the collection day (customer familiarity with collection schedules). Driver familiarity with scheduled routes is another advantage.

Despite its applicability, the PVRP has inherent disadvantages in some settings. Annual routing costs rise with the collection frequency, but the time between two consecutive visits must be short enough to avoid overspill at the collection points. When the variability over time of generated volumes at the collection points is high, a periodic solution will either have a too-high routing cost or will result in overspills. Another strategy is to go for smart drop-off sites that monitor the generation rate and use that information to insert visits just-in-time (to prevent overspill) in a dynamic (non-periodic) planning model. When the time between two consecutive visits is relatively large, e.g. every two months, this situation matches with the distribution model of Dror and Ball (1987). The model works typically with a (rolling) planning horizon of one or two weeks. The model of Bell et al. (1983) is similar but is more appropriate when collection points need typically more than one visit in the planning period.

There seem to be no vehicle routing models available that focus on call services. The problem here is to plan the routes over the planning horizon in anticipation of future calls. Assumptions are needed about the probability that any given customer will call in any given period. There is a lack of multi-period routing models for combined collection in multi-compartment vehicles and for all forms of integration with deliveries.

Minimizing the fixed cost. Most (PVRP) models focus on the minimization of the variable transportation costs. Only some references consider the more strategic objective of minimizing the fleet size, i.e. minimizing the maximum number of vehicles simultaneously deployed over the planning horizon. Gaudioso and Paletta (1992) study the fleet size problem for the PVRP in the context of (soft drink) distribution. Multiple routes can be assigned to the same vehicle on the same day, but the total travel time per day and per vehicle is bounded from above. They compared their bottleneck objective heuristic with results from the literature that used the minimum total distance objective. Computational experiments indicate that there is a trade-off between both objectives, while other differences are attributed to performance differences between the algorithms. Hanafi et al. (1998) address the PVRP fleet size problem by assigning workloads to the edges or arcs of a graph, and use tabu search to improve the current workload plan, aiming for a better balance between the workdays. Eisenstein and Iyer (1997) develop a somewhat special model for the city of Chicago. The arc routing component is reduced to a network of city blocks. Each block is serviced once a week, and residents are not promised a fixed collection day. A vehicle visits several blocks per day, collects for 2.5 hours, and then goes to the landfill; this can be done once or twice per day. A Markov decision process forms the basis to divide the work of a service region among vehicles and to set up a five-day decision plan for each vehicle determining which blocks should be visited per day and whether this should include one or two dump trips. At the end of each day, the vehicle crew checks the work done and then a dynamic probabilistic program proposes the workplan for the next day in order to maximize the service level, i.e. the probability of completing in five days the blocks assigned to the vehicle.

Considering the more strategic question of fleet size naturally leads to the question of what the impact would be of other design parameters, in particular the collection frequencies. Baptista et al. (2002) model the collection of recyclable paper from unmanned drop-off sites with the objective of optimizing total profits. Although the frequency of collection is a decision variable, the impact on fleet size is not investigated.

There are still many other gaps in the multi-period routing literature (see Campbell et al., 1998). It appears that much more work has to be done on lower bounding procedures. Multi-period models addressing the issues discussed in Section 5.3.1 also seem to be unavailable as of yet.

5.3.3 Conclusion

The normative vehicle routing models to date have laid a firm foundation for many of the current vehicle routing approaches for reverse logistics. On the other hand, considering the specific characteristics of reverse logistics, some specific areas deserve increased research attention. Most striking are the lack of lower and upper bounds for multi-period (periodic) models in the context of arc routing, or in the context of node routing and the integration of deliveries and collections (with backhauling, mixing, etc). There is also a need for models that consider co-collection in multi-compartment vehicles.

It's not only important to have good vehicle routing algorithms. Many issues in reverse logistics and vehicle routing (see Section 5.2.) ask for the proper design of the logistics system in which the vehicle routing occurs. The system design fixes the collection infrastructure and policy, the scope and scale of the collection, and the vehicle types. The design determines the constraints and degrees of freedom for the vehicle routing and usually has a large impact on the collection cost. For example, for the PVRP insights are welcome about the impact of the collection frequencies or the variation in demand on the fleet size, on the variable routing cost, and on the optimal periodic schedules for the customers. Normative vehicle routing models are rather intricate by nature and difficult to use for obtaining such insights. The next section therefore delves deeper into an alternative approach.

5.4 System Design Models

Descriptive or system design models reveal the impact of critical spatial, temporal, and vehicle- and demand-related parameters on the total logistics cost. By making some simplifying assumptions and using aggregated data about customer locations, demand patterns, or vehicle routing, closed or near-closed form solutions can be obtained.

In the area of (multi-echelon) distribution, models based on (probabilistic) asymptotic analysis are reviewed in Federgruen and Simchi-Levi (1995) and Anily and Bramel (1999). This line of research is directed at constructing (polynomial) algorithms and analyzing their performance, e.g. proving the asymptotical accuracy in the limit when the size of the problem tends to infinity. Another type of model uses the continuous approximation approach from Newell (1971) and Daganzo (1984). This stream of literature is more focused on developing and understanding solutions by using an analytic approximation for the vehicle routing cost to evaluate the performance of design decisions. Although these models do not assume asymptotic conditions, the approximations become more accurate for larger problems. Currently, continuous approximation models are more and more integrated into mathematical programming models. A review of models is given by Langevin et al. (1996) and Daganzo (1999).

In this section, some system design models are developed or reviewed for the PVRP in a reverse logistics context. Return flows are assumed to arise at a constant rate at a fixed but large number of collection points. The main question is how to find, in various system designs, an optimal periodic collection schedule for each point. A single stream of goods is considered in the first two subsections. The last two subsections analyze and compare various forms of co-collection and integration with distribution, respectively.

5.4.1 Economic Order Quantity Models

A popular approach to finding optimal periodic schedules is the use of economic order quantity (EOQ) models. Two models, adapted from Burns et al. (1985), are examined. These models also serve to introduce the notation and assumptions used in later sections.

Direct Shipping

The standard EOQ for *direct shipping* between a depot and a single customer is widely known. It can, of course, be applied to a collection problem as well. For every collection trip, costs are incurred corresponding to the loading at the customer, the transport, and the unloading. This gives

$$\text{transport cost per item} = (K_s + c_t\, g\, \ell + K_r)/Q_r, \tag{5.1}$$

where K_s is the cost per stop at the collection point, c_t is the transport cost per unit of distance, ℓ is the travel distance between the depot and the collection point, K_r is the setup cost for dispatching a vehicle and unloading at the depot, Q_r is the lot size, and g is equal to one in the case that only the backhaul trip is charged (e.g. when outsourced to a common carrier), and equal to two in the case that line and backhaul trips are charged.

Three relevant waiting times are involved: waiting at the collection point, the recovery lead time, and the time at the depot before the item is further treated. Assuming the schedule of the latter process and the transportation are uncorrelated:

$$\text{average holding cost per item} = h_r(Q_r/2\, u_r + L_r + Q_r/2\, u_r), \tag{5.2}$$

where h_r is the holding cost rate, u_r the return rate at the collection point, and L_r the transit time. The total cost is the sum of (5.1) and (5.2). First order conditions give the optimal shipment size Q_r^* and frequency u_r/Q_r^*, where

$$Q_r^* = min(\sqrt{(K_s + c_t\, g\, \ell + K_r)(u_r/h_r)}, Q_r^{max}, W), \qquad (5.3)$$

and where W is the vehicle capacity and Q_r^{max}, commonly omitted in other texts, represents the accumulation capacity of the collection point.

The problem now is that in many instances of reverse logistics $Q_r^* = Q_r^{max}$. The transport cost largely exceeds the value of the items, and even volume reduction before shipping can often not alleviate the problem. This is a real issue, limiting the success of recycling and reuse; the collection points and the vehicles that fit the existing road infrastructure are not able to deal with the large volumes which would be required.

Multiple Stops per Vehicle

To allow the collection of smaller quantities per point and per trip, a vehicle can also make *multiple stops per tour*. To express the distance travelled by the vehicle in this model, Daganzo's continuous approximation is used, introduced in Section 4.5 of this book. It is a function of the Euclidean distance $\ell(x)$ between the center of the vehicle's delivery region x and the depot, the average number of stops per vehicle $(= W/Q_r)$, and the *local delivery distance* which is a function of the point density in x, $\rho(x)$, a location-dependent function which is slowly varying in x within some overall service area A containing, in total, n collection points. Assuming a full truckload tour, Daganzo's result can be stated as:

$$\text{average distance per stop} = 2\ell(x)/(W/Q_r) + 0.57/\sqrt{\rho(x)}. \qquad (5.4)$$

It follows that $K_s/Q_r + K_r/W + c_t(2\ell(x)/W + 0.57/(Q_r\sqrt{\rho(x)}))$ represents the average transport cost per item in x. Note that this expression reduces to Equation (4.16) for the case of $v = W/Q_r$, $K_s = K_r = 0$, and continuously distributed collection requests. Integration over the whole service area A gives

$$\text{avg. transp. cost per item} = K_s/Q_r + K_r/W + c_t(2I_1/W + 0.57I_2/Q_r), \qquad (5.5)$$

where $I_1 = (1/n) \int_A \rho(x)\ell(x)dx$, $I_2 = (1/n) \int_A \rho(x)^{1/2}dx$. Compared to direct shipping, the waiting time for an item is extended by the transit time on the local collection tour. If it is L_{rl} for the last visit on the route and zero for the first, then on average it is approximately $L_{rl}/2$. The average holding cost per item is thus Equation (5.2), in which L_r is replaced by $(L_r(x)+L_{rl}(x,Q_r)/2)$. For the sake of integration, it is fair to assume that time is related to distance, thus $L_r(x) \sim \ell(x)$ and $L_{rl}(x,Q_r) \sim 0.57/(Q_r\sqrt{\rho(x)})$:

$$\text{average holding cost per item} = h_r(Q_r/u_r + a\,I_1 + 0.57\,b\,I_2/Q_r), \qquad (5.6)$$

where a and b are constants expressing the inverse of the average speed of the vehicle on the backhaul trip and the local collection tour (including loading times), respectively. Adding Equations (5.6) and (5.5) results in the average total cost per item. First order conditions give

$$Q_r^* = min(\sqrt{(K_s + 0.57c_t I_2)(u_r/h_r)}, Q_r^{max}, W). \qquad (5.7)$$

For the case of uniformly distributed collection points, I_2 equals $1/\sqrt{\rho}$, and I_1 equals $\bar{\ell}$. $\bar{\ell}$ denotes the expected distance from a randomly chosen collection point in the service region to the depot. It is easy to show that the average total cost per item is a continuous monotonously decreasing function of point density ρ, ceteris paribus. A higher ρ implies better service for the collection points: Q_r^* will decrease as seen in Equation (5.7), and the optimal frequency will increase. Thus, when the collection is outsourced to more than one collector, it is best for cost *and* service to assign a specific geographic region to each collector, without overlaps between their respective regions. From an overall perspective, it is therefore unfortunate that, in practice, different collectors (vehicles) sometimes operate within the same service area. Descriptive models like the one above easily allow for the assessment of the loss of overall cost and service efficiency from such service region overlaps.

For collecting a given amount δ generated per unit of time in the service region, it is easy to show that the average total cost per item increases with the point density ρ. (Note that dividing ρ by a factor will increase u_r by the same factor, Q_r^* will increase less than by the factor (see Equation (5.7)), and therefore the optimal collection frequency will increase.) This partially explains why curbside collection is more expensive than the collection from drop-off sites or municipal depots (where, obviously, a part of the cost for travelling and (un)loading is on account of the generators). Usually, the cost difference is even worse since, as in the direct shipping model, here too Q_r^* is often limited by Q_r^{max}, and this is especially so for curbside collection. Another cost disadvantage of curbside collection that is not reflected in the above model is related to the flexibility in the visit schedule design. This issue will be clearly explained and demonstrated in Section 5.4.2. Another aspect that sometimes plays a role in this trade-off is that curbside collection typically realizes higher capture rates (higher values of δ). From an overall perspective, higher capture rates may be favorable since this may lead to increased recycling rates.

Finally, note that multiple stops per vehicle often lead to lower collection costs than direct shipping. Neglecting the holding cost, and for the case of uniformly distributed collection points and binding accumulation capacities, it can be easily seen from the comparison of Equations (5.1) and (5.5). It follows that direct shipping is less costly when the ratio Q_r^{max}/W exceeds the threshold value $(K_r + c_t g\bar{\ell} - 0.57\, c_t/\sqrt{\rho})/(K_r + 2\, c_t\, \bar{\ell})$. Direct shipping is more likely to be better the lower ρ, g, K_r, and/or $\bar{\ell}$ are and the more the accumulation capacity is matched with the vehicle capacity.

In conclusion, if the low value of the goods makes the accumulation capacity of the collection points the binding constraint, the design problem reduces to finding a balance between the transport cost and the service aspects, i.e.

assuring the accessibility of the collection points for the generators and providing a sufficient collection frequency to avoid collection point overspills or odor problems. This approach is used in the so-called sector design models, studied in the next section.

5.4.2 Sector Design Models

Sector design models aim to find a set of sectors by which to partition a service region and, for each sector, a periodic schedule with the objective to minimize somehow the investment cost in vehicles and crew as well as to minimize the weekly routing cost, while giving a certain level of service to the generators located in the service region.

Sector design models differ from the above EOQ models in a number of ways: a different objective function, the distinction between holidays and workdays, another way of dealing with the accumulation capacity, and the consideration of strategies that are, in general, so-called 'discriminatory' or 'asymmetric', i.e. all generators are not treated alike. This will become clear by first presenting an example, then a sector design model, and finally an analysis of the example at the end of this section. For the sake of clarity, uniformly distributed collection points are assumed throughout this section. The models are also valid for the general case of a location-dependent collection point density which is slowly varying in the overall service area.

A Sector Design Problem Example

An example, adapted from Mansini and Speranza (1998), serves to state the problem more specifically. Refuse needs to be collected from households in a city. A set of $n = 3,360$ uniformly distributed containers daily collects, in total, $p = 179$ tons. The required collection frequency $f = 3$ times per week. The accumulation time between two consecutive visits to any collection point must fall between $a_{LB} = 2$ and $a_{UB} = 4$ days to avoid either a low pickup volume or an overspill. The collector does not work on Sunday. A fleet of dedicated vehicles has to be purchased, and each vehicle has a capacity of 10 tons and can perform two collection tours a day. The problem is partitioning the service region into a set of fixed sectors and creating the periodic schedule per sector so as to minimize the sum of the weekly fixed fleet investment and the variable routing cost. (Mansini and Speranza only considered the first objective.)

A Sector Design Model

The objective function is an additive cost function with two attributes. In general, it is clear that in an hierarchical approach, suboptimal solutions are found. When first the fleet size cost is minimized, an extra constraint is added to the problem of the route cost minimization, and vice versa. In the first part, we analyze this trade-off by developing an expression for the routing cost as a function of the sector design. In the second part, we present a model for the sector design.

Some notations are introduced. The reference time interval of one week is represented by an ordered set $T = (0, 1, ..., t - 1), t = 7, 0 =$ Monday, $1 =$ Tuesday, etc. Consider any sector design. It partitions the service region of area A into a set S of sectors. Sector $s \in S$, of area A_s, contains n_{A_s} collection points, and has an associated periodic schedule $\mathbf{s} = (s_0, s_1, ..., s_i, ..., s_{kt-1})$. The integer $\lfloor i/t \rfloor$ is called the time interval of period i in $\mathbf{s}$. There are k time intervals in $\mathbf{s}$, and k is the periodicity of the schedule. A schedule has the following properties (necessary and sufficient): 1) it contains $k f$ non-zeros, 2) each non-zero s_i corresponds with a planned visit on weekday $i/modulo(t)$ in the $\lfloor i/t \rfloor^{th}$ time interval, 3) any s_j must be zero if it corresponds to a non-working day, 4) there has to be f non-zeros in every time interval, and 5) let s_i and s_j be two consecutive non-zeros (from left to right, subscripts taken modulo $k t$) with $v - 1$ zeros in between, then v represents the accumulation time between them and $s_j = v$, $a_{LB} \le v \le a_{UB}$.

The routing cost as a function of the sector design will now be analyzed. Take any non-zero s_i from a sector s. In that sector are n_{A_s} points, each having a load of $s_i (p/n)$ in period i. The distance travelled in s on the corresponding workday is approximated by the integration of Daganzo's result (5.4) over A_s, which gives

$$\text{transport cost in } A_s \text{ in period } i =$$
$$(K_r + 2\bar{\ell}_{A_s} c_t) \frac{n_{A_s} s_i (p/n)}{W} + 0.57 c_t \sqrt{A_s n_{A_s}} + K_s n_{A_s}. \qquad (5.8)$$

where $\bar{\ell}_{A_s}$ denotes the average distance from the depot to the collection points in s. This can be repeated for every non-zero element of all sectors. Easy calculus shows that the summation and the division by k results in

$$\text{average transport cost per week} = (K_r + 2\bar{\ell} c_t) \frac{t p}{W} + f(0.57 c_t \sqrt{A n} + K_s n). \qquad (5.9)$$

The continuous approximation theory thus suggests that the average transport cost is not significantly affected by the sector design. Neither the relative size of the sectors, nor their location in the service region, nor their corresponding schedules, nor the accumulation constraints are present in Equation (5.9). Daganzo (1999) reports similar findings in the context of multi-period distribution, where the total cost also appears to be largely dependent on the distribution frequency. The simplicity of the equation is also attractive for decision-makers to quickly get an idea of the impact on the routing cost of a change in density of drop-off locations or in collection frequency. For typical refuse collection data, the formula predicts, for example, savings in distance travelled of around 20% when reducing the collection frequency from twice to once per week, which is also the number measured in empirical studies (see McMillen and Skumatz, 2001).

The above analysis indicates that the most important objective for the sector design is to minimize the fixed cost related to the fleet size (hence supporting the objective formulation of Mansini and Speranza). Beullens (2001) formulates the problem of sector design as follows.

$$\min q,$$
$$subject\ to$$
$$\sum_{(s \in S)} \mathbf{s}^T\ z_s \le \mathbf{q}^T$$
$$\sum_{(s \in S)} z_s = 1$$
$$z_s \ge 0 \qquad \text{for } s \in S$$
$$q \ge 0$$

where q is the peak load, i.e. the maximum workload over all workdays; $z_s \in [0,1]$ indicates the amount of refuse collected in sector s as a fraction of p; $\mathbf{q} = (q, ..., q) \in R_+^{kt}$. The surrogate objective function minimizes q. The first constraint set expresses that the amounts collected on any workday should be less than q. The last constraint states that no part of the service region can be left unserviced. Note that the outcome of the model is independent of p. Therefore, a slow (steady or seasonal) variation in the value of p will also not change the optimal sector design. A lower bound for q is $t/(t - h)$, where h is the number of days in T the collector does not work. The problem is trivial if there are no holidays ($h = 0$), or if refuse is only produced on workdays, or if $n = 1$. Indeed, divide the service region into $t - h$ sectors so that in each $1/(t - h)$ of refuse is generated daily, and visit one sector per workday and collect each time $p\,t/(t-h)$ (thus obtaining the lower bound). It is also trivial when $n = (t - h)$: there can only be one sector, the service region itself, that is visited every workday. The peak load is equal to p times the largest number of consecutive holidays in (modulo) T.

In the model, either all feasible schedules from S are implemented from the start, or it can be seen as a master problem to which improving sectors are added by column generation. Feasible (and improved) schedules can be found as shortest paths in $(k\,f + 1)$ staged networks (as shown in Beullens, 2001). Tests show that typical instances can be solved by column generation in seconds on a Pentium II, 300 Mhz.

Analysis of the Sector Design Problem

Table 5.1 summarizes the model output for the problem instance given at the beginning of this section. There exists no sector design that can achieve a perfectly balanced workload. Plan 1 in the table is the optimal plan for $k = 1$. The service region is divided into four sectors (column 1) of different size, indicated as a percentage of p (column 2). The days on which to collect in

Table 5.1. Impact of sector design on workload balance

Plan	Sector size(%)	Schedule
1		
Sector 1	32.3	(3, 0, 2, 0, 2, 0, 0)
Sector 2	15.5	(2, 0, 0, 3, 0, 2, 0)
Sector 3	46.6	(0, 3, 0, 2, 0, 2, 0)
Sector 4	5.23	(2, 0, 2, 0, 0, 3, 0)
Workload(ton)		(250, 250, 134, 250, 116, 250)
2: week 1		
Sector 1	2.5	(3, 0, 2, 0, 2, 0, 0)
Sector 2	22.5	(2, 0, 0, 3, 0, 2, 0)
Sector 3	7.5	(2, 0, 2, 0, 0, 3, 0)
Sector 4	33.75	(2, 0, 2, 0, 2, 0, 0)
Sector 5	33.75	(0, 4, 0, 2, 0, 2, 0)
Workload(ton)		(241, 241, 157, 241, 130, 241)
2: week 2		
Sector 1	2.5	(3, 0, 2, 0, 2, 0, 0)
Sector 2	22.5	(2, 0, 0, 3, 0, 2, 0)
Sector 3	7.5	(2, 0, 2, 0, 0, 3, 0)
Sector 4	33.75	(0, 4, 0, 2, 0, 2, 0)
Sector 5	33.75	(2, 0, 2, 0, 2, 0, 0)
Workload(ton)		(241, 241, 157, 241, 130, 241)

a sector correspond to the non-negative numbers in the schedule (column 3). There is a peak workload (reported in tons) on Monday, Tuesday, Thursday, and Saturday, and a much lower workload on the other workdays. Plan 2 is the optimal solution. It comprises five sectors. The workload is still not balanced and the plan involves two sectors in which the visit schedule repeats itself every two weeks ($k = 2$).

A general insight from this analysis is that balancing the workload is not always possible but is more likely when increasing the allowed periodicity of the schedules. In these circumstances, it may be better to implement solutions using unmanned drop-off points instead of curbside collection, since curbside collection usually demands schedules that are repeated weekly, as in Plan 1, and thus faces a higher peak load than unmanned drop-off points which may be serviced with Plan 2. Another general insight is that although generators have all the same characteristics, the optimal strategy may be discriminatory or asymmetric, i.e. all generators are not treated alike. The main reason for this is the presence of the holidays on which generation of refuse continues, leading to non-equal demands on the workdays which need to be balanced by proper sector and schedule design.

5.4.3 Co-collection Models

Intuitively, the combined collection of different flows of goods seems likely to further reduce the collection cost. The issue also arises naturally when a source separation program (see Section 5.2.1) is introduced. Mansini and Speranza (1998) were the first in analyzing this, but they only considered the impact on resource use. Beullens (2001) conducted a more extensive analysis, comparing different forms of combination, and considering both resource use and routing cost. This approach is further clarified in this section.

We remark that the focus here is on the determination of the best combination level. The actual benefits of source separation also depend on the further process, in particular potential sorting costs. The models here presented could be easily extended to include such issues.

A Co-collection Problem Example

Returning to the example of Section 5.4.2, suppose the city council has decided to introduce source separation. On the same locations as the existing $n = 3,360$ refuse containers, an equal amount of recycling containers is added. The two classes of flows are to be collected and transported to the municipal depot. The city council still has to decide on the collection frequencies for both classes, and on the contracting of collection companies. Three companies are interested. They are not allowed to work on Saturday and Sunday. Company 1 and 2 have single-compartment vehicles, while Company 3 has dual-compartment vehicles. All vehicles have a capacity of 10 tons and can perform two collection tours a day. Three possible ways are considered.

- Independent collection: Company 1 collects the refuse, Company 2 the recyclables (or vice versa).
- Shared resources: Either Company 1 or 2 will deploy its fleet to collect both classes. A vehicle can only collect one class at the time.
- Co-collection: Company 3 will collect both classes. A vehicle is capable of collecting both classes simultaneously.

(Note that in practice it is always possible to contract several companies at the same time by assigning them each to their own part of the service region.) The city council wishes to estimate the impact of these design options on the weekly collection cost.

A Co-collection Model

For the independent collection, the model introduced in Section 5.4.2 can be used for each class to estimate the number of vehicles required, and Equation (5.9) for estimating the associated transport cost.

For the shared resource case, note that the continuous approximation analysis in Section 5.4.2 showed that the routing cost was largely unaffected by the sector design. Therefore, Equation (5.9) can still be used for the transport cost estimation of each class in the shared resources case, and the sector design model can be easily adapted as follows.

$$
\begin{aligned}
&\min q, \\
&subject\ to \\
&\sum_{S^c} \sum_{(s \in S^c)} \mathbf{s^c}^T\ z_s^c \le \mathbf{q}^T \\
&\sum_{(s \in S^c)} z_s^c = 1 \qquad \text{for all } S^c \\
&z_s^c \ge 0 \qquad\qquad \text{for all } s \in S^c, S^c \\
&q \ge 0
\end{aligned}
$$

where c is the index specifying a class, and the non-zero elements of the schedules $\mathbf{s^c}$ are defined as $s_i^c = p^c\,v$, i.e. they are the accumulation time v between two consecutive visits multiplied by $p^c \in [0,1]$, the amount of class c expressed as a fraction of the total amount generated per day p. This model is not more difficult than the single-class model of Section 5.4.2 and can be solved with the same approach.

For the co-collection scenario, though, the routing cost is not independent of the sector design. Co-collection involves the simultaneous pickup at a location of both classes, hence it is impossible if there are no 'overlaps', i.e. if these two classes are never scheduled for collection in the same period in some area of the service region. Label the classes a and b, and consider two of their sectors $s^a \in S^a$ and $s^b \in S^b$, respectively. Co-collection is possible in period i of the first time interval if $s_i^a \ne 0$ and $s_i^b \ne 0$ and if there is some overlap $y_{s^a s^b} \ne 0$ ($y_{s^a s^b} \in [0,1]$) between the sectors s^a and s^b; $A_{s^a} \cap A_{s^b} = y_{s^a s^b} A$. In that area of overlap $y_{s^a s^b}\, n$ points are visited, and the load of class c ($c = a$ or b) collected in that period is $y_{s^a s^b} n(s_i^c/p^c)(p\,p^c/n) = y_{s^a s^b}\, s_i^c\, p$. The associated distance travelled is

$$
2\bar{\ell}_{A_{s^a} \cap A_{s^b}} \frac{y_{s^a s^b}\, p(s_i^a + s_i^b)}{W} + 0.57 y_{s^a s^b} \sqrt{A\,n}. \tag{5.10}
$$

The distance to collect class c separately in the considered area and period is also given by Equation (5.10) if the factor $(s_i^a + s_i^b)$ is changed to s_i^c. Taking the sum for both classes and subtracting the co-collection distance (5.10) thus results in a net savings of $0.57\, y_{s^a s^b} \sqrt{A\,n}$. Straightforward calculus shows that the summation over all schedules and the k time intervals leads to the following optimization problem:

$$
max_{(S^a,S^b)}\left(\triangle(S^a, S^b) \right), \tag{5.11}
$$

where

$$
\triangle(S^a, S^b) = \sum_i \sum_{(s^a \in S^a:s_i^a \ne 0)} \sum_{(s^b \in S^b:s_i^b \ne 0)} y_{s^a s^b}. \tag{5.12}
$$

This optimization problem shows that the routing cost savings are not a function of the particular overlapping sectors, of how far overlaps are located from the depot, of the periods in which co-collection occurs in the overlapping areas, or of the amount of accumulation that occurred before the co-collection. Note that the impact of the collection frequency is implicitly incorporated, since the total overlap is also determined by the number of times two sectors of both classes have to be visited in the same time period. The routing cost for co-collection can thus be calculated as the sum of the separate transport cost (Equation (5.9)) for the two classes, and, given the sector designs S^a and S^b and assuming that the cost of a co-collection stop is the sum of the two stop costs for the separate collection, subtracting a savings of

$$0.57\,c_t\,\sqrt{A\,n}\,\triangle\,(S^a, S^b). \qquad (5.13)$$

To obtain S^a and S^b, the sector design model for co-collection can be formulated as a pre-emptive goal program with weights C_f and C_t ($c \in \{a, b\}$, $d \in \{a, b\}$, $c \neq d$):

$$
\begin{aligned}
&\min\left(C_f\,q - C_t\,\triangle\,(S^c, S^d)\right),\\
&subject\ to\\
&\sum_{S^c}\sum_{(s\,\in\,S^c)}\mathbf{s^{c}}^{T}\,z_s^c \leq \mathbf{q}^T \text{ for all } c\\
&\sum_{(s\,\in\,S^c)} z_s^c = 1 && \text{for all } S^c\\
&\sum_{(s^d\,\in\,S^d)} y_{s^c s^d} \leq z_s^c && \text{for all } s^c \in S^c, S^c, S^d\\
&z_s^c \geq 0 && \text{for all } s \in S^c, S^c\\
&y_{s^c s^d} \geq 0 && \text{for all } S^c, S^d\\
&q \geq 0
\end{aligned}
$$

When $C_f \gg C_t$, the peak load is the most important feature (and vice versa). The third set of constraints is new and specifies that the overlap of two sectors can never be larger than the part of the region assigned to one of the sectors, z_s^c.

Analysis of the Co-collection Problem

The different design options available to the city council can now be compared. Table 5.2 summarizes the results. Recall that the decision variables are the collection frequencies f^c (per week) for both classes, refuse and separated recyclables, as well as the combination level used in the transportation (independent collection, shared resources, and co-collection). Scenario 1 serves as the base case (no source separation, as in the example of Section 5.4.2). The other scenarios consider several situations in which source separation receives a different response from the generators, indicated by different levels in which recyclables have been separated from the refuse stream. The corresponding parameters in the table are the fraction of class c in the total waste stream p^c. The performance of a given collection method is the number of vehicles needed #V and the distance travelled per week d (km). For these measurements, the service area is assumed to be 320 km^2 and the average distance

Table 5.2. Fleet size and route cost for different design options

Scenario	Refuse (p^c, f^c)	Recyclables (p^c, f^c)	Ind.Coll. $(\#V, d)$	Shared.Res. $(\#V, d)$	Co-Coll. $(C_f \gg C_t)$ $(\#V, d)$	Co-Coll $(C_f \ll C_t)$ $(\#V, d)$
1	(1.0, 3)	(0.0, -)	(27, 3445)	(-, -)	(-, -)	(-, -)
2	(0.8, 3)	(0.2, 2)	(26, 4612)	(22, 4626)	(22, 5035)	(22, 5035)
3	(0.5, 3)	(0.5, 2)	(23, 4607)	(19, 4627)	(19, 4036)	(19, 4036)
4	(0.2, 3)	(0.8, 2)	(20, 4601)	(17, 4634)	(17, 4039)	(17, 4039)
5	(0.2, 3)	(0.8, 1)	(16, 4038)	(13, 4038)	(13, 4039)	(21, 3444) (*)
6	(0.2, 2)	(0.8, 1)	(14, 3432)	(13, 3444)	(13, 3000)	(16, 2853)
7	$(x, 1)$	$(1-x, 1)$	(14, 2855)	(13, 2855)	(13, 2265)	(13, 2265)
8	$(x, 1)$	$(1-x, 0.5)$	(14, 2558)	(13, 2558)	(13, 2265)	(13, 2265)

to the depot to be 6.67 km. For the co-collection method, both the minimum peak load solution and the minimum distance solution are reported.

Bad scenarios in terms of resource use are those in which the workload cannot be balanced. This is more likely to occur with higher collection frequencies. Therefore, when the frequency for recyclables is lower than for refuse, the number of vehicles reduces with the percentage of the captured recyclables. Higher peak loads are also more likely for independent collection. Once the workload is balanced, the additional reduction of collection frequency has no effect on the resource use.

In terms of routing cost, source separation is in general only beneficial if the sum of the collection frequencies of both classes is at least as low as the initial frequency (the exception is indicated (*) in the table). A change in the capture rate of recyclables has no significant effect on the routing cost for independent or separate collection. This can be understood since, although the optimal sector design can change with the capture rate, the sector design has a limited impact on the routing cost of a single class (as seen in Section 5.4.2). For the same reason, the routing costs of these scenarios are comparable. Although co-collection routing is more sensitive to the sector design, and thus to the capture rate, it becomes insensitive once frequencies are once a week or lower. Co-collection typically has a much lower routing cost compared to separate routing. Weekly co-collection is still about 11% cheaper in routing cost than weekly collection of refuse and separately collecting recyclables once every two weeks. It is therefore somehow surprising to see the latter scenario occurring in practice. Reasons for this can be 1) many single-compartment vehicles are still in use and a change to dual-compartment vehicles is expensive; 2) a dual-compartment vehicle needs to be flexible to adjust the relative compartment size to the local capture rate observed during its tour, otherwise one compartment might be full sooner than the other, decreasing collection efficiency; and 3) both collection tasks are outsourced to different organizations. The latter, combined with the consideration of a specific depot for each organization, will tend to decrease the value of co-collection, as indicated by some preliminary analysis in Beullens et al. (1999).

The example illustrates that source separation can decrease or increase the routing cost (and the environmental cost related to the transportation): it depends on the initial situation *without* source separation versus how it is organized *with* source separation. If the collection frequency initially was more than once a week, source separation will generally reduce the collection cost, e.g. going from scenario 1 to 6 or 7 in Table 5.2. Starting from weekly collection, however, source separation can increase the cost as well (for example, when going from no separation, which would give a routing cost equal to the co-collection scenario 7, to the separate collection scenario 8).

5.4.4 Integration Models

Another way to combine several flows of goods is to integrate the collection and distribution activities, by means of backhauling or mixing (introduced in Section 5.2.2). The main issues in this section are the estimation of how much integration reduces the transportation cost, the comparison between backhauling and mixing performance, and the exploitation of the lower time pressure in the reverse channel.

General Bound

Compared to separately organizing the delivery and collection routes, integration can, at most, reduce the total distance travelled by 50%.

Jacobs-Blecha and Goetschalckx (1992) proved this for the single-period VRPB, and Mosheiov (1994) for the single-period VRPM. However, it is valid for every form of integration in many types of vehicle routing problems with delivery and collection requests, such as multi-period problems, problems with time windows, upper bounds on the route duration, multiple depots, vehicle-specific capacities, vehicle-site compatibility constraints, or problems with a mixture of these characteristics.

This is based on two observations (Beullens, 2001). 1) *Optimality principle*: For every instance of a vehicle routing problem, mixing is at least as good as partial mixing, partial mixing as good as backhauling, and backhauling as good as no integration. Indeed, going from the latter to the first, the set of allowed solutions can never become smaller, so the optimal solution is at least as good. Other forms of integration, e.g. using mixing for some vehicles and backhauling for others, will have their optimal solution falling somewhere in the interval given by the optimal solutions of mixing and no integration. 2) *Tight upper bound*: The optimal route plan with no integration can never be longer than twice the optimal route plan for whatever form of integration, and this bound is tight. The claim is easily verified as follows. Consider the vehicles and their capacity, and construct for each an 'adapted' vehicle which is equal to the original, except for having two compartments, each with a capacity equal to the original vehicle capacity. One compartment of each adapted vehicle is for deliveries, the other for collections. Construct, using the adapted vehicles, an optimal solution for the vehicle routing problem instance at hand so that on each vehicle route the total load of deliveries and collections does not exceed the compartment capacity, while respecting all other constraints. This solution is a lower bound for mixing given that in mixing, using the original vehicles, an additional constraint states that total load may never violate vehicle capacity. Travelling twice along these routes with the original vehicles — one vehicle for the deliveries, another vehicle for the collections — provides an upper bound when there is no integration. Examples can be easily constructed to show the bound is tight. Given the first observation, the statement must hold for whatever form of integration.

The savings from integration thus depend on the particular instance; special instances can be easily constructed where the savings are either 0 or 50% for mixing and/or backhauling. Clearly, this general bound provides not much insight. More specific estimates are obtainable by considering a probabilistic approach, discussed next.

Asymptotic Behavior

Probabilistic analysis allows the characterization of the optimal solution (total distance) when the number of customers tends to infinity. Performing such an analysis under different routing strategies gives insight into their relative asymptotic performance. In particular, we will compare separating delivery and pickup tours, mixing, and backhauling.

The analysis in Beullens (2001) of the CVRP shows that the relative performance depends on the assumptions being made about the vehicle capacity. When the vehicle capacity grows with the number of customers such that all delivery demands and all collection demands can be serviced separately by a single vehicle, backhauling is asympotically as 'inefficient' as separate routing, while mixing clearly performs better. How much better depends on how delivery and collection requests are geographically distributed. More specifically, let $x_1, x_2, ..., x_n$ be a set of points (customers) which are independently drawn from an identical distribution μ over a bounded region of the Euclidean plane (area A). Each point x_i is chosen independently as a delivery location with probability p_d, as a collection location with probability p_c, and as a location with both a delivery and collection request (denoted as exchange locations) with probability $p_e = 1 - p_d - p_c$. The depot housing the vehicle is arbitrarily located at point x_0 inside or at a finite distance from the region. Let T^p denote the total distance travelled in the optimal solution under policy p ($p \in sepr, back, mix$). Then it is shown that with probability one,

$$\lim_{n \to \infty} (T^{mix}/n^{1/2}) = I_3, \tag{5.14}$$

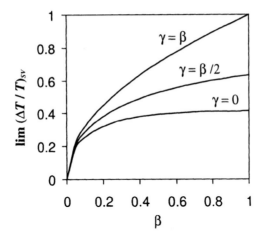

Fig. 5.1. Asymptotic gap between mixing and backhauling or separating delivery and pickup tours, single-vehicle case

$$\lim_{n \to \infty} (T^{back}/n^{1/2}) = \lim_{n \to \infty} (T^{sepr}/n^{1/2})$$
$$= ((p_d + p_e)^{1/2} + (p_c + p_e)^{1/2})I_3 , \qquad (5.15)$$

where $I_3 = \zeta \int_A \mu^{1/2} dv$, i.e. the well-known result of Beardwood et al. (1959) for the asymptotic value of the optimal travelling salesman tour, in which ζ is a constant. Thus, with probability one, the gap between backhauling and mixing is $\lim_{n \to \infty} ((T^{back} - T^{mix})/T^{mix}) = (p_d + p_e)^{1/2} + (p_c + p_e)^{1/2} - 1$. The gap is largest (100%) when $(p_d + p_e) = (p_c + p_e)$, thus when $p_e = 1$. Figure 5.1 visualizes the gap using two parameters: the ratio of exchange locations to delivery requests, $\gamma = p_e/(p_d + p_e)$, and the ratio of collection requests to delivery requests, $\beta = (p_c + p_e)/(p_d + p_e)$. Then, with probability one,

$$\lim(\Delta T/T)_{SV} \equiv \lim_{n \to \infty} ((T^{back} - T^{mix})/T^{mix})$$
$$= (1 + \beta^{1/2} - (1 + \beta - \gamma)^{1/2})/(1 + \beta - \gamma)^{1/2} . \quad (5.16)$$

Hence, the benefit of mixing compared to backhauling becomes larger with the similarity of the number of delivery requests and collection requests, and the more delivery and collection requests geographically coincide (or occur at exchange locations).

The situation is different when the vehicle capacity is fixed or does not grow sufficiently with the number of points, so that the number of vehicles needed also tends to infinity with the number of points. Then backhauling becomes as efficient as mixing. The benefit of integration compared to separating delivery and collection tours depends on β, and on the relative size of

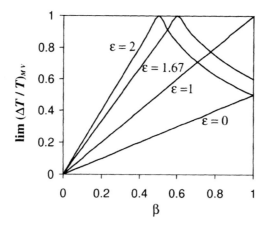

Fig. 5.2. Asymptotic gap between integrating and separating delivery and pickup tours, multi-vehicle case

delivery and collection loads. The γ parameter, which specifies how the collection requests are distributed among collection and exchange points, as well as μ, the distribution of points in the region, appear to become irrelevant. In particular, associate with each delivery (collection) demand a random variable drawn from an identical distribution ϕ_d (ϕ_c) with support on $[0, C]$. Let b_n^d (b_n^c) be the minimum number of bins (vehicles) of capacity C needed to pack the delivery (collection) demands associated with n points. Let $\lim_{n\to\infty} b_n^x/n = \varphi_x$ (a.s.), where $x \in d, c$. φ_d (φ_c) is called the bin packing constant of delivery (collection) demands. Let $E(r)$ denote the expected distance between x_i and the depot. Then almost surely,

$$\lim_{n\to\infty} (T^{mix}/n) = \lim_{n\to\infty} (T^{back}/n)$$
$$= 2E(r)max((p_d + p_e)\varphi_d, (p_c + p_e)\varphi_c) , \qquad (5.17)$$

$$\lim_{n\to\infty} (T^{sepr}/n) = 2E(r)((p_d + p_c)\varphi_d + (p_c + p_e)\varphi_c). \qquad (5.18)$$

It follows that, almost surely,

$$\lim_{n\to\infty} ((T^{back} - T^{mix})/T^{mix}) = 0, \qquad (5.19)$$

and, with $\epsilon \equiv \varphi_c/\varphi_d$, in case $\beta\epsilon \leq 1$:

$$\lim(\Delta T/T)_{MV} \equiv \lim_{n\to\infty} ((T^{sepr} - T^{mix})/T^{mix}) = \beta\epsilon, \qquad (5.20)$$

or $\lim(\Delta T/T)_{MV} = 1/\beta\epsilon$ in case $\beta\epsilon \geq 1$.

The latter gap is displayed in Figure 5.2. Note that $(p_d + p_e)\varphi_d$, for sufficiently large numbers of n, represents the number of bins needed to meet all delivery demands (ignoring collection demands). Hence $\beta\epsilon$ can be interpreted as the ratio of the number of vehicles needed for satisfying collection demands to, separately, satisfying delivery demands. Since the gap grows as $\beta\epsilon$ goes to 1, the benefit of integration (in percent) is larger in situations where these numbers are more similar.

We note that delivery, collection, and exchange locations are drawn from the same distribution μ. Anily (1996) provides an asymptotic value for the case without exchange locations where delivery and collection locations are drawn from different distributions μ_d and μ_c. Mixing and backhauling have again the same asymptotic behavior, but the result is not in closed form. The characterization of the length of the optimal solution requires solving a minimum cost bipartite matching problem between the set of delivery and collection customers. The cost of matching becomes asymptotically a negligible fraction of the total cost when $\mu_d = \mu_c$, leading of course to the formulas presented in this section.

Continuous Approximation

The asymptotic probabilistic analysis gives little insight into the relative performance of integration when the number of vehicles is approaching neither one nor infinity. Continuous models have the benefit of being able to predict behavior in these circumstances. In the remaining part of this section, we show how continuous models may help in determining the performance of several forms of integration in a context of periodic vehicle routing. The basic question is: how can the lower time pressure in the reverse channel be exploited when aiming at efficiently integrating the collection requests with the distribution activities?

An Integration Problem

Consider a service region in which three types of points are all uniformly distributed. One type corresponds to customers with delivery requests, another type to customers with collection requests, and the last type to customers with both requests (exchange points). The number of points is denoted by n_d, n_c, and n_e, respectively. A depot is arbitrarily located in or outside the service region. It houses the vehicles, all goods to be delivered, and all goods collected. On every workday, the daily delivery (collection) request is of size Q_d (Q_c)(the same for all points). Delivery requests should be satisfied on a daily basis, but collection requests can be postponed up to the accumulation capacity Q_c^{max}. Can postponement reduce the transport cost?

To start with the simplest case, only symmetric models will be developed here. That is, we wish to specify one optimal value of postponement. We will indicate that better postponement strategies will exist by allowing the postponement value to vary based on customer type and location.

The main decision variable of interest is thus v, which denotes, as in Section 5.4.2, the accumulation time between two consecutive visits to a collection point. To facilitate the analysis, let v be a continuous variable, i.e. $v \in [1, \lfloor Q_c^{max}/Q_c \rfloor]$, although obviously only integer values are feasible.

A Separate Collection Model

Consider no integration first. Since there is no generation of requests nor accumulation on holidays, an optimal sector design for the separate collection will, as seen in Section 5.4.2, consist of partitioning the service region into v sectors of equal size, visiting one sector per workday. Each period, an area of size A/v is visited, containing $(n_c + n_e)/v$ points, each with a load of vQ_c. Taking the average over v consecutive periods gives an average collection transport cost per period of

$$(K_r + 2\bar{\ell}c_t)\frac{(n_c + n_e)Q_c}{W} + \frac{1}{v}(0.57c_t\sqrt{A(n_c + n_e)} + K_s(n_c + n_e)). \quad (5.21)$$

Since this continuous function in v is monotonously decreasing, we have the following result.

Proposition 5.1. *For separate collection,* $v^* = \lfloor Q_c^{max}/Q_c \rfloor$.

A Backhauling Model

Consider first backhauling without postponement (i.e. $v = 1$). To find an expression for the distance, based on the continuous approximation approach, the following partitioning scheme is constructed. Partition the service region into a number of zones, each zone corresponding to a vehicle's territory of operation, and containing W/Q_d delivery requests in case $(n_d + n_e)Q_d \geq (n_c + n_e)Q_c$ or W/Q_c collection requests in case $(n_d + n_e)Q_d < (n_c + n_e)Q_c$; except maybe a single zone close to the depot containing fewer points. Each vehicle travels to its zone, delivers according to a minimum-length travelling salesman (TSP) tour, then collects according to a TSP tour, and finally returns to the depot. The total distance travelled by all vehicles can be approximated by

$$\frac{2\bar{\ell}}{W}\,max((n_d + n_e)Q_d, (n_c + n_e)Q_c) + 0.57(\sqrt{A(n_d + n_e)} + \sqrt{A(n_c + n_e)}).$$
$$(5.22)$$

One can show the routing scheme is asymptotically optimal and the above cost approximation asymptotically accurate (see Beullens, 2001). When postponement is now applied in the sector design scheme, backhauling is only performed in one sector A_s, of size A/v, containing $(n_d + n_e)/v$ delivery requests of size Q_d and $(n_c + n_e)/v$ collection requests of size vQ_c. The average cost for the backhauling vehicles per period is

$$\frac{K_r + 2\bar{\ell}\,c_t}{W}\,\max(\frac{(n_d + n_e)Q_d}{v}, (n_c + n_e)Q_c)$$
$$+ \frac{1}{v}(0.57c_t(\sqrt{A(n_d + n_e)} + \sqrt{A(n_c + n_e)}) + K_s(n_d + n_c + 2\,n_e)), \quad (5.23)$$

in which the potential but small reduction in visiting cost is ignored, associated with the possible single visit of a point requiring both a delivery and collection in the backhauling tour of a vehicle. The average cost for delivery in each period in the $(v - 1)$ other sectors is

$$\frac{v - 1}{v}((K_r + 2\bar{\ell}\,c_t)\frac{(n_d + n_e)Q_d}{W} + 0.57c_t\sqrt{A(n_d + n_e)} + K_s(n_d + n_e)).$$
$$(5.24)$$

The total average cost per period, $C_{back}(v)$, is the sum of (5.23) and (5.24). In the interval $v \in [1, a/b)$ where $a = (n_d + n_e)Q_d$ and $b = (n_c + n_e)Q_c$ if $a/b > 1$, else $a/b = 1$, the function $C_{back}(v)$ is continuous and $\partial C_{back}(v)/\partial v = -C_1 v^{-2}$, $C_1 \in R_+$. Therefore there are no local roots and the function is monotonously decreasing. In the interval $v \in (a/b, \infty)$, $C_{back}(v)$ is continuous. There, $\partial C_{back}(v)/\partial v = +C_2 v^{-2}$, $C_2 \in R_+$, if and only if

$$(K_r + 2\bar{\ell}\,c_t)\frac{(n_d + n_e)Q_d}{W} \geq 0.57c_t\sqrt{A(n_c + n_e)} + K_s 2 n_e, \quad (5.25)$$

and $C_2 \in R_-$ otherwise. If Condition (5.25) holds, there are no local roots and the function is monotonously increasing, and decreasing if the condition is false. This gives the following policy.

Proposition 5.2. *For backhauling, if Condition (5.25) is not met, then $v^* = \lfloor Q_c^{max}/Q_c \rfloor$. If Condition (5.25) is true, then*

$$v^* = min(\,max(1, \frac{(n_d + n_e)Q_d}{(n_c + n_e)Q_c}), \lfloor Q_c^{max}/Q_c \rfloor). \quad (5.26)$$

In words, when Condition (5.25) is true and the total daily volume of goods to be delivered is smaller or equal to the total daily volume of goods to be collected, it is optimal not to postpone and to backhaul every day in the complete service region. If the total volume to be delivered is larger, then the ratio in these total volumes will determine the optimal periodic schedules for the collection, except when this would violate the accumulation capacities of the collection points, in which case $v^* = \lfloor Q_c^{max}/Q_c \rfloor$. When Condition (5.25) is not met then, whatever the volume ratio, a maximum postponement strategy is always best.

The backhauling scheme and associated cost approximation (5.22) may overestimate the cost when dealing with small values of $(n_c + n_e)Q_c$ relative to $(n_d + n_e)Q_d$. Indeed, for the case with no capacity restrictions on the collections, Daganzo and Hall (1993) developed another backhauling approach based on performing the pickups on routes perpendicular to the line-haul paths of the vehicles. Applying this model requires more effort but leads to lower backhauling costs. Its validity in a postponement strategy is, however, rather limited, since postponement tends to balance delivery and collection loads and, as they also point out, the assumption of unrestricted capacity then tends to make their backhauling scheme and cost approximation unrealistic.

A Mixing Model

The difficulty in obtaining a closed form solution for mixing is the constraint specifying that the total load in the vehicle may never exceed its capacity. But a continuous approximation for a lower bound on the distance travelled can be obtained: the constraint is relaxed and replaced by the constraints that separately the sum of the total delivery and the sum of the total collection load in the vehicle do not exceed vehicle capacity. When there is no postponement, the same partitioning scheme can be constructed as for backhauling. But now a vehicle delivers and collects in its zone according to a single minimum-length travelling salesman (TSP) tour. The total distance travelled by all vehicles can be approximated by

$$2\,\frac{\bar{\ell}}{W}\,max((n_d + n_e)Q_d, (n_c + n_e)Q_c) + 0.57\sqrt{A(n_d + n_c + n_e)}. \qquad (5.27)$$

Although a lower bound, it can be proven that the expression is asymptotically accurate and a fair approximation for the real distance under mixing when there are many points. Postponement can now be applied in the sector design scheme for mixing as was done for backhauling. Without demonstrating it here, this strategy leads to a similar condition as the Equation (5.25), the same policy for postponement, but a total cost which is at least as small and often smaller (for every value of v). Another and even smarter approach to using mixing is as follows. The collection points requiring both a delivery and a collection are serviced daily, and the points requiring only a collection are postponed and then mixed with the other requests in the sector of the day. It can be shown that this special form of mixing will give a total cost at least as small as with the mixing strategy for whatever value of v, but a different v^* if $n_c \neq 0$. The cost associated with this policy is derived as follows. Per period, mixing the collection-only requests is done in one sector of size A/v, containing $(n_d + n_e)/v$ delivery requests of size Q_d and $(n_c + n_e)/v$ collection requests of which the size of the load for the first type of points is $v Q_c$ and for the second type is Q_c. The average cost per period for the vehicles mixing with the collection-only points is thus

$$\frac{(K_r + 2\,\bar{\ell}\,c_t)}{W}\,max(\frac{(n_d + n_e)Q_d}{v}, (n_c + \frac{n_e}{v})Q_c)$$
$$+ \frac{1}{v}(0.57c_t\sqrt{A(n_d + n_c + n_e)} + K_s(n_d + n_c + n_e)). \qquad (5.28)$$

The average cost per period for the delivery and mixing of the requests from the points requiring both delivery and collection in the $(v-1)$ other sectors is

$$\frac{v-1}{vW}((K_r + 2\bar{\ell}c_t)\,max((n_d+n_e)Q_d,\, n_e\,Q_c)$$

$$+ 0.57c_t\sqrt{A(n_d+n_e)} + K_s(n_d+n_e)). \tag{5.29}$$

The total average cost per period, $C_{mix}(v)$, is the sum of (5.28) and (5.29). Similar analysis as for the backhauling case leads to the following condition:

$$\frac{K_r + 2\bar{\ell}c_t}{W}((n_d+n_e)Q_d - n_eQ_C)$$

$$\geq 0.57c_t\sqrt{A}(\sqrt{n_d+n_c+n_e} - \sqrt{n_d+n_e}) + K_s\,n_e, \tag{5.30}$$

which determines the policy for this type of 'special' mixing.

Proposition 5.3. *For special mixing, if $(n_d+n_e)Q_d \leq n_eQ_c$, or if Condition (5.30) is not met, $v^* = \lfloor Q_c^{max}/Q_c \rfloor$. Else, if Condition (5.30) is true,*

$$v^* = min(\,max(1,\, \frac{(n_d+n_e)Q_d - n_eQ_c}{n_c\,Q_c}),\, \lfloor Q_c^{max}/Q_c \rfloor). \tag{5.31}$$

Similar models can be derived for other forms of integration. For example, daily servicing the points requiring both a delivery and a collection, as in the previous strategy, but postponing and backhauling the points requiring only a collection.

Finally, as pointed out before, the cost approximation for mixing (5.27) was actually a lower bound since it did not take into account the capacity restriction for 'mixed' loads. An argument of Mosheiov (1994) (Proposition 2) can be used to show that the local distance cost of a vehicle, to make its route feasible with respect to its capacity constraint, will increase no more than by twice the longest distance between any two points it services. These lengths are clearly negligible for numerous points and hence our approximation is then quite accurate. When there are fewer points to visit, the probability that a vehicle will have to use a tour which is significantly longer than a simple TSP will increase. For this case, the model may therefore need fine-tuning.

Analysis of the Value of Postponement

Figures 5.3 and 5.4 provide numerical examples for the different integration models, displaying three graphs each for a total of six different parameter settings. In each of these graphs, the average cost per day is displayed as a function of the number of days between consecutive collections, v, for backhauling (B), mixing (M), and special mixing (SM). The fixed parameters and their values are: $K_r = 50$, $K_s = 2$, $c_t = 1$, $\bar{\ell} = 6.67\,km$, $A = 320\,km^2$, $W = 50$,

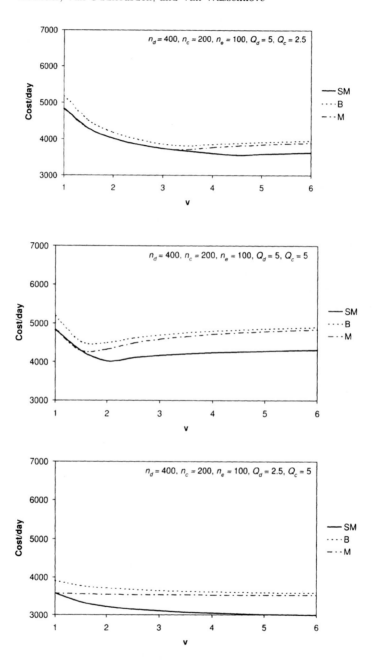

Fig. 5.3. Average cost as a function of v, many delivery points

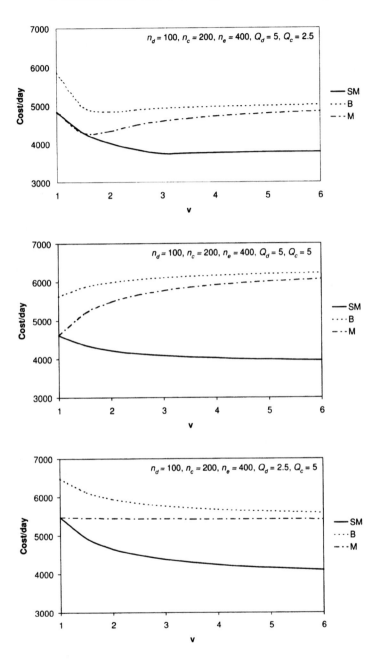

Fig. 5.4. Average cost as a function of v, many exchange points

and a maximum value for v of 6 (i.e. $Q_c^{max} = 6\,Q_c$). In each figure, the top graph corresponds to $Q_d = 5$ and $Q_c = 2.5$, the graph in the middle considers values of respectively 5 and 5, and the bottom graph of 2.5 and 5. Furthermore, Figure 5.3 has $n_d = 400$, $n_c = 200$, and $n_e = 100$, while Figure 5.4 considers values of respectively 100, 200, and 400.

It is no surprise that special mixing always outperforms the other integration methods, and that backhauling is always the worst. It is interesting how the optimal value of postponement, i.e. v^*, appears to be a function of both the parameter settings and the integration method. Analysis shows that there are circumstances where the simpler policy of backhauling with just a little postponement can beat (or has at least an equal performance to) the more difficult mixing policy without postponement (such as in the first two graphs of Figure 5.3, and the first graph of Figure 5.4).

The general superiority of asymmetric postponement strategies, as pointed out before, can be derived when looking closer at the meaning of the Conditions (5.25) and (5.30). They actually compare the costs of dispatching vehicles and linehaul travelling toward their service subareas with the local travelling and stop costs. The optimal value of postponement can thus largely depend on the underlying parameters that determine these costs. One of the important consequences is that the optimal postponement value may differ depending on the relative location of the depot to the service (sub)area. For example, for the case depicted in the third graph of Figure 5.4, maximum postponement was optimal. However, when the depot is more than $\bar{\ell} = 28$ km away, ceteris paribus, then no postponement is optimal. This indicates the general superiority of asymmetric models.

Although these system design models assume some particular problem characteristics, they analytically show how the efficiency of a particular way of integration is a function of some critical parameters, and they provide insight into the value of postponement. The convexity of the cost curve tells how important it is in the specific context to specify the right value of postponement. These models certainly reveal the circumstances where there is a lot to gain from *multi-period* vehicle routing models with integration, models which are currently lacking in the literature.

The system design models presented in Section 5.4 only tried to capture the essential elements of reverse logistics in comparison to typical distribution logistics. In particular, we considered the lower value of the items and the greater degree of freedom for the collector to organize efficient routing strategies, including the option of postponement. Some of the examples given were situated in a context of waste collection, however the models are valid for many other product streams. For many applications, postponing the collection for a few days does not increase costs for processes further down the reverse logistics network. Valuable items or products with a high depreciation rate, and valuable reusable distribution items, are the exceptions for which the models need to be adjusted. The explicit incorporation of holding costs in the models may make them more applicable for valuable return flows. The issue of holding costs for reverse logistics is, however, a topic unto itself (see Chapter 7 and Chapter 11). Differentiating more between customers is another issue which may be needed to fit such models to this application area. As a starting point, we refer to Daganzo (1999), a book which contains more elaborate models in the context of distribution.

5.5 Conclusions

Increasing reverse logistics flows from the disposer (and (re)use) markets introduce new issues in the areas of collection and vehicle routing. Two prominent differences with 'traditional' distribution logistics are the usually low value of the goods and the large degree of freedom in deciding the moment and method of collection. The search for the lowest-cost approach leads to multi-period vehicle routing models which may need to include strategies that combine the transport of multiple flows of goods.

There are algorithms in demand for problems dealing with, for example, a mixed fleet, arc- and general routing, simultaneous delivery and pickup, and partial mixing. For arc routing, it seems worthwhile to investigate the impact on efficiency of relaxing the requirement to service edges or arcs in one go. The major gap, however, appears to be in the area of multi-period routing models dealing with arc routing, the co-collection in multi-compartment vehicles, and the integration with distribution activities. It is particularly striking to see that there is a lack of multi-period routing models for all forms of integration (backhauling, mixing, etc).

System design models have been used to analyze such strategies for the periodic vehicle routing problem in the case of infinite horizon constant deterministic demand and return rates. The basic tools are the continuous approximation approach and small-sized mathematical programming models. The advantage of models using concise summaries instead of detailed numerical data is the insight into the broad properties of optimal design approaches, even if the system is not yet in place.

The sector design models in this chapter determine the number, size, location, and visit schedule of the sectors in which the overall service area needs to be divided in order to minimize the fleet size and routing cost. The continuous approximation theory shows that the routing cost, although a function of, for example, the collection frequency, is not significantly affected by sector design. This insensitivity to sector design supports a hierarchical solution approach. In a first stage, sector design models need to look at minimizing the fleet size by determining an optimal set of sectors and their periodic schedules, and the locations of sectors can be chosen rather arbitrarily. The problem can be modelled as a linear program. The solutions imply, in general, asymmetric strategies, even for collection points with identical characteristics (in contrast with EOQ solutions). In a second stage, single-period vehicle routing models determine the individual vehicle routes within each sector.

The routing cost, however, becomes an important factor for sector design when planning for the co-collection of two flows toward the same depot. We have shown how to estimate the effect on total cost when a source separation program is introduced, and how to design sectors. Using optimal sector designs for both situations, before and after the introduction of source separation, the cost can either increase or decrease. This largely depends on the choice of collection frequencies and the level of combination (independent routing in single-compartment vehicles, sharing single-compartment vehicles, or co-collection in two-compartment vehicles). In all investigated cases, a good sector design makes co-collection the lowest cost approach.

The chapter ends with analyzing the value of postponing pickups when they can be integrated with non-postponable deliveries. Models with symmetric backhauling and (special) mixing policies are developed. Conditions are derived that determine when postponement will generate savings and how to calculate the optimal postponement level. The value of postponement demonstrated in these models underlines the observed approaches in practice and stresses the importance of developing multi-period routing models with integration. It was also indicated that asymmetric policies will, in general, lead to even better performance.

It should be clear that this field shows many opportunities for research. One promising line of research that is imminent, is to develop normative multi-period vehicle routing models incorporating the issues mentioned. Another road for research is to fine-tune the system design models. For co-collection, for example, other conclusions may appear when using multi-compartment vehicles with rigid compartments, or when relaxing the assumption that flows have to be brought to the same depot. Considering non-uniformly distributed return rates may also lead to guidelines for locating the sectors in which co-collection is desired the most. For integration, the search for a characterization of optimal asymmetric strategies may be undertaken. Holding costs may be explicitly incorporated. We hope this overview of vehicle routing issues will stimulate new and interesting developments that will ultimately contribute to good implementations of reverse logistics in practice.

Product and Material Returns: Handling and Warehousing Issues

Marisa P. de Brito[1] and M.B.M.(René) de Koster[2]

[1] Rotterdam School of Economics, Erasmus University Rotterdam, P.O. Box 1738, 3000 DR Rotterdam, The Netherlands, `debrito@few.eur.nl`

[2] Rotterdam School of Management / Faculteit Bedrijfskunde, Erasmus University Rotterdam, P.O. Box 1738, 3000 DR Rotterdam, The Netherlands, `rkoster@fbk.eur.nl`

6.1 Introduction

According to the Material Handling Industry of America (see MHIA.org), material handling is the movement, storage, control, and protection of material, goods, and products throughout the process of manufacturing, warehousing, consumption, and disposal. The focus is on the methods, mechanical equipments, systems, and related controls used to achieve these functions, usually internal, within the company (see, for example, Tompkins et al., 1996). In this chapter, we focus primarily on warehousing activities. According to research of the ELA (1999), 7.7% of the revenues of the 500 European companies interviewed consists of logistics costs, of which 2% is in warehousing and 3.1% is in transport (with 1.2% in administration and 1.4% in inventory costs). Return handling is even more costly. In some businesses, return rates can be over 20% (for example, fashion in the catalog industry) and returns can be especially costly when not handled properly (see Meyer, 1999, and Morphy, 2001). The Reverse Logistics Executive Council has announced that U.S. firms bear losses of the order of billions of dollars on account of return handling (see Rogers and Tibben-Lembke, 1999).

In spite of the aforementioned numbers, many managers have disregarded product returns or they have handled returns extemporarily (see Meyer, 1999). The fact that quantitative methods to support return handling decisions barely exist adds to this problem. In this chapter, we bridge those issues by first going over the key decisions related with return handling and then identifying quantitative models to support those decisions. Due to the lack of quantitative methods dedicated to return handling, we sometimes depart from existent quantitative models for general material handling. We review the main findings and discuss how such models could be adjusted to support return handling. In this way, we are able to provide insight on directions for future research.

The remainder of the chapter is organized as follows. We clarify the operations involved with return handling and then provide an illustrative example. Subsequently, we consider the decisions to be taken when return handling is at stake. After that, we review the existent models for warehousing and material handling and identify promising research matters within the topic. We finish with overall conclusions.

6.2 Return Handling

Different actors in the supply chain face different return flow types (see De Brito and Dekker, 2002). Independent of the return flow type, the following warehousing processes can be distinguished (see Figure 6.1).

- Receipt
- Inspection, sorting & other handling
- (Interim/stock) storage
- Internal transport

Operations details may differ per return type. For instance, products that come back in as-good-as-new state can be restocked to be sold again, while end-of-life products may only need interim storage until they are sold to a third party recycler. We come back to this in the following sections.

Material handling is among the most recent research issues within RL, which is a relatively new topic as a whole. Therefore, material handling aspects (e.g. handling, storage) have not yet been dealt with in great detail. Yet the coercion of reuse and recycling quotas in Europe (see EUROPA online) has served as impetus for specific research on subsequent consequences for return handling. For instance, Anderson et al. (1999) have investigated how firms are coping with EU legislation for packaging recovery, and Fernie and Hart (2001) have considered the required assets to carry out the European packaging regulations. Very recently, De Koster et al. (2002) discussed in detail retail operations when a diversity of reverse flows are present. In this exploratory study, the handling of product and material returns of nine retailer

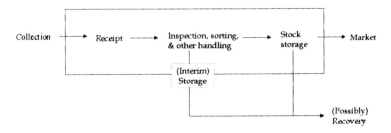

Fig. 6.1. Return handling in the warehouse

warehouses are compared, and receipt and storage of returns, in particular, are analyzed. Furthermore, the aforementioned paper identifies bothersome aspects and slackening procedures for return handling. We come back to specific findings throughout the chapter.

6.3 Return Handling: An Illustrative Case

In this section, we present the Wehkamp case, one of the largest mail-order (business-to-consumer) companies in the Netherlands. Wehkamp has about 3.5 million customers, of which 1.5 million place about 6 million orders per year (daily average about 24,000 orders), with an average 1.8 lines per order. The assortment consists mainly of fashion and hardware, totaling about 50,000 articles. Sales vary greatly over the days of the week, but also particularly over the season. There are two selling seasons of 26 weeks per year, with different assortments. Sales are largest in the first weeks of the season, after a new catalogue has appeared. Customers are supplied from three warehouses, one for hanging and boxed fashion and small appliances, one for dry groceries, and one for furniture and large appliances. The dry groceries flow is fully integrated with the flow of boxed fashion and small appliances. Return rates in mail-order businesses are large; for Wehkamp they vary from a few percent (mainly warranty returns for furniture and large appliances) to 15% for small appliances, and, on average, 40% for fashion. The reason for these returns is (besides warranty returns) partially due to legislation (consumers are permitted to return products at no cost within 10 days of purchase), but accepting returns at no cost to the customer is also seen as an important service element. Too, telephone operators sometimes stimulate customers to buy more fashion products than they really need, thereby creating a sure return. However, as return transportation and handling costs are free to the customer (this also holds for the delivery cost, unless the customer buys only one item), this may still lead to lower cost for Wehkamp, as compared with the customer placing multiple orders for one unit, of which only the last product is bought. Charging a return fee would, of course, reduce the number of returns, but it would probably reduce the net sales as well.

The home delivery of customer orders is outsourced to a Third Party Logistics Provider (3PL) with a very dense network. Orders are delivered within 24 hours. The 3PL makes use of a hub-and-spoke network to achieve this. When a customer wants to return an item, Wehkamp informs the 3PL, who schedules the pick-up on a route carried out in the customer's area. If there are no such routes yet, the pick-up is delayed, or the customer can decide to bring it to the post office herself. On many occasions, pick-ups are within 24 hours, within a time window agreed upon with the customer. Pick-ups go via the two sorting centers back to Wehkamp in large trucks. The same trucks are used for picking up the new orders at Wehkamp's warehouse, thereby achieving efficient use of the trucks and vans in all stages of the network.

In the following description, we focus on the following elements: the receipt of commercial (reimbursement) returns, handling and storage of the returns in the fashion and small appliances warehouse (where the majority of the stock keeping units are stored). We consider those aspects of reverse flows in relation to forward flows. (The average return percentage in this facility is 28%, which corresponds to about 10.000 returns every day.)

Return Receipt, Inspection, and Sorting

Returned items are handled in a large separate area (separated from the receiving area of purchased goods). About 25 people per shift (three shifts per day) sort and recondition the products. The receiving and handling process is illustrated in Figure 6.2.

Unsorted returns are received in roll cages and buffered, after which they are pre-sorted by product category (shoes, media, boxed fashion, etc.). People working in small teams process returns by product category, unpacking the fashion products, checking them for stains, reconditioning (for example, ironing or steaming) the product, and grading and repacking it. After this, products are generally graded as 'new quality' and fit for sale at the original price. The receipt is confirmed in the information system, where it is decided whether the client should be credited or not, based on whether or not the product is returnable (e.g. the condition of the product [unsealed cds may not be returned], payments made, and date of original purchase). The products are marked with a label that can be used to trace the number of returns of this particular item, so that items can be monitored individually. Returned items are sorted per storage zone and put in rolling bins, which are regularly brought to the storage area.

Storage

Bins with returned items are regularly picked up and distributed over the forward pick locations. This is done on a route, visiting multiple storage locations. At the forward location where the product is stored, the returned item's bar code and the location bar code are scanned to confirm the storage and to update the inventory record. In the picking process, no explicit priority is given to returned items over new stock.

Considering the decisions made by Wehkamp to control the returns, we can conclude the following.

- Collection of returns at customers has been integrated with the distribution of products to customers. The same third party logistics company carries out both processes and the route planning is based on both types of orders, although return collection has a slightly lower priority. Even the return drop-offs at Wehkamp are integrated: the truck that brings returned items also collects new orders for distribution. Vehicle routing for collection and distribution is addressed in Chapter 5.

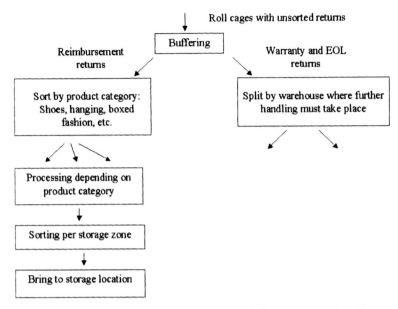

Fig. 6.2. Flow scheme of return handling for fashion and small appliances

- The return handling process is completely separated from the regular receipts of suppliers, with dedicated workstations and handling equipment like conveyors. For administrative handling, a tailor-made software module is used.
- The inbound storage process differs from the regular receipts. For regular receipts, one product carrier (usually a pallet) consists of one product. Returns are consolidated for inbound internal transport in rolling bins.
- Returned products with a 'good' status are consolidated with new products on a location in a forward pick area, whereas newly purchased items are stored in a reserve area. Other than 'good' products are either returned to vendors or to brokers.
- In the picking process, no explicit priority is given to returned items.

6.4 Return Handling: A Framework of Decisions

The description of the case in the last section served to illustrate some of the decisions to be made regarding return handling. We now formally address warehousing decision-making according to long-, medium-, and short-term decisions (see also Ganeshan et al., 1999). Table 6.1 provides an overview. One should notice that the decision topics have many interdependencies. In practice, when deliberating over a decision, the whole set of decisions is to be kept in mind.

Long-term Decisions

Facility Layout and Design

Facility layout and design is essentially a long-term decision regarding warehousing. Companies have to ensure that sufficient storage and handling capacity is going to be allowed for return handling and, additionally, how the space in the facility is going to be organized. For instance, whether a facility dedicated to return handling is preferable or, if not, whether returns should be handled in a separate area of the facility. By integrating return and forward handling, resources can be shared but handling complexity increases. In the illustrative example, returns were handled in the same facility, but in a separate area. Actually, the same was observed in nine retailers in several industries, in a comparative study by De Koster et al. (2002). From the analysis, a threshold on return volume seems to determine whether returns are allocated to a separate area or not: above the threshold it is more efficient to handle returns in a dedicated area. In practice, one even finds companies with a separate facility dedicated for returns, such as Quelle in Germany, and Sears and Kmart in the U.S., which deal with huge amount of returns (*Chain Store Age*, 2002). Chapter 4 of this book, which deals with network design, further explains how the decision of settling a separate facility for return handling is affected by the amount and type of returns. An obvious trade-off is between investments and the capacity for return handling. In addition, the location of a return handling area within a facility affects both the internal handling systems that can be used and the associated internal transport costs. In the design process, future return handling must therefore be taken into account.

Medium-term Decisions

Medium-term decisions concern outsourcing, integrating operations (return policy and type of packaging), inventory management, internal (return) transportation, and information systems. In the following sections, we briefly address each of them.

Outsourcing

In the illustrative case, Wehkamp keeps the warehousing operations in-house. In this case, outsourcing would be difficult as the activities take place within the warehouse and are closely related with order picking and storage activities. Provided that the return process can be clearly separated from the forward process, return handling may be outsourced. Examples include Albert Heijn (the return warehouse in Pijnacker) and Sears and Kmart in the U.S., where the returns are handled by Genco (De Koster and Neuteboom, 2001; *Chain Store Age*, 2002). Other determinant factors are the value and type of product as well as the availability of experienced third parties. At this moment such companies are emerging. Examples include Genco and Universal Solutions Incorporated (see *Chain Store Age*, 2002).

Table 6.1. A framework of decisions for return handling warehousing

Long-term decisions

- Facility layout and design

  Are returns handled in separate facilities?
  Are returns handled in a separate area of the same facility?
  How should these facilities be laid out?

Medium-term decisions

- Outsourcing

  Are warehouse (return handling) operations going to be
  totally/partially outsourced?

- Integrating operations (return policy and reusable packaging)

  Under which conditions are returns accepted at the warehouse
  (related with return policy/responsibility)?
  Which returns are to be totally/partially credited by the warehouse?
  What responsibility does the warehouse have with respect to returns
  (e.g. collection, sorting)?
  Specific decisions on reusable vs. one-way packaging;
  waste reduction/disposal

- Inventory management

  Where are products to be stored?
  Are product returns in condition as-good-as-new
  to be stored together with new products?

- Internal (return) transport

  Which type of product carriers should be used?
  To which extent should the operation be mechanized or automated?
  Which type of vehicles should be used?

- Information Systems

  Which IT systems are to support the return handling?
  Which information is to be kept and for how long?
  Will the warehouse make use of dedicated software for return handling?

Short-term decisions

- Inventory, storage, and order picking

  Controlling return storage
  Planning returns-storage vs. order picking

- Vehicle planning and scheduling

  Route selection taking into account reverse and forward flows.

Integrating Operations

When products are returned, responsibilities for related processes have to be settled. In the case of chain stores, for example, either a central facility (often the distribution center) collects and processes the products and waste materials that have been returned from the stores, or the stores take care of it themselves. According to IGD (2002), in the U.K. all major food retailers recycle packaging materials in a central facility, either a distribution center or a separate recycling unit. Actors in the chain also have to establish whether different types of returns incur different levels of accountability. For instance, the manufacturer takes the responsibility for all the processes related with end-of-life returns but no responsibility for reimbursement returns independent of the motive. Another issue to decide upon is the explicit return policy. For instance, stores have to know which products can or must be returned to the warehouse in which situations, or how to grant permission for this. In the illustrative example, the mail-order company does not have much freedom with respect to the return policy, as it is a legal requirement to allow for returns within a specified period of time.

The urge to reduce packaging materials leads to the decision of whether packaging materials can be replaced by reusable packages. Reusable packages will demand a higher initial investment, but have the inherent benefit of repetitive use. Many industries choose to reduce packaging waste by using less material and by using standardized boxes (Schiffeleers, 2001), or by replacing packaging materials with other materials that involve less material handling. Some retailers opt to start recycling their own packaging materials. Albert Heijn, for example, recycles wrap foil into plastic bags, integrating the forward and reverse flows for a practical purpose. Yet, this demands changes in its operations. Stores have to separate the wrap foil, put it in roll cages, and store it until collection. After recycling by a third party, bags are resent to the warehouse to be integrated with the distribution.

Inventory Management

Returned products may have a quality status that differs from new. Depending on this status and the timing, they may be sold in the same market (e.g. early in the season reimbursement returns, as in the case of Wehkamp) or not (namely leftovers or end-of-life returns). Both newly supplied and (as-good-as-new) returned merchandise will be stored for future sales as long as the season runs. The two streams can be stored separately or the merchandize can be consolidated in the same location. In the case of Wehkamp, returned items are stored together with new merchandise. De Koster et al. (2002) have also found mixed and hybrid storage policies among the operations of nine different retailers. For instance, one retailer would first store returns at a separate location and consolidate them only when the location's capacity would be exceeded. The analysis led to the conclusion that separate storage policies are mainly found among retailers wishing a high degree of control over returns.

Returns can be consolidated if the future market is the same as for newly supplied merchandise. The 'future market' is also a critical factor for returns other than the ones 'as-good-as-new.' Leftovers are usually consolidated in interim storage locations per vendor or potential 'customer'. A well-known example is that of books that go back to the respective publishers. End-of-life returns are also separated and kept together by the potential broker.

Internal (Return) Transport

An important issue in internal transport is the choice of reusable carriers. Choosing the type depends on the willingness of the parties in the chains to adopt one of the available standards. This is usually a complex process, where power plays a major role. Once a standard is in place, the decision that is left is the amount needed. Product carriers require collection, transportation back to the warehouse, checking, storage, and possibly cleaning before being used again. In order to limit the amount of product carriers needed, they should rotate rapidly. Since the timing and quality of such returns are difficult to anticipate, many companies have searched for ways to reduce such uncertainty. De Brito et al. (2003) review real applications of incentives to persuade parties to behave in a desired way. A common incentive is to charge a deposit fee between the different parties in the chain, especially if the material has some intrinsic value. Each receiver must pay this fee to its supplier. Often, when multiple companies participate in such networks, there is a central organization that tracks the ownership of carriers and registers financial transactions. The deposit fee not only prevents items from being lost but it also provides a natural mechanism to motivate careful handling and therefore a minimum quality is ensured per returned item, as transactions have to be tracked. Several international pools for pallets exist in Europe, like Deutsche Bahn (DB) and Chep. Many manufacturers supply their products on such pallets. If the reusable packaging can be shared between different users, the benefits can sometimes be huge, as large savings can be achieved as the numbers needed decrease (Koehorst et al., 1999). For this reason, in the Netherlands and in the U.K. food retailers have switched to reusable container systems (IGD, 2002). In food retailing, reusable crates are mostly used for product categories such as produce and chilled, where warehouse stocks are small and sales volumes are high. This means the crates are used intensively, reducing the cost per trip. However, wholesalers and retailers with many suppliers and different pallet types have considerable work in handling, sorting, and storing pallets returned from their customers (retail organizations or stores). In case of large return volumes, mechanization or automation may become economically attractive. Choices need to be made regarding the degree of mechanization and the appropriate equipment for internal transport and storage.

Information Systems

Finally, at this level, one has to determine which information system is going to be used to register product returns. Chapter 16 examines the overall needs and existing supporting technology for managing information in the reverse logistics context. Commercial software particularly designed for supporting return handling is, however, lacking (Caldwell, 1999). The commonly used ERP packages generally lack the ability to properly deal with returns (De Kool, 2002). At this stage, decision-makers may consider the in-house development of dedicated software. This has been the case at Wehkamp and Estée Lauder (see Meyer, 1999). The specialized software system checks returns for expiration date and damages, speeding up return handling. Besides this, the software is linked to an automatic sorting system, which has lowered labor costs. In general, the type and the horizon of information to be registered has to be determined, as well as how decisions with respect to returns can be supported. Also attention can be given to potential abusive returns (see Schmidt et al, 1999) and how this affects warehousing operations. In the case of Wehkamp, though law enforces accepting returns, returned merchandise is first checked and only then it is decided whether the client should be credited for it.

Short-term Decisions

Inventory, Storage, and Order Picking; Vehicle Planning and Scheduling

When storing returns, the work has to be planned in coordination with forward operations such as order picking and internal transport. In the case of Wehkamp, to put returns back into inventory in a timely fashion is crucial, to prevent stock outs. Another aspect is the organization of the picking process. In the case of Wehkamp, no explicit priority is given to returned items over regular stock. There are, however, cases where priority to returns might be given. This was explicitly observed at a mail-order company as reported by De Koster et al. (2002). The aim is to have quick feedback on returned items, and thus items that are being returned repeatedly can be identified and proper action can be taken.

We have presented here a framework of long-, medium-, and short-term warehousing decisions with respect to return handling. In the remainder of this chapter, we bring into focus the available and potential models to support those types of decisions.

6.5 Models and Research Opportunities

Quantitative models on return handling are practically non-existent. Therefore in this section we sometimes depart from quantitative models on general material handling. We indicate how they can be adapted to accommodate returns and in which areas there are research opportunities for developing quantitative models that aid decision-making. An early overview of the use of OR tools in material handling can be found in Matson and White (1982).

The main research areas in materials handling and warehousing are (see also Van den Berg and Zijm, 1999) the following.

- Facility layout and design
- Outsourcing
- Integrating operations: return policy
- Integrating operations: reusable packaging
- Inventory management (see also next-to-last item)
- Information systems for return handling (see Chapter 16)
- Internal (return) transport
- Inventory, storage, and order picking
- Vehicle planning and scheduling

Facility Layout and Design

One successful research area is facility layout. A well-known approach is Muther's (1973) Systematic Layout Planning concept, a structured approach leading to a layout. The formulation of the problem usually leads to a nonlinear objective function (for example, a quadratic assignment problem). Many improvement algorithms, not necessarily leading to an optimal solution, have been produced that may support designers in coming to efficient layouts. Classic algorithms such as CRAFT, ALDEP, and CORELAP are described in Tompkins et al. (1996). There also algorithms for multi-story buildings (MULTIPLE, Bozer et al, 1994) and algorithms for areas which are restricted in length/width ratios (Goetschalckx, 1992). The major objective in the underlying models is to minimize the total cost of daily transport between the different areas in the facility. For some facility types, special layout models have been developed with the reduction of travel time as the primary objective. These facilities include cross-dock centers, where goods have to be transported from receiving dock doors to shipping dock doors (Bartholdi and Gue, 2000) and picker-to-parts warehouses where the objective is to find the layout for a given storage capacity that minimizes the order picking travel time (Roodbergen, 2001).

Based on these methods, commercial software has been developed to graphically aid designers in interactively developing designs. An important shortcoming in the models is that, in reality, many other restrictions and objectives are important as well, such as congestion reduction. The models and solution methods can be applied straightforwardly to facilities with return flows and handling areas. Departments where returns have to be received or sorted do not change the models fundamentally. However, it is important to incorporate returns operations from the start in layout decisions, as the solutions may change dramatically when compared to the situation without returns (think of crossing flows or travel distances). It may be interesting to study how return flows impact such changes in layout. This is a topic that has not received attention from researchers so far.

Outsourcing

Literature on process outsourcing is mostly of a qualitative or quantitative empirical nature. Although much literature deals with outsourcing of warehousing and transport to logistics service providers (see for example Rabinovich et al., 1999; Van Laarhoven et al., 2000), academics have not specifically addressed the outsourcing decision of return handling so far. One exception is Krumwiede and Sheu (2002), where a model is presented to aid third-party logistics providers to decide whether to enter the reverse logistics market.

Integrating Operations: Return Policy

Distant sellers and other retailers usually accept merchandise back up to a number of weeks after delivery or purchase. Wekhamp puts the threshold at about two weeks and does not charge the client in case of return. In practice, we find various charging schemes including partial refundable costs. Schemes to charge returns and some of the associated dilemmas have been studied in the literature (for example, how to control opportunistic returns; see Hess et al., 1996). Researchers have established relations between return policies and a number of elements, among which are salvage value (Davis et al., 1995 and 1998; Emmons et al., 1998), mismatching probabilities (Davis et al., 1995; Hess et al., 1996), speed of consumption (Davis et al., 1998), the product's value to the consumer (Davis et al., 1995; Hess et al., 1996), and the product's quality (Moorthy and Srinivasan, 1995; Wood, 2001).

However, return handling costs have been kept out of the discussion. We present here a model to explicitly include return handling costs. The model below is, in fact, a holistic approach to return policies as it gives plenty of room for the incorporation of multiple critical factors. Let U_r be the utility function of a retailer. U_r is a function of factors like

- t, maximum time until return,
- p, price of the product,
- x, quantity sold/(bought),
- y, quantity returned,
- c_h, handling/warehousing costs,
- c_{rh}, return handling costs, and
- c_{rc}, return charges.

Customers $i=1,...,\ m$ have, respectively, utility functions u_i. Each u_i is a function of factors like

- p, price,
- t, maximum time until return,
- x, quantity sold/(bought),
- y, quantity returned,
- q, quality of the product, and

- c_{rc}, return charges.

All the actors involved intend to maximize their utility function. The retailer is the leader as (s)he sets a priori some of the parameters, like t and p. In turn, customers will react to the values of these parameters and maximize their utilities. The problem can be written as follows:

$$\max_{p,t,c_{rc}} U_r(p, t, c_{rc}; y^*, x^*, c_h, c_{rh})$$

s.t.
$$(x^*, y^*) = \sum_{i=1}^{m} \arg\max u_i(x, y; p, t, c_{rc}, q)$$

For a given strategy that the retailer may choose, customers will have a 'best' strategy to follow. The retailer has to anticipate the reactions of its customers and only then set the return strategy.

The problem can be put in the form of a Mathematical Programming problem with Equilibrium Constraints and it can be solved as such (MPEC problem, see Luo et al, 1996). The MPEC is closely linked to the Stackelberg game-theoretical models. For more on closed-loop supply chain coordination, please see Chapter 12.

The previously mentioned literature on return policies can be used to estimate the utility functions. Furthermore, experiences similar to the ones conducted by Wood (2001) can be applied to tune the relations between the parameters.

Integrating Operations: Reusable Packaging

Several authors have paid attention to the evaluation of durable versus one-way packaging. Organizational, technical, economical, and environmental aspects play a role in this evaluation. Organizational aspects concern responsibilities of different actors in the supply chain with respect to the handling, return, and storage of the packaging, including the financial setup. Technical aspects involve investments needed in equipment to properly handle the packaging.

In order to determine economic benefits, the needed number of reusable packages and the distribution over the network has to be determined. In order to do this, quantitative models can be used. Kroon and Vrijens (1995) use a location-allocation model to determine the locations of empty container depots and to allocate stocks of empty containers in the different depots to the users of the containers. The objective is to minimize empty container transport cost and the fixed costs of the container depots. Duhaime et al. (2001) use a minimum-cost-flow model to determine the number of empty containers that should be returned to the central hub of Canada Post each month and the distribution over the network. To determine the environmental impact of reusable crates, Life Cycle Analysis (LCA) can be used (see Chapters 14 and 15). For every process (production, handling, transport) or material usage, the impact on the environment (for example, global warming or acidification) can be established by calculating the emissions of different components. It is still necessary to trade-off the different impacts. Furthermore, different methods exist to calculate these emissions. The best method depends on the production and transport technologies used, on the country involved, etc. Gradually, standard tools will become available to do such analysis.

Inventory Management

If return rates are high, such as in distant selling business, it becomes necessary to manage return stocks explicitly. One can also look at this from a completely different perspective: remote sellers are overselling to customers that return merchandise (and therefore underselling to customers that do not return merchandise). One can actually put customers that (systematically) return merchandise versus not into two classes: less- and more-profitable customers. Mail-order companies are a good example as these companies carry large amounts of historical customer data, which can be employed to draw the profile of each customer class. Return-handling costs can be explicitly utilized to draw the line between the two classes. Existent inventory models with priority customers can be stretched to this new application (see Kleijn and Dekker, 1999). Research has shown that '...in some cases it is optimal to...reserve inventory for possible orders from higher-margin customers' (Cattani and Souza, 2003). In other words, it may be optimal to reserve inventory for customers that do not systematically return merchandise. This nurtures confidence in the opportunities that this sort of model can offer to retailers like mail-order companies. Moreover, return-handling costs can be plugged in explicitly, bringing realism to inventory management with product returns.

Internal (Return) Transport

OR models (stochastic models, mixed-integer linear programming models or simulation) can be applied successfully in the evaluation of material handling systems, in particular estimating the number and type of vehicles needed. By comparing multiple scenarios, with multiple types of material handling systems, an evaluation of the best system can be made. The underlying models try to determine the number of vehicles of a certain type in a certain facility. Examples are the evaluation of single-load versus multiple-load vehicles (Van der Meer, 2000), or lifting versus non-lifting vehicles (Vis, 2002). Other related design areas are vehicle transport track design, choice of pick-up and delivery points, design of deadlock-free tracks, track claim design, and design of battery loading areas. The paper by Goetz and Egbelu (1990) is an example. Material handling systems used for both forward and return flows may have to meet different requirements. Depending on the return volume, separate systems that are better fit for return handling may be needed.

Inventory, Storage, and Order Picking

This topic includes particular problems within warehouses, such as order batching, order picking, routing pickers in a warehouse, warehouse zoning (dividing a picking area in zones to achieve certain objectives), product storage allocation, and forward versus reserve storage area decisions.

Many papers deal with warehouse planning and control. There are several recent overview papers in this area, including Van den Berg (1999), Rouwenhorst et al. (2000), and Wäscher (2002). The following topics have been addressed in OR literature in particular.

- Product to storage allocation
- Order batching and wave picking
- Routing order pickers in a warehouse
- Worker balancing and warehouse zoning
- Forward-reserve problem
- Pallet or container loading

Product to storage allocation tries to allocate products to locations such that the order picking process (and sometimes also the storage process) is optimized. The objective is usually to minimize the worker's travel time. Well-known methods for storage allocation are closest open location storage, the cube-per-order index method (Heskett, 1963), class-based storage, random storage, full-turnover based storage, and family grouping.

Order batching and wave picking is concerned with grouping customer orders and releasing them to pickers as a group, in order to reduce the processing time. By combining orders in a pick route, the routes may become more efficient. This is often at the expense of the picker having to sort the orders while picking. De Koster et al. (1999) present a performance evaluation of different order batching methods.

Routing order pickers is the problem of finding efficient routes within a warehouse to pick all orders, the objective being the minimization of travel time. The models depend on restrictions that play a role, the warehouse layout (for example, the length and number of aisles or the presence of cross aisles), start and finish locations (these may be fixed or variable), and the type of material handling equipment used. Roodbergen (2001) presents an extensive literature overview on this topic.

The problems of *worker balancing and warehouse zoning* have been addressed by, among others, De Koster (1994), Bartholdi and Eisenstein (1996), and Bartholdi et al. (1999). The latter papers show that the throughput can be increased substantially if workers have free, rather than fixed, working zones.

The *forward-reserve* problem is concerned with the storage of small (daily or weekly) pick quantities of some items in a forward pick zone and the storage of bulk quantities in a larger reserve zone. The forward zone is replenished from the reserve zone. Some items may be stored exclusively in the reserve zone and have to be picked there. The objective is usually minimization of cost or of total work. Literature on this topic includes Hackman and Platzman (1990).

The *pallet or container loading* problem is closely related to the bin-packing problem (Coffman et al., 2000). One tactical problem is to determine the right container sizes to be used. At the operational level, a choice must be made from the available sizes, and a stacking scheme must be developed.

All of these topics are strongly interdependent in their impact on warehouse performance (e.g. order through time). There is some literature on combinations of these above-mentioned problems, but due to the complexity of such problems, it is not abundant. Gademann et al. (2001) combine batching and routing, and Roodbergen (2001) combines routing and layout design. Recently, storage and routing have received some more attention (see, for example, Dekker et al., 2002).

Currently, no models exist that explicitly include return flows. In our view, creating models is most interesting for the following decision areas.

- One important decision is setting up separate storage areas for returns (for example, storage of multiple products per location), or consolidating them with existing stock. The trade-off is in the size of the storage area and order pick travel time, but also in the cost of a warehouse of increased size. Storing returns in separate locations may save time in storage, but potentially requires more space, which in turn increases pick times.

- With respect to putting returned products back in stock, a further point to investigate is how long to buffer product returns before consolidating them in stock. Important factors that play a role in such a decision are the quantity, variety, and timing of the returned items. The larger the stored batch size is, the more efficient the storage process becomes, while on the other hand the buffer space may become overcrowded and the stock will not be available in time.

- Another decision area is in the sequencing and routing of pickers that have to store returns in a location on a route, but also have to pick orders on routes. On many occasions, return job scheduling issues play a role, since picks can only be carried out when the (returned) products are at their location (this is the case at Wehkamp, see Section 6.3). On the other hand, picks are much more urgent, because of due times for shipment. This problem has received no attention from researchers so far.

Vehicle Planning and Scheduling

A large part of the literature on material handling equipment covers planning and control issues of unit-load automated storage and retrieval systems (AS/RS). This has yielded some important results.

- The optimal length-height ratio of storage systems, with respect to crane cycle times (Bozer and White, 1984).

- Calculation of crane cycle times using different storage or retrieval strategies. (Initial papers are by Bozer and White, 1984, 1990, and Hausman et al., 1976.)

These initial papers have led to a large number of papers on cycle time calculation for AS/RS systems with slightly different layout, behavior, location usage, interleaving policies, and storage strategies.

Another large group of papers are on planning and control of automated guided vehicle (AGV) systems. The nature of such internal transport systems is that the environment they work in is highly dynamic, that is the horizon over which information is available regarding loads that need transportation is usually short. On a slightly longer horizon, information may be known in advance, but is uncertain. Both load arrivals and transportation times are stochastic, due to congestion and intersection control policies. The following activities have to be carried out by a controller of the system.

- Dispatching of vehicles to pick up certain loads
- Route selection
- Vehicle scheduling
- Dispatching vehicles to parking positions

Most literature addresses one or at most two topics simultaneously. Many heuristic rules have been evaluated for vehicle dispatching, with or without prior load arrival information. Overviews can be found in Van der Meer (2000) and Vis (2002). Vehicle routing and scheduling problems are often modelled as mixed integer programming problems, which may have mixed pick-ups and deliveries, time windows for the deliveries, dynamic versions of the problem and different objective functions (like makespan and average load waiting time). Overviews are given by Desrochers et al. (1988), Solomon and Desrosiers (1988), and Savelsbergh and Sol (1995).

Quantitative research on planning and control of other material handling equipment is not abundant. There is some literature on conveyors and sorting systems (Meller, 1997; De Koster 1994), carousels (Rouwenhorst et al., 1996), compact storage, and other systems.

When returns are involved, the planning and scheduling of such material handling equipment will change. Returns can be modelled as another type of (storage or transport) job that has to be carried out. However, particular-sequencing restrictions will apply. For example, storage jobs resulting from returns must be carried out prior to retrieval jobs. Or, for unit load handling machines, it must be attempted to combine jobs in a double play to increase machine utilization. The above type of model can be applied to situations with returns as well.

In conclusion, there is a shortage of quantitative models to support return handling decisions. At the same time, there are many opportunities to extend existent forward models or to initiate new research paradigms. Presumably, these quantitative models will not be straightforward and more has to be learned on return processes before they can be developed and implemented successfully.

6.6 Summary and Conclusions

The handling of return flows is not the same as the usual handling of forward flows coming into the warehouse. As illustrated in Section 6.3 and discussed in Section 6.4, handling return flows additionally involves collection, inspection, grouping, splitting, and recovery. This demands that decisions regarding a larger number of issues are taken.

We have presented a decision-making framework for return handling for long-, medium-, and short-term decisions: 1) facility layout and design, 2) outsourcing, 3) integrating operations (return policy and reusable packaging), 4) inventory management, 5) internal (return) transport, 6) information systems, 7) inventory, storage, and order picking, and 8) vehicle planning and scheduling. We have stressed that the whole set of decisions has to be taken into account for every single decision because of the many interdependencies. Some of the known results on efficient material handling were reviewed. In short, when high volume of returns is present, receipt and sorting is likely to be in a dedicated area of the warehouse. With respect to storing of returned products, the future market of returns and the desired degree of control on returns are the influencing factors (see De Koster et al., 2002).

Though available quantitative models can be adapted to support warehousing return handling, this has been, up to now, largely ignored. For a vast number of research areas, we have in Section 6.5 highlighted how forward models can be extended or new models can be launched to include return handling. We brought to attention several research holes, which can be turned into a research agenda: What impact do return flows have on 1) the warehouse layout 2) material handling systems (e.g. whether dedicated return handling systems do pay off), 3) vehicle planning and scheduling (mind sequencing restrictions), and 4) storage and picking procedures (e.g. route combination of returns storage and pick-ups). Furthermore, we suggested the innovative use of MPEC formulation to help retailers decide on a return policy. A new direction to inventory modelling was also put forward by categorizing customers in two classes, less- and more-profitable, depending on whether or not the customer (systematically) returns merchandise.

Most likely some of the ideas for future research laid down here are not straightforwardly carried out. We believe that learning more about the practice of return handling can subtract some of the latent modelling difficulties. Not only quantitative models are needed, but in order to implement it successfully more qualitative or empirical approaches are necessary. For example, research towards best practices may help decision makers to come up with solutions for sorting, buffering, and storage. By conducting simultaneous desk and field research, quantitative models will plausibly aid on real return handling decision-making.

Acknowledgements

We would like to thank Ilker Birbil for his helpful comments on the MPEC formulation. Marisa P. de Brito was financially supported by the Portuguese Foundation for the Development of Science and Technology, i.e. "Fundação Para a Ciência e a Tecnologia".

Part III

Inventory Control and Production Planning
for Product Recovery

7

Lot Sizing Decisions in Product Recovery Management

Stefan Minner and Gerd Lindner

Faculty of Economics and Management, Otto-von-Guericke University Magdeburg,
P.O.Box 4120, 39016 Magdeburg, Germany,
minner@ww.uni-magdeburg.de, gerd.lindner@ww.uni-magdeburg.de

7.1 Introduction

Inventories play a central role in operations and supply chain management, and there exist several motives for storing units instead of following a pure just-in-time strategy with zero inventories (see, for example, Silver et al., 1998). A first motive is that of transactions, i.e. the presence of setup costs, setup times, and economies of scale. This results in cycle stocks, the underlying problem of which is lot sizing. This will be addressed in this chapter. Incorporating uncertainty, the safety motive results in safety stocks, an issue that plays a central role in Chapter 8. The presence of dynamic aspects, e.g. changing prices and costs may result in speculative stocks, an issue that is addressed in Chapter 9. The presence of product returns can be incorporated by a different category of opportunity stocks (see Fleischmann and Minner, 2003). In the following, we will focus on lot sizing issues and refer to the inventory-related chapters for combined issues.

Lot sizing addresses the problem of how many units should be aggregated into a single order. This phenomenon occurs in procurement (how many units should be purchased at once), in manufacturing (how many units should be processed between consecutive setups of the equipment), and in transportation (how many units should be shipped in one lot). This problem has been studied extensively for production systems without product recovery (for example, see Kuik et al., 1994, for an overview). The incentives of batching, in contrast to just-in-time operations, where the extreme, a processing of every single unit, takes place, are economies of scale. For example, the fixed cost of a procurement batch is allocated to several units, quantity discounts may be achieved, or the number of setups between production batches for different products may be reduced. On the other hand, the disincentives of batching are resulting inventories that tie up capital and occupy space. Lot sizing aims to trade-off the costs associated with process transactions (ordering, setups, transportation) and the costs of inventory holding (see Chapter 11 for details) which consist of opportunity costs for capital and out of pocket costs.

The two well-known approaches towards these problems are the economic order quantity (EOQ) model and the discrete time, dynamic lot sizing problem (Wagner-Whitin model). Both models are covered in operations textbooks and have resulted in numerous extensions to make the approaches more suitable for practical application (for example, see Silver et al., 1998). In this chapter, we discuss the additional problems that arise in lot sizing for reverse logistics processes and outline the progress that has been made in this regard. The main differences in contrast to the classical EOQ and the Wagner-Whitin model are 1) the presence of two alternative supply modes by having the opportunity to remanufacture returned items instead of procuring all requirements, one mode being capacitated by the availability of returned items and 2) having to decide whether and when a returned item is remanufactured or disposed. Where the cost trade-off in the EOQ model is between costs for each setup or replenishment on the one hand and inventory holding costs on the other, these trade-offs in reverse logistics lot sizing models are complicated by having two modes of supply, where the classical trade-off has to be considered for each mode together with the selection between the two modes. Additionally, when returned items are kept in a separate recoverables inventory before being processed and are then stored together with procured or newly manufactured items in a serviceables inventory, there arises another cost trade-off between the (generally cheaper) storing of units upstream (in recoverables inventory) rather than downstream (in serviceables inventory).

The practical need to incorporate product returns into lot sizing considerations was emphasized in several case studies. Ashayeri et al. (1996) discuss the problem of service parts in computer electronics where products are returned decentrally and can be used as service parts. Procurement and manufacturing are used to replace defective units and are carried out in batches. The remanufacturing processes, mainly disassembly and component recovery, are carried out in batches as well, which requires prescribed lot sizes. Van der Laan (1997) describes the problem of automotive spare parts at Volkswagen (see also the DaimlerChrysler case presented in Chapter 9). In Teunter et al. (2000), the case of solvent recycling in a pharmaceutical company is described. The main processes in this multi-product problem are the manufacturing of different active ingredients and the recycling of different solvents being used in the manufacturing processes. Cleaning requirements cause significant changeover times, both between batches of different ingredients, as well as in the case of solvent recycling, which requires distillery setup time. To minimize this non-productive time, there is a clear incentive to have recovery batches. The manufacturing processes require different solvents as major input resources. During processing, impure solvents are retained and stored. Subsequently, they can be partly recovered through distillation and then serve as a substitute for new pure solvents. Moreover, demand for and returns of a given solvent are directly linked to the manufacturing of associated main products. As a result, the planning problem involves two interacting lot sizing and scheduling problems.

The material in this chapter is organized as follows. In Section 7.2, we present extensions of the EOQ model considering reverse logistics environments. First, we examine approaches that assume a certain production/procurement and recovery policy and find the optimal batch sizes for this given strategy. Then we discuss why this proposed batching strategy with identical lot sizes excludes the optimal strategy and review an improved approach that allows for non-identical batches. Thereafter, we address extensions with finite production rates and approaches that explicitly take important manufacturing issues like capacity and scheduling issues into account. In Section 7.3, we consider the problem of discrete time, dynamic lot sizing model extensions with return flows. Section 7.4 reviews lot sizing issues in rework operations (see also Chapter 10) with imperfect yield before the main results are summarized and lines for future research are outlined in Section 7.5.

7.2 Lot Sizing for Constant Demand and Returns

7.2.1 Constant Production and Recovery Lot Sizes

If returns are ignored, i.e. demands are completely satisfied from procured or manufactured parts, traditional lot sizing models apply. However, if returns have an economic value and recovered parts can be used as substitutes for procured or manufactured parts, leaving the lot sizes unchanged and using uncoordinated production and recovery batch sizes results in additional inventories and associated costs for the returned items. If one setup cost, either associated with manufacturing or remanufacturing, is negligible and processing rates are infinite, a netting approach can be applied. If the remanufacturing setup cost is negligible, we can remanufacture incoming returns for demand and determine the relevant manufacturing batches using the EOQ approach with a net demand rate. In the opposite case of negligible manufacturing setup costs, we can first determine the remanufacturing batches and then manufacture for the remaining net demands just-in-time. If both setup costs are significant, a simultaneous consideration of the involved trade-offs is required, which will be illustrated in the following.

As an extension of the EOQ environment that includes a recovery option, the following assumptions are made. A constant, deterministic demand rate of d units per unit of time has to be satisfied, and backorders are not permitted. Used products are returned with a constant and deterministic rate u_r, measured in units per unit of time. The rate u_r does not exceed d in order to guarantee a stable system behavior. This assumption seems to be realistic since usually not all products consumed by customers come back to the manufacturing system or, equivalently, some items are in such a damaged condition that it is impossible to remanufacture them. In the following, u_r/d represents the return and, in the case of no disposal, the reuse fraction. There are two inventories: serviceable parts that are used to satisfy customer demands and recoverable items that have been returned but are not yet ready for use by customers. Serviceable parts can either be replenished by manufacturing or procuring new items or by remanufacturing parts from recoverables inventory. It is assumed that both processes have negligible lead times, which is not a limitation since respective replenishments can be scheduled in advance to account for fixed delays. Remanufactured parts are assumed to be as good as new and all returns have the same quality. A special case of this model was first studied by Schrady (1967) and later extended by Richter (1996a, 1996b) and Teunter (2001). In order to facilitate the understanding of the following presentation, we do not include a disposal option; all reusable items are recovered (for the respective extension, see Richter, 1996a, 1996b, and Teunter, 2001). Note that because of this assumption, variable manufacturing and remanufacturing costs are not relevant for the decision. Each manufacturing setup is associated with a cost K_p, and each remanufacturing batch with a remanufacturing setup cost K_r. Recoverable items are subject to holding costs h_u per unit and unit of time. For reasons of simplicity, we presume that for each item at the serviceables inventory, irrespective of being remanufactured or procured (newly produced), a corresponding holding cost rate h_s is charged. Regarding the more general case where different holding cost rates were quoted for units stemming from procurement (production) or remanufacturing, we refer to Chapter 11 of this book for an in-depth discussion. Because more capital is tied up in serviceable parts and less in recoverable items, we assume $h_s \geq h_u$. These assumptions are a straightforward extension of the EOQ model. Therefore, all criticism on the EOQ assumptions are, of course, valid here as well and call for respective model extensions.

Subsequently, two cases are analyzed. In the first case (referring to Schrady, 1967), a single manufacturing batch of size Q_p is followed by R remanufacturing batches of identical size Q_r. Note that having identical batch sizes is an additional assumption that will be relaxed in Section 7.2.2. Therefore, the following lot sizes are only optimal within this class of policies with identical batch sizes for each process type. The manufacturing batch serves to satisfy the difference between the demand and return rate whereas, with respect to one inventory cycle, the relation between remanufactured and newly produced items can be expressed via $u_r(RQ_r+Q_p)/d = RQ_r$, i.e. all units being returned in a cycle are remanufactured. For illustrative purposes, Figure 7.1 shows an exemplary behavior of system-wide inventories in an inventory cycle that is composed of one production lot as well as $R = 6$ remanufacturing batches. The upper part displays the development of recoverables inventory whereas the lower part depicts the corresponding serviceables inventory pattern.

In order to determine the optimal solution, we use the criterion of minimization of total relevant costs per unit of time. The total relevant costs as a function of the manufacturing and remanufacturing batch sizes are

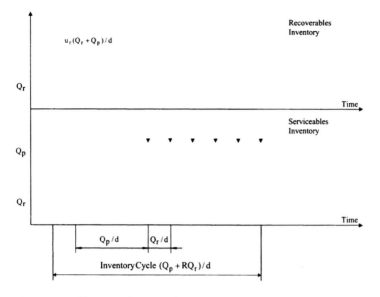

Fig. 7.1. Recoverables and serviceables inventory

$$TC = \frac{K_p(d - u_r)}{Q_p} + \frac{K_r u_r}{Q_r} + \frac{h_s}{2}\left[\left(1 - \frac{u_r}{d}\right)Q_p + \frac{u_r}{d}Q_r\right] + \frac{h_u}{2}\frac{u_r}{d}(Q_r + Q_p),$$

(7.1)

i.e. the sum of production setup costs per unit of time, remanufacturing setup costs per unit of time, and serviceable as well as recoverable holding costs per unit of time. The evaluation of the first order conditions of this convex objective function reveals the following optimal lot sizes Q_p^*, Q_r^*, and recovery batch number R^*

$$Q_p^* = \sqrt{\frac{2(d - u_r)K_p}{h_s\left(1 - \frac{u_r}{d}\right) + h_u\frac{u_r}{d}}}$$

(7.2)

$$Q_r^* = \sqrt{\frac{2dK_r}{h_u + h_s}}$$

(7.3)

$$R^* = \frac{\frac{u_r}{d}}{1 - \frac{u_r}{d}}\frac{Q_p^*}{Q_r^*} = \frac{\frac{u_r}{d}}{1 - \frac{u_r}{d}}\sqrt{\frac{K_p(h_s + h_u)}{K_r\left(h_s + \frac{u_r}{1-\frac{u_r}{d}}h_u\right)}}.$$

(7.4)

These results are simple adjustments of the EOQ formula. The production lot size takes into account the net demand rate $d - u_r$ and uses a weighted holding cost rate, whereas the remanufacturing lot size uses the gross demand rate and an adjusted holding cost rate which equals the sum of the serviceables and recoverables holding cost rate. It should be noted at this point that the above result is only an approximation. In reality, the number of remanufacturing batches R^* has to be an integer which is, in general, not given by the above analysis. When it comes to an implementation of the above lot sizes, the realized costs will be larger than predicted by the analysis, and adjustments, e.g. rounding of R^*, are required (for a detailed discussion and a numerical study on the errors and adjustments of the integrality issue see Minner, 2002a). Inserting the above results into the cost function, the total relevant costs for an optimal solution are

$$TC^* = \frac{u_r}{d}\sqrt{2dK_r(h_s + h_u)} + \left(1 - \frac{u_r}{d}\right)\sqrt{2dK_p\left(h_s + \frac{\frac{u_r}{d}}{1 - \frac{u_r}{d}}h_u\right)}. \quad (7.5)$$

In contrast to Schrady (1967), Teunter (2001) analyzes the opposite case where a single remanufacturing batch Q_r is followed by M identical manufacturing batches Q_p. Note that these lot sizes may be different from the ones determined by the previous approach. The total relevant costs per unit of time are

$$TC = \frac{(d - u_r)K_p}{Q_p} + \frac{u_r K_r}{Q_r} + \frac{h_s}{2}\left[\frac{u_r}{d}Q_r + \left(1 - \frac{u_r}{d}\right)Q_p\right] + \frac{h_u}{2}Q_r \quad (7.6)$$

with resulting optimal lot sizes and manufacturing batch number

$$Q_p^* = \sqrt{\frac{2dK_p}{h_s}} \quad (7.7)$$

$$Q_r^* = \sqrt{\frac{2u_r K_r}{\frac{u_r}{d}h_s + h_u}} \quad (7.8)$$

$$M^* = \frac{1 - \frac{u_r}{d}}{\frac{u_r}{d}}\frac{Q_r^*}{Q_p^*} = \frac{1 - \frac{u_r}{d}}{\frac{u_r}{d}}\sqrt{\frac{K_r h_s}{K_p\left(h_s + \frac{d}{u_r}h_u\right)}} \quad (7.9)$$

and optimal total relevant costs

$$TC^* = \left(1 - \frac{u_r}{d}\right)\sqrt{2dK_p h_s} + \frac{u_r}{d}\sqrt{2dK_r\left(h_s + \frac{d}{u_r}h_u\right)}. \quad (7.10)$$

In this case, the production lot size remains unadjusted, whereas the recovery lot size takes into account the return rate and uses a weighted holding-cost rate. Again, the integrality of the optimal value of M is ignored and the same problems discussed earlier apply. A comparison of both cost functions reveals that the $1 : R$ model dominates the $M : 1$ model if

$$\sqrt{K_p\left(h_s + \frac{\frac{u_r}{d}}{1 - \frac{u_r}{d}}h_u\right)} - \sqrt{K_p h_s} \quad (7.11)$$

$$< \frac{\frac{u_r}{d}}{1 - \frac{u_r}{d}}\left(\sqrt{K_r\left(h_s + \frac{d}{u_r}h_u\right)} - \sqrt{K_r(h_s + h_u)}\right).$$

For $\frac{u_r}{d} \to 1$ this tends to the $1 : R$ model, whereas in the opposite case of $\frac{u_r}{d} \to 0$ the $M : 1$ model is preferable. Based on the assumption that R as well as M can be real-valued, Teunter (2001) proves that it is sufficient to analyze $1 : R$ and $M : 1$ lot size policies in order to determine the optimal solution, because one can show that other policies are dominated by one of the above two.

Mabini et al. (1992) have suggested extensions of the approach proposed by Schrady (1967), allowing for backorders, and are presenting the analysis of a multi-item case where only a limited repair capacity is available for the remanufacturing process. Richter and Dobos (1999) and Minner (2002a) analyze integer-valued setup numbers, an important requirement that is relaxed by almost all the EOQ extensions to lot sizing issues in product recovery. All approaches discussed up to now assume the property that each production batch or remanufacturing lot has the same size. However, this is only valid for consecutive manufacturing batches and not for consecutive remanufacturing batches, as we will show in the following.

7.2.2 A Policy with Non-identical Batch Sizes

All models presented in Section 7.2.1 assume that batch sizes are identical, a property that holds in the EOQ model. However, as we will show in the following, this property does not necessarily hold for the recovery model. We relax this assumption by assuming that an integer number of n_p consecutive manufacturing/procurement batches is followed by an integer number of n_r remanufacturing batches in an inventory cycle of arbitrary length T. Then, we determine the optimal timing of the batch releases for any given cycle length and any given numbers n_p and n_r, use this result to obtain the optimal cycle length for any given batch numbers, and finally obtain the optimal numbers n_p and n_r. However, even this extended approach will not necessarily guarantee an overall optimal solution since it might well happen that in subsequent cycles, different batch numbers and timings are optimal.

Besides the cost parameters introduced in 7.2.1, we introduce the following additional notation. A number of n_p production batches are released at times $\theta_0 < \theta_1 < ... < \theta_{n_p-1}$ with sizes $Q_i^p = (\theta_{i+1} - \theta_i)d$, $i = 0, ..., n_p - 1$. Thereafter, n_r remanufacturing batches follow at times $\theta_{n_p} < ... < \theta_{n_p+n_r-1}$ with sizes $Q_i^r = (\theta_{i+1} - \theta_i)d$, $i = n_p, ..., n_p + n_r - 1$. For simplicity of notation, let $n = n_p + n_r - 1$. At the beginning of each inventory cycle and with each placement of an order of any type, serviceables inventory is zero. Otherwise, further postponement of the next batch saves costs since $h_s \geq h_u$. Under stationary conditions, as a consequence of restricting the analysis to a stationary manufacturing-remanufacturing pattern, initial recoverables inventory is given by $I_u(0) = (T - \theta_n)u_r$, assuming that the final recovery batch depletes recoverables inventory to zero. These are the returns delivered during the consumption of items from the final remanufacturing batch.

The costs per inventory cycle consist of setup costs for production and remanufacturing, holding costs for serviceable items, and holding costs for recoverable items.

$$C(T, n_p, n_r, \theta_1, ..., \theta_n) = n_p K_p + n_r K_r + \frac{h_s}{2} d \left(\sum_{i=1}^{n} (\theta_i - \theta_{i-1})^2 + (T - \theta_n)^2 \right)$$

$$+ \frac{h_u}{2} u_r \left(\theta_{n_p}^2 + \sum_{i=n_p+1}^{n} (\theta_i - \theta_{i-1})^2 + (T - \theta_n)^2 \right)$$

$$+ h_u \left((T - \theta_n) u_r \theta_{n_p} + \sum_{i=n_p+1}^{n} (\theta_i - \theta_{i-1})(d(T - \theta_i) - u_r(\theta_n - \theta_{i-1})) \right) \quad (7.12)$$

The cost function is convex with respect to the timing of batches. For the optimal timing of production batches $i = 1, ..., n_p - 1$, we find the following first order conditions.

$$\frac{\partial C}{\partial \theta_i} = d h_s(-\theta_{i+1} + 2\theta_i - \theta_{i-1}) = 0 \Leftrightarrow \theta_i - \theta_{i-1} = \theta_{i+1} - \theta_i \quad (7.13)$$

This result implies that the time distance between consecutive production batches has equal length, therefore we find identical production batch sizes. For the first $n_r - 1$ recovery batches $i = n_p + 1, ..., n - 1$ we find

$$\frac{\partial C}{\partial \theta_i} = d(h_s - h_u)(-\theta_{i+1} + 2\theta_i - \theta_{i-1}) = 0$$

$$\Leftrightarrow \theta_i - \theta_{i-1} = \theta_{i+1} - \theta_i, \quad (7.14)$$

which again implies that identical batch sizes are desired (given that a sufficient amount of recoverables is available). For the final remanufacturing batch $i = n$, we find

$$\frac{\partial C}{\partial \theta_n} = d(h_s - h_u)(-\theta_{n-1} + 2\theta_n - T) - h_u u_r T = 0$$

$$\Leftrightarrow \theta_n - \theta_{n-1} = T - \theta_n + \frac{h_u u_r}{(h_s - h_u)d} T. \quad (7.15)$$

This implies that the final remanufacturing batch is smaller than the previous ones, which states the first important result: it is never optimal to have identical remanufacturing batch sizes. The reason for this result is the following. Items that are returned after the final remanufacturing batch processing have to be stored over the entire production interval until the next remanufacturing batch is started. Therefore, it becomes attractive to reduce this amount of returns by implementing a smaller final recovery batch.

The results so far are only valid for the unconstrained problem. Taking into account that the remanufacturing batch size is limited by the available number of recoverables, the following constraints have to be taken into account for each recovery batch.

$$Q_i^r \leq I_u(0) + \theta_i u_r - \sum_{j=1}^{i-1} Q_j^r$$

$$\Leftrightarrow \theta_{i+1} d \leq (\theta_i - \theta_n) u_r + T d \qquad i = n_p, ..., n-1 \qquad (7.16)$$

The available amount of recoverables is given by the amount $I_u(0)$ returned during the previous cycle's final batch consumption plus the amount returned since the release of the first manufacturing batch less the amount used for the previous recovery batches of the current cycle. For the final recovery batch, the condition that all returns of a cycle have to be remanufactured yields the following equality which also determines the total length (in terms of a fraction of the inventory cycle length) of the production and the remanufacturing time interval.

$$\theta_{n_p} = \left(1 - \frac{u_r}{d}\right) T \qquad (7.17)$$

Incorporating these linear constraints into the analysis by a Lagrange-multiplier approach results in three different cases. The first case is the unconstrained solution discussed above, where a number of $n_r - 1$ batches of identical size is followed by a single batch of smaller size. The second case is represented by the other extreme, all recovery batches are limited by the respective constraints, that is that each remanufacturing batch equals the number of available recoverable items. Therefore, the size ratio of all recovery batches and their immediate successors is d/u_r. The third case covers all intermediate cases. Note that once a remanufacturing batch depletes recoverables inventory to zero, the same holds for all following recovery batches in the same cycle (see Minner, 2001). Here, a number of identically sized recovery batches is followed by a number of geometrically decreasing batches.

Reinserting the optimal timing of production and remanufacturing batches into the objective function results in a cost expression with a setup cost and a holding cost term $H(n_p, n_r)$ where only the latter depends on the three cases (for details of the analysis, see Minner, 2002b). Using the case-specific definition of the time-weighted holding costs, the objective function becomes

$$C(T, n_p, n_r) = \frac{n_p K_p + n_r K_r}{T} + H(n_p, n_r) T. \qquad (7.18)$$

Taking the first derivative with respect to T provides the optimal length of an inventory cycle.

$$T^* = \sqrt{\frac{n_p K_p + n_r K_r}{H(n_p, n_r)}} \qquad (7.19)$$

Then, the cost function only depends on the batch numbers and is given by

$$C(n_p, n_r) = 2 \cdot \sqrt{(n_p K_p + n_r K_r) H(n_p, n_r)}. \qquad (7.20)$$

Finally, the optimal numbers of production and remanufacturing batches have to be determined by integer search. A further property similar to the results of the models reviewed in 7.2.1 can be used to simplify this, that is that the optimal combination is either to have a single manufacturing batch followed by several remanufacturing batches or to have several manufacturing batches followed by a single remanufacturing batch, of course, including the special case of a 1:1 policy (see Minner, 2002b).

7.2.3 Noninstantaneous Production and Remanufacturing Processes

The models we have reviewed so far reflect a planning situation which is characterized by the main feature that the remanufacturing process is carried out without being time consuming. Such an assumption is justified whenever the time needed to fabricate a remanufacturing lot is only a very small fraction of the time period during which demand can be satisfied from this lot. However, in several practical situations this condition definitely does not hold. As a consequence, application of the models presented in the previous sections in such environments would generate unnecessary costs since they do not properly emulate the behavior of system-wide inventories in the case of a finite remanufacturing rate. Due to this fact, we will subsequently devote our attention to approaches which allow for determining appropriate production (procurement) lot sizes as well as suitable remanufacturing quantities in the presence of a finite remanufacturing rate.

The model suggested by Nahmias and Rivera (1979) was the first one considering a finite remanufacturing rate r, measured in units per unit of time. Except for this difference, Nahmias and Rivera (1979) examine the same planning situation as well as the same operating policy as the Schrady approach (see Section 7.2.1 of this chapter). However, in contrast to Schrady (1967), who analyzes the case where the number of shipments between the remanufacturing area and the serviceables inventory is given by a so-called sequential movement of units, they study a parallel movement. Sequential movement means that the entire lot is conveyed in one go between two consecutive stages. When parallel movement is used, each item, or, to be exact, each infinitesimal unit of a lot, flows immediately after completion from the remanufacturing department to the serviceables inventory (see Szendrovits and Truscott, 1989, p. 334, for a detailed description of both transportation alternatives). The cost function which we have to minimize is given by

$$C(Q_p, Q_r) = \frac{K_p(d - u_r)}{Q_p} + \frac{K_r u_r}{Q_r} + \frac{h_u u_r}{2d} \left[Q_p + Q_r \left(1 - \frac{d}{r} \right) \right]$$
$$+ \frac{h_s u_r}{2d} \left[Q_p \left(\frac{d}{u_r} - 1 \right) + Q_r \left(1 - \frac{d}{r} \right) \right], \qquad (7.21)$$

which represents a straightforward extension of Equation (7.1). Since Equation (7.21) is convex for all positive values of its variables, we obtain the optimal procurement lot size Q_p^* and the optimal remanufacturing lot size Q_r^* by solving the extremal equations $\partial C(Q_p, Q_r)/\partial Q_p = 0$ and $\partial C(Q_p, Q_r)/\partial Q_r = 0$. The solution is:

$$Q_p^* = \sqrt{\frac{2(d - u_r)K_p}{h_s \left(1 - \frac{u_r}{d}\right) + h_u \frac{u_r}{d}}}, \tag{7.22}$$

$$Q_r^* = \sqrt{\frac{2dK_r}{(h_u + h_s)\left(1 - \frac{d}{r}\right)}}. \tag{7.23}$$

A comparison of Equation (7.22) with the expression for Q_p^* in the Schrady approach given by Equation (7.2) reveals the interesting result that the optimal procurement lot size is the same in both situations. Hence, this quantity remains unchanged for systems with an infinite as well as a finite remanufacturing rate. Of course, like in the traditional Economic Production Quantity (EPQ) approach (for example, see Silver et al., 1998 , pp. 170-171), the optimal remanufacturing lot size is larger than its counterpart in the case of an instantaneous remanufacturing process (see Equation (7.3)). Recently, Koh et al. (2002) generalized the model by studying not only 1:R lot size policies, as done by Nahmias and Rivera (1979), but also M:1 coordination rules (see Section 7.2.1 for a detailed explanation of these two control policies). They proposed a search algorithm in order to ascertain the optimal solution from both lot size policies considered.

Finally, returning to Nahmias and Rivera (1979), we compare the cost consequences when using a parallel movement of units instead of a sequential one. It can be shown that the first alternative never results in costs higher than that of the second variant (see Nahmias and Rivera, 1979, p. 219). Numerically the cost advantage is given by

$$(h_u + h_s)\frac{u_r}{2r}Q_r. \tag{7.24}$$

From the perspective of total costs, the outcome suggests that individual units should be conveyed between the remanufacturing area and the serviceables inventory. In practice, however, due to the presence of transportation costs, it is not economically suitable or even possible (e.g. caused by constraints on the transportation equipment availability) in every planning situation to transfer item by item from one stage to the next (see Szendrovits, 1976, p. 334). As a result, in order to make a flexible choice of shipment quantities possible, the alternatives of transporting those items incorporated in one lot should not be limited to either parallel or sequential movement of units. It seems to be more realistic to allow the conveyance of partial lots larger than one and smaller than the entire lot, as well. In the literature, this is known as combined movement of units (see Szendrovits and Truscott, 1989, pp. 335-337).

A model which integrates the determination of appropriate transportation quantities into the planning of production and remanufacturing lot sizes was presented by Lindner and Buscher (2002). Although the system they analyze contains the same elements as the models discussed in Section 7.2.1, their approach differs from them in many aspects. First of all, Lindner and Buscher assume a finite remanufacturing rate r as well as a finite manufacturing rate p. Since there is only one system available to perform the production and remanufacturing operations, it has to be ensured that this facility is used either for production activities *or* remanufacturing operations at one time. Thereby, all demand which occurs at a constant rate d must be met without deliberate shortages, since it is presumed that the relations $r > d$ and $p > d$ do always hold.

The production and remanufacturing activities are carried out at the facility in lots. Each changeover from production to rework and vice versa causes a setup cost (setup time) of K_r or K_p (t_r or t_p), respectively, as the machine must be prepared within a certain time span according to the requirements of the subsequent lot. Lindner and Buscher (2002) restrict their attention to a coordination policy where each production lot is followed by a remanufacturing lot and then a production lot is fabricated again. Therefore, one inventory cycle consists of one remanufacturing and one production run.

Used products arrive at the recoverables inventory with a constant and deterministic rate u_r. There they can be stored at a holding cost rate h_u per unit per unit of time. It is not necessary to wait until all items are collected which are needed for one remanufacturing lot. In order to allow an overlap between the collection process and the remanufacturing activity equally sized batches q_r can be shipped from the recoverables inventory to the manufacturing area at a fixed cost of $B_{r,1}$ per transport. Because of the close proximity of both locations, the corresponding transportation time can be neglected. The number of shipments per remanufacturing lot is given by m. Hence, each sublot has a size of $q_r = Q_r/m$. Moreover, we have to be careful in selecting an appropriate batch size, since each transport consumes a portion $e_{r,1}$ of the total handling capacity E_r available at the recoverables inventory per unit of time.

Once the production or remanufacturing process has started at the machine, it must be continued without interruption until all units of the actual lot are completed. However, before the whole production (remanufacturing) lot is finished, equally sized batches $q_p = Q_p/m$ (q_r) may be conveyed to the serviceables inventory at a fixed cost B_p ($B_{r,2}$) per shipment. Consequently, the shipment frequency m is assumed to be the same between any pair of consecutive stages within the considered system. Furthermore, for every item at the serviceables inventory, a holding cost rate h_s per unit per unit of time is charged. Moreover, each batch q_p (q_r) consumes a fraction e_p ($e_{r,2}$) of the total handling capacity E_s available at the serviceables inventory. Once again, the transportation time between the manufacturing area and the serviceables inventory is negligible. For determining the optimal lot and batch size policy with respect to the considered planning situation, the following constrained minimization problem has to be solved. It can be shown that Q_p can be expressed in terms of Q_r via the relation $Q_p = Q_r(\frac{d}{u_r} - 1)$.

$$\min\ C(Q_r, m)$$

$$= \left[\frac{Q_r h_u u_r}{mr} + \frac{Q_r}{2}\left(\frac{1}{u_r} - \frac{1}{r}\right) h_u u_r + \frac{Q_r h_s u_r}{mr} + \frac{Q_r}{2}\left(\frac{1}{d} - \frac{1}{r}\right) h_s u_r \right]$$

$$+ \left[\frac{Q_r h_s u_r}{mp}\left(\frac{d}{u_r} - 1\right)^2 + \frac{Q_r}{2}\left(\frac{1}{d} - \frac{1}{p}\right) h_s u_r \left(\frac{d}{u_r} - 1\right)^2 \right]$$

$$+ \left[K_r + K_p + (B_{r,1} + B_{r,2} + B_p)m\right] \frac{u_r}{Q_r} \tag{7.25}$$

subject to:

$$Q_r \left\{ \left(\frac{1}{d} - \frac{1}{r}\right) + \frac{1}{mr} - \frac{1}{mp}\left(\frac{d}{u_r} - 1\right) \right\} \geq t_p \tag{7.26}$$

$$Q_r \left\{ \left(\frac{1}{d} - \frac{1}{p}\right)\left(\frac{d}{u_r} - 1\right) - \frac{1}{mr} + \frac{1}{mp}\left(\frac{d}{u_r} - 1\right) \right\} \geq t_r \tag{7.27}$$

$$e_{r,1} m \frac{u_r}{Q_r} \leq E_r \tag{7.28}$$

$$e_{r,2} m \frac{u_r}{Q_r} + e_p m \frac{u_r}{Q_r} \leq E_s \tag{7.29}$$

$$Q_r \geq 0; m \text{ being a positive integer} \tag{7.30}$$

The objective function contains the recoverables and the serviceables inventory holding costs per unit of time related to the remanufacturing process (first term in square brackets), the serviceables inventory holding costs per unit of time regarding the production process (second term in square brackets), and the last term in square brackets represents the fixed cost (setup and transportation of m batches) per unit of time. Restrictions (7.26) and (7.27) guarantee that the time between finishing the manufacturing operation for the current remanufacturing (production) lot and exactly that moment when the corresponding serviceables inventory level hits zero is sufficiently large in order to prepare the facility for the production (remanufacturing) run and to fabricate the first batch of newly made (remanufactured) items. Constraints (7.28) and (7.29) ensure the adherence of the constraints on the handling capacity regarding the recoverables or serviceables inventory, respectively (a detailed derivation of (7.25)-(7.29) can be found in Lindner and Buscher, 2002).

If we relax Constraints (7.26)-(7.29), the optimal remanufacturing lot size for a given value of m can be obtained by solving the first-order-condition $\partial C(Q_r, m)/\partial Q_r = 0$ for Q_r, since it can be shown that (7.25) is a convex function for all positive values of its variables.

$$Q_r^*(m) = [K_r + K_p + (B_{r,1} + B_{r,2} + B_p)m]^{1/2}$$

$$\times \left[\frac{h_u}{mr} + \frac{1}{2}\left(\frac{1}{u_r} - \frac{1}{r}\right)h_u + \frac{h_s}{mr} + \frac{1}{2}\left(\frac{1}{d} - \frac{1}{r}\right)h_s \right.$$

$$\left. + \frac{h_s}{mp}\left(\frac{d}{u_r} - 1\right)^2 + \frac{1}{2}\left(\frac{1}{d} - \frac{1}{p}\right)h_s\left(\frac{d}{u_r} - 1\right)^2 \right]^{-1/2} \tag{7.31}$$

Now, we describe the logic of a relatively simple algorithm for determining the optimal solution (for a more sophisticated, faster procedure see Lindner and Buscher, 2002). First of all, starting with $m = 1$, we determine the smallest integer shipment frequency for which the terms in rambled brackets on the left sides of Restrictions (7.26) and (7.27) are, for the first time, larger than zero. This value, denoted by m^+, represents an initial solution for m. Then, for this transportation frequency, we calculate $Q_r^*(m^+)$ as well as the smallest lot size which fulfills (7.26)-(7.29). We select the larger one of the two previously mentioned Q_r values as the optimal lot size regarding m^+ and determine the corresponding total cost. Then we search for a better cost by generating the best feasible solution for $m^+ + 1, m^+ + 2, \ldots$ etc., and retain that solution which gives the lowest cost so far as our temporary optimal solution. Note, $C(Q_r^*(m), m)$ is a lower bound on the cost obtainable for a given m. Therefore, we compute $C(Q_r^*(m), m)$ for each considered m and end the search if this cost value is for the first time larger than the cost of our best feasible solution. Since the total cost function is convex, this procedure guarantees an optimal solution.

7.3 Dynamic Lot Sizing

In a dynamic environment, parameters, especially demand and returns, may vary over time, and the static models presented in Section 7.2 no longer apply. A straightforward extension would be to replace the constant demand and return rates by continuous functions and to model the problem as an optimal control problem with impulses (see Chapter 9). Many applications, especially MRP-type concepts, utilize a different approach where time is divided into a finite number of buckets (time periods). Then, dynamic lot sizing models for the simple product recovery model are extensions of the model by Wagner and Whitin (1958) for the single-stage, dynamic lot sizing model. We first present the model assumptions, a formulation as a mixed integer linear programming model, and discuss the properties of an optimal solution in general and then for several special cases. We conclude with a discussion of principles for the development of lot sizing heuristics for the recovery model.

7.3.1 Model Formulation

Assume a finite planning horizon of T discrete time periods $\tau = 1, 2, ..., T$. For each period τ, known (or, in a rolling horizon framework, forecasted) and time-varying customer demands d_τ have to be satisfied. Backordering is not permitted. Additionally, customers return u_τ used products in each period τ. There exist two types of inventories, serviceables and recoverables. Let I_τ^o denote the serviceables inventory at the end of period τ and I_τ^u the recoverables inventory at the end of period τ. Demands can only be satisfied from serviceable items which are replenished by manufacturing a batch of size Q_τ^p and/or remanufacturing a batch of size Q_τ^r (assuming that remanufactured units are as good as manufactured ones). Excessive returns can be disposed of, and the respective quantity is Q_τ^w. All processes are instantaneous, i.e. have zero lead times. There is no loss of generality by imposing this assumption because fixed time lags can easily be incorporated into the model formulation. We assume the following sequence of events within a period: 1) decisions are made, 2) items become available in serviceables inventory, 3) demands and returns take place, and 4) inventories are measured and holding costs incurred. Then, the inventory balance equation for serviceable items is given by

$$I_\tau^o = I_{\tau-1}^o - d_\tau + Q_\tau^p + Q_\tau^r \qquad \tau = 1, 2, ..., T.$$

Recoverables inventory changes due to returns u_τ, remanufacturing Q_τ^r, and disposal Q_τ^w. The respective balance equation is

$$I_\tau^u = I_{\tau-1}^u + u_\tau - Q_\tau^r - Q_\tau^w \qquad \tau = 1, 2, ..., T.$$

The initial inventory levels I_0^o and I_0^u for both types are given.

The decisions on manufacturing, remanufacturing, and disposal are associated with setup costs K_p for each manufacturing, K_r for each remanufacturing, and K_w for each disposal batch. The release of a batch is indicated by binary variables γ_τ^p, γ_τ^r, and γ_τ^w, which are equal to one if a batch of the respective type is released in period τ and zero otherwise. Further, there are the following variable costs: c_p manufacturing cost per unit, c_r remanufacturing cost per unit, and c_w disposal cost per unit. Inventories are subject to inventory holding costs h_s and h_u per unit and unit of time for serviceables and recoverables respectively.

This leads to the following mixed-integer, linear optimization problem that minimizes total relevant costs over the planning horizon.

$$\min C = \sum_{\tau=1}^{T} (K_p \gamma_\tau^p + K_r \gamma_\tau^r + K_w \gamma_\tau^w + c_p Q_\tau^p$$
$$+ c_r Q_\tau^r + c_w Q_\tau^w + h_s I_\tau^o + h_u I_\tau^u) \qquad (7.32)$$

$$\text{s.t. } I_\tau^o = I_{\tau-1}^o - d_\tau + Q_\tau^p + Q_\tau^r \qquad \tau = 1, 2, ..., T \quad (7.33)$$
$$I_\tau^u = I_{\tau-1}^u + u_\tau - Q_\tau^r - Q_\tau^w \qquad \tau = 1, 2, ..., T \quad (7.34)$$

$$Q_\tau^p \leq M\gamma_\tau^p, \ Q_\tau^r \leq M\gamma_\tau^r, \ Q_\tau^w \leq M\gamma_\tau^w \qquad \tau = 1, 2, ..., T \ (7.35)$$
$$Q_\tau^p, Q_\tau^r, Q_\tau^w, I_\tau^o, I_\tau^u \geq 0. \ \gamma_\tau^p, \gamma_\tau^r, \gamma_\tau^w \in \{0, 1\} \quad \tau = 1, 2, ..., T \ (7.36)$$

where M is a number large enough to allow all relevant lot sizes (for example, the sum of all demands and returns over the planning horizon). The constraints represent the inventory balance equations and logic constraints that ensure that a manufacturing, remanufacturing, or disposal process can only take place if the respective setup is present. The difference to the Wagner-Whitin model is the presence of two supply modes and a disposal option with respective setup and variable costs, and two inventory states with respective balance equations.

7.3.2 Solution Properties and Algorithms

Golany et al. (2001) consider a generalized version of the above model with concave instead of linear cost functions. They give an interpretation of the problem as a network flow problem and show that its computational complexity is NP-complete. For the special case that all setup costs are zero and the remaining cost function is linear, the problem reduces to the well-known transportation problem. In Yang et al. (2001), this research is extended by stating conditions under which the problem can be solved in polynomial time.

In general, the above linear problem can either be solved by commercial packages for mixed integer linear programming or, in light of the Wagner-Whitin tradition, by dynamic programming. However, the computational complexity cannot be reduced in the same way known from the Wagner-Whitin model since the zero inventory property, that is a lot is only released when the serviceables inventory is zero and it contains a consecutive number of period demands, no longer holds. To see this, let us consider the following problem with three time periods. The demands are $d_\tau = 10$ for all periods, and the returns are $u_1 = u_2 = 3$ and $u_3 = 4$. There are zero initial serviceables and 11 recoverable units. Inventory holding costs are $h_s = 1.5$ and $h_u = 1.0$ per unit and unit of time and the setup costs are $K_p = 100$ and $K_r = 2$. There exists no disposal option. The optimal policy is to remanufacture 11 units in the first period and 10 units in the third period and to produce 9 units in the second period. Therefore, serviceables inventory is positive at the beginning of the second period when a production batch is released. The reason is that it is cheaper to remanufacture the single excessive unit in advance instead of keeping it in recoverables inventory until period three. A reason for having positive serviceables inventory when a remanufacturing batch is released is that the available returns for satisfying the period's demand may only be close to be sufficient. In order to permit a remanufacturing batch without having backorders, the missing items have to be manufactured in advance. To see this, let us consider the above example and change the return in period three from 4 to 2 units. Then, the optimal policy becomes to remanufacture

10 and 9 units in the first and third period and to manufacture 11 units in the second period in order to compensate for the unit missing in period three.

Richter and Sombrutzki (2000) investigate special cases where the zero inventory property of the Wagner-Whitin model is retained for serviceables inventory (under the assumption that holding recoverables inventory is cheaper than serviceables). This is the case under the following restrictive conditions: either 1) no consideration of demands and serviceables inventory, where only the trade-off between setup costs for remanufacturing and holding costs for recoverables is included, or 2) two inventories for serviceables and recoverables and positive demands, but only the remanufacturing option exists, or 3) there exists a large quantity of initial (and no intermediate) returns and the manufacturing setup cost is smaller than the remanufacturing setup cost, where the result is an alternate manufacturing and remanufacturing policy. The analysis of these special cases is extended in Richter and Weber (2001) to include variable costs for manufacturing and remanufacturing.

Kelle and Silver (1989a) analyze a stochastic Wagner-Whitin type of model where containers are returned into serviceables inventory. Since net demands for serviceables may become negative, they prescribe a chance constraint such that the inventory level at the end of each period is positive with some predetermined probability. Beltran and Krass (2002) analyze the somewhat different special case of merchandise returns. Returned products have the same quality as new products and therefore no repair/remanufacturing is required. The returns either immediately enter serviceables inventory and can be used to satisfy customer demands or are disposed of. The authors derive properties of an optimal solution, i.e. a generalized version of the zero inventory property where there is at least a period with zero initial inventory between two actions that allows for a reduction of the state space and an efficient dynamic programming algorithm.

7.3.3 Heuristics

In order to provide decision support, heuristics in analogy to the single-stage, single product, dynamic lot sizing problem (for example, Silver-Meal, Part Period Balancing, Least Unit Cost, Groff, etc.) can be developed (see Zoller and Robrade, 1988a, 1988b, for an overview). In general, an heuristic has to resolve the main trade-offs of the product recovery problem discussed earlier in this chapter. Like with other heuristic methods, we can distinguish between constructive and improvement techniques. The functionality of a solution construction heuristic involves the following choices (see Wojanowski, 1999).

(1) Select a replenishment mode (manufacturing, remanufacturing, or both modes simultaneously) in a period where initial serviceables inventory does not cover the demand.
(2) Provide increments for lot extension:
  a) manufacturing batch: next demand, assuming that the properties of the Wagner-Whitin model would still hold, and

b) remanufacturing batch: next demand or clear recoverables inventory.
(3) Determine a stopping criterion for lot-extensions, for example, if the cost per period increases, as used in the Silver-Meal-heuristic.

However, since these choices do not explicitly address some of the differences of the recovery model compared to the Wagner-Whitin model, the error of such methods is larger compared to similar heuristics applied to the simple, single-stage, dynamic lot sizing problem (Wojanowski, 1999). Therefore, solution improvement procedures that combine or re-group batches may be desirable. For the improvement of a given solution, problem-specific (for example, a method based on savings from combining consecutive batches) and general purpose local search methods like steepest descent, simulated annealing, tabu search, or genetic algorithms (see Michalewicz and Fogel, 2000) can be used.

7.4 Lot Sizing Models with Imperfect Yield and Rework

Nowadays companies face an ever-increasing pressure to fulfill customer needs regarding high quality products at low costs. For manufacturing purposes, most often quite sophisticated production equipment is used. Nevertheless, some defective units are fabricated even by production systems comprised of high quality facilities. These defective items can either be reworked, scrapped, sold at reduced prices, or be subject to other corrective actions. In the overwhelming majority of cases, situations with rework also represent situations which are characterized by variations with respect to the output (yield) of the considered manufacturing system. However, although the last mentioned topic has received a lot of attention in the literature, nearly all approaches dealing with this problem assume (implicitly) that defective items have to be disposed of (see Yano and Lee, 1995, for an excellent overview on this area of investigation). Subsequently, we describe the research on lot sizing in the presence of imperfect yield and rework. Following Yano and Lee (1995), we base our review on the distinction between single-stage and multi-stage approaches. Regarding the manifold planning problems which could arise in situations with rework and imperfect yield, the interested reader is referred to Flapper and Jensen (2003).

7.4.1 Single-stage Models

Porteus (1986) considers a manufacturing system producing a single item. Although the system operates in a nearly stable environment, the process used for fabricating units can go 'out of control' with a given probability each time it produces another item belonging to the actual lot, and starts to make defective parts. Once the process is out of control, it remains in this state until the remainder of the lot is manufactured. Due to the inspection policy the firm uses, the manufacturing facility is in an 'in control' state when a new production run is started. All defective items of a particular lot have to be reworked at a cost. The production as well as remanufacturing activities are carried out instantaneously. Since demand is constant and total cost per unit of time consists of setup and holding costs, as well as rework costs, this approach represents a modification of the classical EOQ model. In an extension of this basic model, Porteus (1986) allows the investment of capital in improving the process quality and in reducing setup costs. Here the task is to simultaneously determine the optimal investment level and lot size.

Regarding the inspection policy followed in Porteus (1986), monitoring the state of the manufacturing equipment is restricted to the end of a production cycle and, if necessary, the facility can be put back into control. Relaxing this constraint, Lee and Rosenblatt (1987) and Porteus (1990) allow multiple inspections during the manufacturing cycle time of one lot. Once the production process is found to be out of control via these inspections, it will be restored to the 'in control' state at a fixed cost. Lee and Rosenblatt (1987) consider the case of a finite production rate and a parallel movement of units, and assume that inspection and restoration times are negligible. In contrast to this approach, Porteus (1990) presumes that there is a delay in obtaining the results of a certain inspection. Thus, if the process is out of control at the time of an inspection, then some more defective units will be produced before the status of the process is known. As in Porteus (1986), one feature of Lee and Rosenblatt (1987) and Porteus (1990) is that all defective items can be remanufactured in negligible time.

Within a setting which incorporates more stochastic elements, a comparable planning problem was studied by Lee (1992). He assumes that production and remanufacturing activities are carried out on the same machine with different expected manufacturing times per unit. After finishing the production of a lot, a time-consuming setup is required in order to be able to rework all defective items. Afterwards, again within a particular time span, the equipment is prepared for the subsequent lot. The entire lot leaves the system only when all incorporated items have finished production and reprocessing. With respect to the inspection policy, Lee (1992) distinguishes two different situations. In the first case, there is an imperfect monitoring during processing, which means that there is only a constant probability that the 'out of control' state is detected and adjustments are made within a so-called correction time prior to the manufacturing of the next unit. A final inspection is used in order to determine whether items have to be reworked or not. In the second case, each item of a lot is inspected perfectly immediately after its completion, and hence the 'out of control' state can be corrected without delay. Based on the objective of minimizing the average processing time per unit, the aim is to find the optimal lot size regarding both planning situations.

The models discussed so far share the common feature that all units produced prior to the point where the process goes 'out of control' are presumed to be of acceptable quality and all subsequent items are defective. Another interesting planning problem arises when there is a priori knowledge about the probability that an item may be defective. Such an assumption is justified when the quality characteristics of the process output exhibit stable expected values representing the environmental manufacturing conditions. Modeling yield uncertainty in this way is appropriate for systems that, in terms of statistical process control, are in control for a long period of time (see Yano and Lee, 1995, p. 313).

In order to control single product manufacturing systems of this type, So and Tang (1995a, 1995b), Liu and Yang (1996), and Chern and Yang (1999) suggest the use of a so-called threshold policy. Without going into detail, a threshold policy works as follows. The manufacturing cycle starts with processing a lot consisting of new units. Each completed item or lot is then fed into a perfectly reliable inspection process. Here, an incoming job is classified as good, reworkable, or non-reworkable with known probabilities. Whereas good items or non-reworkable units leave the system immediately, a reworkable product or lot is routed back to a rework buffer to wait for remanufacturing. After finishing the actual lot of new items, the manufacturing system is prepared within a certain time span for processing the accumulated reworkable items if their amount has reached or exceeded some prespecified threshold. Subsequently, the manufacturing cycle repeats with setting up the production equipment. If the available inventory level of reworkable jobs is below the threshold, then the manufacturing cycle continues with producing equally sized lots until the threshold condition is satisfied. Obviously, a threshold policy can be applied to a manufacturing system if one has specified the lot size for new units as well as the threshold value itself. Regarding this optimization problem, all four models make use of a profit-maximization approach. Based on revenues via sales, fixed costs related to switching between regular production and rework and vice versa, and inventory holding costs related to the inventory of new items until they enter processing as well as holding costs related to the inventory of reworkable defectives, the optimal solution is determined without considering constraints on the quantity which can be sold.

Although the last mentioned assumption is a reasonable approximation for certain classes of real systems, most often a manufacturer faces market-driven restrictions on sales quantities. In an approach suggested by Lindner (2002), the case of a single item facing a constant and continuous demand d which has to be satisfied without deliberate shortages is analyzed. Production as well as rework takes place on the same machine with finite rates p (r). Due to an imperfect production process, each lot of new items contains only a constant portion of α good items. After finishing the actual manufacturing run, these units are shipped using sequential movement to the serviceables inventory. During processing, all reworkable defectives are transported continuously to a so-called recoverables inventory and wait there until the perfectly reliable remanufacturing process is started. Since recoverables inventory holding costs h_u are assumed to be smaller than serviceables inventory holding costs h_s, it is on principle economically desirable to delay the reprocessing activity in such a way that all reworked items arrive at the serviceables inventory exactly at that moment when the previously delivered good items are used up. Furthermore, if all items of the actual lot are consumed by the demand process, the entire amount of good units out of the next production lot must enter the serviceables inventory. Regarding each switchover from production to rework and vice versa, a setup cost (setup time) K_r or K_p (t_r or t_p) has to be taken into account. For ascertaining the optimal production (rework) lot size Q_p^* [$Q_r^* = (1 - \alpha)Q_p^*$], we have to solve the constrained minimization problem as stated below. Note, we presume that the relations $\alpha(r + d) > d$ and $(1 - \alpha)p > d$ always hold.

$$\min \ C(Q_p) = \left[\frac{\alpha^2 Q_p}{2} \left(\frac{1}{\alpha p} + \frac{1}{d} \right) h_s d \right] + \left[\frac{(1 - \alpha)^2 Q_p}{2} \left(\frac{1}{r} + \frac{1}{d} \right) h_s d \right]$$
$$+ \left\{ Q_p h_u d \left[\frac{1 - \alpha}{2p} + (1 - \alpha)\alpha \left(\frac{1}{d} - \frac{1 - \alpha}{\alpha r} \right) + \frac{(1 - \alpha)^2}{2r} \right] \right\}$$
$$+ (K_r + K_p) \frac{d}{Q_p} \tag{7.37}$$

subject to:

$$Q_p \left[\frac{\alpha}{d} - \frac{1 - \alpha}{r} \right] \geq t_r \tag{7.38}$$

$$Q_p \left[\frac{1 - \alpha}{d} - \frac{1}{p} \right] \geq t_p \tag{7.39}$$

$$Q_p \geq 0, Q_r \geq 0 \tag{7.40}$$

The objective function is comprised of serviceables inventory holding costs related to good as well as reworked items (first and second terms, respectively, in square brackets), recoverables inventory holding costs (third term in rambled brackets), as well as fixed setup costs. Moreover, constraints (7.38) and (7.39) ensure that the time period during which demand can be satisfied via all good (reworked) items incurred in the actual production (rework) lot is sufficiently large in order to prepare the facility for reprocessing (production) and to rework (manufacture) the entire amount of defectives (a production lot).

If we first ignore Constraints (7.38) and (7.39), the optimal production lot size can be obtained by solving the extremal equation $dC(Q_p)/dQ_p = 0$ for Q_p.

$$\hat{Q}_p = (K_r + K_p)^{1/2} \left\{ \frac{\alpha^2}{2} \left(\frac{1}{\alpha p} + \frac{1}{d} \right) h_s + \frac{(1 - \alpha)^2}{2} \left(\frac{1}{r} + \frac{1}{d} \right) h_s \right.$$
$$\left. + h_u \left[\frac{(1 - \alpha)}{2} \frac{1}{p} + (1 - \alpha)\alpha \left(\frac{1}{d} - \frac{1 - \alpha}{\alpha r} \right) + \frac{(1 - \alpha)^2}{2r} \right] \right\}^{-1/2} \tag{7.41}$$

If the production quantity determined using (7.41) is not feasible, the optimal solution is represented by the smallest lot size which fulfills (7.38) as well as (7.39).

The models reviewed so far in this section presume (implicitly) that the state of defective items does not change while they wait to be reworked. However, in several practical situations (for example in the metal processing industry; see Flapper et al., 2002), the state of non-conforming units worsens (deteriorates) over the course of time. As a consequence, the cost and time of rework operations necessary to bring defective products back into an acceptable state will increase (in general) the longer these items have to wait for being reprocessed. Thereby, if a significant rate of deterioration exists, its impact on ascertaining the optimal lot size policy must not be neglected. This topic was recently studied by Teunter and Flapper (2003) as well as Inderfurth, Lindner, and Rahaniotis (2003).

7.4.2 Multi-stage Models

Relatively little research has been done on multi-stage approaches dealing with the topic of imperfect yield and rework. A typical situation with respect to a make-to-order environment where production is in small quantities and items are custom made was considered by Spence Wein (1992), Barad and Braha (1996), and Grosfeld-Nir and Gerchak (2002). Once again, without describing differences in details, these three models analyze a planning problem where a single known order needs to be satisfied in its entirety. However, since the yield at each stage of a serial manufacturing process is uncertain, several manufacturing runs may be required until the quantity of items ready for sale is sufficient. The trade-off at a particular stage is (in general) between using small lots, possibly resulting in multiple production (rework) setups, and large lots which may correspond to costly over-production. Therefore, the decision problem is to determine the production and rework lot sizes which should be used at the different stages during the manufacturing cycle time of a certain order. Whereas Spence Wein (1992) allows only the processing of one manufacturing and recovery lot at each stage, in Grosfeld-Nir and Gerchak (2002), multiple runs of both types could be initiated if necessary. Moreover, all processing and reprocessing activities are carried out on the same equipment. The last mentioned aspect is known in the literature as in-line rework (see Flapper and Jensen, 2003). The situation of off-line rework was modeled by Barad and Braha (1996). In their model, the manufacturing line is dedicated to the processing of production lots. At each inventory point between two subsequent stages, remanufactured items may enter the system fabricated by using an outside rework shop. The optimization problem involved in specifying stage- and run-specific lot sizes is represented by minimizing the expected total costs associated with an order. The cost elements influencing this set of decisions are variable raw material and manufacturing and rework costs, as well as fixed setup costs per stage.

7.5 Conclusions and Recommendation for Further Research

In this chapter, we reviewed the problem of lot sizing in reverse logistics systems. The main conclusion is that the presence of two supply modes, with one being capacitated and under limited control by the independent customer product returns, and the other being two-location storage (serviceables or recoverables inventory), heavily complicates planning and control of batching activities. The same holds for dependent returns from imperfect production. Well-known properties of simple inventory systems without return flows, the property of identical batch sizes in the EOQ model, and the zero inventory property when an order is placed in the dynamic lot sizing model, are no longer valid and can therefore not be applied to construct simple solution methods. Therefore, there is a need for well-performing heuristic methods and principles that overcome the complexity of the problem but nevertheless address the main trade-offs of the underlying problem.

The most simple models are already getting very complex and practical problems with more complicated lot sizing problems and interactions with other areas like production scheduling are far from being adequately covered by the existing research. Though there are a few extensions of traditional lot sizing models available to address these challenges, there remain many open research issues to be tackled in the future. In light of the severe complexity, one question addresses the robustness and sensitivity of the performance of simple models and approximate solutions with respect to the real recovery problem. For the NP-complete problem of dynamic lot sizing, a proper development and testing of simple lot sizing techniques is a prerequisite of the successful implementation of lot sizing methods in MRP-type systems with discrete planning time buckets. Additionally, more research directed towards a simultaneous analysis of lot sizing and scheduling problems, as pointed out for the pharmaceutical case and as started in Lindner and Buscher (2002), multi-level lot sizing with a proper incorporation of the effects of lot sizes on upstream and downstream planning problems is needed. Regarding the problems to be discussed in the following chapters of this book, lot sizing has a strong interaction with other areas: in inventory management, where a combined optimization of cycle and safety inventories offers further potential for cost improvement, lot sizes may vary over time as a result of changes in cost parameters and as a result of learning; and in strategic operations management, where the impact of batching can be reduced by implementing other technologies with less setup requirements (for example, flexible manufacturing and remanufacturing technologies).

Stochastic Inventory Control for Product Recovery Management

Erwin A. van der Laan[1], Gudrun Kiesmüller[2], Roelof Kuik[1],
Dimitrios Vlachos[3], and Rommert Dekker[4]

[1] Rotterdam School of Management / Faculteit Bedrijfskunde,
   Erasmus University Rotterdam, P.O. Box 1738, 3000 DR Rotterdam,
   The Netherlands, elaan@fbk.eur.nl, rkuik@fbk.eur.nl
[2] Faculty of Technology Management, Eindhoven University of Technology,
   P.O. Box 513, 5600 MB Eindhoven, The Netherlands
   G.P.Kiesmueller@tm.tue.nl
[3] Department of Mechanical Engineering, Aristotle University of Thessaloniki,
   P.O. Box 461, 54124 Thessaloniki, Greece, vlachos1@auth.gr
[4] Rotterdam School of Economics, Erasmus University Rotterdam, P.O. Box 1738,
   3000 DR Rotterdam, The Netherlands, rdekker@few.eur.nl

8.1 Introduction

Essentially, inventory management concerns the process of deciding on 1) how often to review stocks, 2) when to replenish stocks, and 3) how much to replenish. This basic focus of inventory management persists in the presence of item returns that can be recovered and then used for servicing demand. However, the details and complexities with which the three basic decisions manifest themselves can, and usually do, differ greatly due to the presence of recoverable-item flows. This, and the practical relevance of inventory management with recoverables, warrants the development of inventory theory that explicitly includes flows of recoverable items.

A first categorization lists the foundations of the differences between recoverable-item inventory management (RIIM) and traditional inventory management under three headings.

- *Multiple sourcing.* When items can be recovered, requirements of items can be met from multiple sources: the source(s) with newly manufactured or ordered items and the source(s) with recovered items. This extends the three basic decisions with a fourth one: 4) where to replenish from. Added complexities arise since the source of recoverables is capacitated: at any point in time the number of items available from recovery is limited. The sources may not only differ in item-availability, but also in per unit price and supply reliability.

- *Absence of monotonicity.* Return streams are often difficult to control. This means that between regular reviews and replenishments, inventories can go up because of product returns even when inventory levels are sufficiently high to maintain a targeted service level. Therefore (heuristic) analysis based on observation of the inventory just prior to the epochs that a replenishment is due may be highly inaccurate. A further consequence of the loss of monotonicity is that the total inventory in the system is not bounded from above, unless an appropriate disposal policy is implemented.
- *Unreliable sources.* The added complexity of multiple sourcing is frequently compounded by the circumstance that sources of recovered items are unreliable concerning availability, as the item returns may be uncertain both in timing and quality, and thus in suitability for recovery.

This chapter gives an overview of the different approaches that have been presented in the scientific literature to account for the above peculiarities of stochastic inventory control for product recovery. Before discussing the literature in more detail, we first present two examples from practice, each representing a different planning problem.

Case A: Commercial Returns at a Mail-order Company

A large mail-order company in Western Europe (see Mostard and Teunter, 2002) faces a difficult inventory problem for their fashion products. Lead times are long, so well before the selling season starts a replenishment order is placed to accommodate the demand for the whole season. Return rates are usually around 40%, but can be as high as 75%. Therefore, returns really need to be taken into account when determining the order size. To obtain more accurate demand forecasts, preview catalogs are sent out to a selected group of customers. On the basis of customer orders placed in response to the preview mailing, another replenishment order is placed just before the start of the selling season. Additionally, an (expensive) emergency order can be placed some three weeks into the season that makes use of the (limited) information regarding sales and returns during this beginning of the season. At the end of the season, shortages result in lost sales and overages in obsolete products that have to be disposed of. Since fashion products are highly seasonal, sales and return volume forecasts are very crude. Since the last opportunity to order is just three weeks after the season's start, there is little opportunity to adapt to realized demand.

Case B: Product Remanufacturing at Volkswagen

Car parts that have failed during operation on the road, varying from injection pumps to complete engines, are collected by Volkswagen via the car dealers (see Van der Laan, 1997). These parts are subsequently remanufactured by a third party and eventually resold as spare parts for approximately half of the

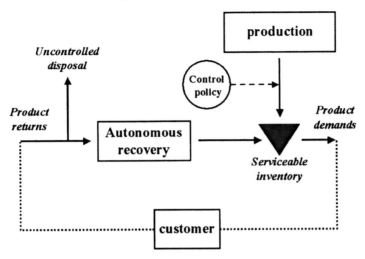

Fig. 8.1. Autonomous recovery

price of newly manufactured parts. Due to normal fluctuations in both demand and supply, and due to a lack of control of the return stream, there may be periods in which the number of collected car parts does not match the number of demands. Then, in times of supply shortage, new parts are ordered from the Volkswagen factories, or, in times of supply overage, recoverable products are disposed of. Lead times for production and remanufacturing may differ. As is the case here, it occurs often that it is this interaction between the supply of new and remanufactured products that makes inventory control more difficult than traditional, single-source inventory control.

Case A admits only a limited number of replenishment decisions, whereas Case B asks for recurrent replenishment decisions. Moreover, in the mail-order company case, returns can be recovered with little management intervention. In the Volkswagen case, the company can pursue an active recovery strategy involving the batching and timing of remanufacturing. This distinction, between cases in which returned items can be made serviceable with little control and cases where management pursues an active recovery policy also seems reflected in the status of the theory of models for these cases. The first type of cases, to which we will refer as the cases of *autonomous recovery*, seems to be more amenable to analysis than the second type of cases, to which we will refer as the cases of *managed recovery*. Note that the cases of autonomous recovery do away with the need for answering the fourth basic question: where to replenish from?

So, the autonomous recovery cases are often those situations in which recoverable products only need to undergo minor operations, such as cleaning and repackaging. These operations are relatively cheap so that it is not really

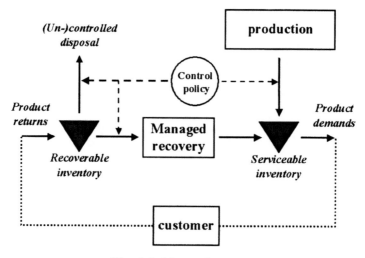

Fig. 8.2. Managed recovery

necessary to delay these operations until actually needed: upon return, items are immediately prepared for reuse and there is no further decision-making involved, hence the provision 'autonomous'. For example, commercial returns are often in very good condition and many of them can be reused directly without much recovery.

As Case B shows, the remanufacture of end-of-life returns calls for extensive recovery operations. The costs involved with these value-adding operations may be such that it is more economical to stock the returns until they are really needed. Thus, cash outflows are postponed and there is less risk of loss of investment due to obsolescence. In this case, decision-making on recovery operations occurs on a continuous basis, hence the provision 'managed'.

Since in the managed recovery cases one recovers items when deemed necessary, these cases are also referred to as *pull strategies* for inventory. In this vein, the autonomous cases are referred to as *push strategies*. Sometimes recovery management is extended with disposal management. An appropriate disposal policy attempts to bound the total inventory in the system. In practice, this is primarily relevant when the return rate is close to or in excess of the demand rate.

Models for inventory management can be further categorized on several dimensions. Below, we will discuss several of these dimensions and discuss the relationship to the dichotomy 'autonomous-managed'.

8.1.1 Problem and Model Dimensions

Nodes and Flows

Inventory systems often consist of multiple stocking points or facilities connected by flows. The structure of the open network consisting of facilities (nodes) and flows (arcs) to a large extent determines the difficulty of the analysis. Divergent (or distribution) systems, are more intricate than serial or convergent (assembly) systems. In divergent systems, the added complication stems from the occurrence of an allocation decision in case of scarcity of items. (This is reminiscent of the 'where' question above.) In the case of managed recovery, we automatically find ourselves in a situation with multiple stocking points: there is (at least) a stocking point for the recoverable items and a stocking point for the serviceable items. Adding to this situation economies of scale in the form of set-up costs for both recovery and new production brings the analysis into a domain equally complex as that of multi-level inventory management with fixed costs. As, in general, no rigorous results on the optimality of policies in the latter situation are known, it is no surprise that the same holds for the case of managed recovery. Even if fixed recovery costs are absent, the situation already seems to be so complex that one needs to resort to heuristics to find inventory strategies (see Section 8.3).

The case of autonomous recovery seems to be best understood. This is especially true in the case of an inventory system consisting of just one inventory facility from which demand is served (see Section 8.2).

The Model of Time

In the literature, there is a clear distinction between a discrete modelling approach and a continuous modelling approach of time. However, the choice for the time model seems mostly to rest on pragmatic grounds. Use of a continuous time model sometimes presents advantages in computational issues. For example, the continuous time approach seems to be more flexible with respect to cost structure and lead time assumptions. So when it comes to numerical analysis, as will be the case for managed returns, continuous time models are more suitable. Section 8.3.2 reports on a general framework for carrying out numerical analysis in a continuous-time setting.

Costs

As far as costs modelling is concerned, the literature on inventory theory with recoverables recognizes the same costs structures as traditional inventory theory. So, besides linear inventory related costs and variable costs of operation, some models consider fixed production and/or recovery costs and some do not. The discussion on how to set the holding cost parameters in recovery models is less straightforward than in traditional single source models, but this discussion is left to Chapter 11.

Decisions

Also in the domain of decision modelling, the inventory theory with recoverables follows largely traditional inventory theory. So most models incorporate the opportunity to release (re-)manufacturing or replenishment orders. There is one additional decision, though, that is sometimes considered in inventory theory with recoverables (and not in traditional inventory theory). This additional decision is the disposal decision.

The distinction between autonomous and managed recovery (or push and pull strategy) seems to be the most fundamental modelling choice in inventory theory including recoverable returns. To honor this, the following two sections of this chapter follow this dichotomy. Note that the distinction is fundamentally based on a choice related to the way models deal with the issue of multiple sourcing.

8.2 Autonomous Recovery

In the case of commercial returns, often only minor actions like inspection or cleaning suffice to render the returned products suitable for direct reuse. This section deals with the situation that no separate stocking facilities of returned products are necessary. If more complex processing is needed and/or the decision maker has the option to delay these activities, then we are in the case of managed recovery. That is the theme of Section 8.3.

The discussion will be restricted to the situation that the inventory system consists of only one stock facility serving end demand. The evaluation of base stock policies (in the absence of fixed costs) for autonomous product returns in a series or assembly system can be found in the working papers DeCroix and Zipkin (2002a) and DeCroix and Zipkin (2002b), the discussion of which is outside the scope of this chapter.

8.2.1 Naive Netting

The naive way of dealing with product returns is the so-called 'netting on averages'. In this approach, one does not take the returns process into account explicitly. Instead, one assumes that returns are deducted from future demand thus cancelling out part of that demand. The remainder of the *expected* demand is treated with traditional methods for single source inventory control. In principle, this method works and yields fair results in case of small return rates (see Van der Laan et al., 1996a)) or high correlation between demands and returns. However, for high return rates and low correlation between demand and returns, the result can be way off the correct one. The reason for this is that netting on averages blatantly ignores the additional variability that is introduced by the return process. The method correctly reduces the

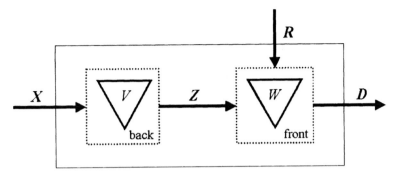

Fig. 8.3. Split inventory model

average demand with the expected returns, but deals incorrectly with the
variability of the processes.

8.2.2 Sophisticated Netting: Split Inventory

Naive netting reduces the problem under returns to the problem without
returns at the cost of inaccuracy. Perhaps surprisingly, we can do much better
by using a netting of the processes and yet cast the problem in one with
only non-negative demand. For ease of exposition let us assume that time is
discrete. Demand in period τ is written as $D(\tau)$ and returns in period τ as
$R(\tau)$. Let $X(\tau)$ be the supply of products from the production facility. Thus,
the inventory balance equation for the facility's inventory at the end of period
τ, $I(\tau)$ reads

$$I(\tau) = I(\tau - 1) + X(\tau) - (D(\tau) - R(\tau))$$

We now split the inventory $I(\tau)$ into two components, $V(\tau)$ and $W(\tau)$, the
sum of which will give back the inventory, through the recursion

$$\begin{cases} V(\tau) = V(\tau - 1) + X(\tau) - Z(\tau) \\ W(\tau) = W(\tau - 1) + Z(\tau) - (D(\tau) - R(\tau)) \end{cases} \tag{8.1}$$

where

$$Z(\tau) \equiv \max\{0, D(\tau) - R(\tau) - W(\tau - 1)\} \tag{8.2}$$

and the initial condition $(V_0, W_0) = (I_0, 0)$. The meaning of these recursive
equations is best illustrated through Figure 8.3.

The figure shows that the inventory system is divided into two subsystems:
a front-end system and a back-end system. It illustrates that (8.1) serves as
the balance equation for the front-end system with inventory level W and the
balance equation for the back-end system with inventory level V. Expression

(8.2) shows that $Z(\tau)$ is determined such that if the joint effect of demands $(D(\tau))$ and returns $(R(\tau))$ would make the inventory in the front-end system, W. negative, then $Z(\tau)$ balances that inventory such that it becomes precisely zero. There is no restriction on the back-end system, so that its inventory V may become negative. The process Z will be referred to as the censored demand, the demand 'cleaned up' for returns.

Three observations now demonstrate the value of the inventory splitting.

1. The front-end system's behavior is independent of the replenishment policy, that is, the inventory pattern of the front-end system is independent of the external supply X. The recursion $W(\tau) = W(\tau-1) + Z(\tau) - (D(\tau) - R(\tau)) = \max\{0, W(\tau-1) - (D(\tau) - R(\tau))\}$ shows that its dynamics are in fact a random walk on a discrete half-line with random step size $R - D$.
2. The flow Z is independent of production input X. The recursion $Z(\tau) = \max\{0, D(\tau) - R(\tau) - W(\tau-1)\}$ tells us that Z measures the virtual undershoot when the random walk hits the end point of the half-line at 0. In short, the front-end system's behavior, inclusive of its input Z, can be described independently of the replenishment flow X, that is, it can be described and analyzed independently of the replenishment policy. The only system affected by the inventory policy is the back-end system.
3. The back-end system's behavior is that of a traditional inventory system: it faces non-negative demand (represented by Z) and has the opportunity to replenish through the flow X.

Before turning to applications of the inventory splitting framework we make two remarks.

Remark 1. Not only are the behaviors of the processes interesting but so are the costs involved. Suppose the inventory related costs take the form $G(I)$. How do such costs translate into the model with split inventory? In particular, how can we translate these costs into costs for the back-end system, especially the inventory V, which still needs to be optimized? In the case in which the processes are stationary the answer is simple. Introduce the cost function H as

$$H(v) = \mathrm{E}_{(W|v)}(G(v + W)),$$

where $\mathrm{E}_{(W|v)}$ is the expectation with respect to $(W|v)$, that is, with respect to the equilibrium of the process $\boldsymbol{W}$ conditioned on $V = v$.

Remark 2. Clearly, the inventory splitting can be used to transform the deterministic lot-sizing problem with time-varying signed demand (Wagner-Whitin problem with signed demand) into one with only nonnegative demand. Such use of the splitting was made in Kelle and Silver (1989a).

8.2.3 Single-decision Models

In this section, we discuss situations in which a single replenishment decision needs to be taken. This problem occurs in two typical cases. In the case of

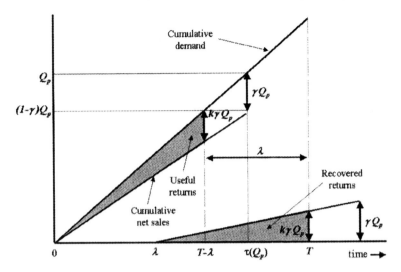

Fig. 8.4. Cumulative orders, net sales, and returns for a constant demand and return rate

fashion products (see Case A in the introduction of this chapter), one needs to determine the initial order quantity before the start of the selling season. Another case is the final order problem, where a manufacturer makes a final production run for his parts/finished products, because he switches over, or already has switched over, to new models. In both cases, the selling period is known in advance. These cases seem similar to the 'news-vendor' problem, but there is a major difference: product returns. Because of the returns, the filled customer orders (transactions) differ from the customer orders filled and not returned by the customer (net sales).

The operation of a single-decision inventory system with returns is explained through Figure 8.4, in which cumulative flows of orders, net sales, and returns are depicted. In Figure 8.4, the solid black line represents the cumulative demand, which attains its maximum value at the end of the selling period, denoted by T.

Suppose one replenishes by ordering (or producing) Q_p units at the beginning of the period over which demand needs to be met. This quantity covers demand until epoch $\tau(Q_p)$. If it were not for recoveries, any demand beyond this epoch would be lost. The total net sales from the initial quantity Q_p are $(1 - \gamma)Q_p$ at time $\tau(Q_p)$. Here γ is the return fraction: the fraction of sales that is returned. The surface between cumulative demand and cumulative net sales represents the cumulative returns of the initial order. The maximum value of these returns is uQ_p.

The returns, so it is assumed, may be used to satisfy further demand occurring after $\tau(Q_p)$ and thus can be used to increase (net) sales. In practical situations, there is a time lag between the time of placement of a customer order and, in case of returns, the time the returned product re-enters the serviceable inventory. This delay is denoted by λ and includes collection and recovery (inspection, packing, minor repair, etc.) times, if necessary. So, the returns become available with a time lag λ with respect to the customer orders from which they are returned (see serviceable returns in figure 8.4). We assume that there is a fixed time T beyond which sales of the product stop. Returns that become serviceable after time T cannot be used to satisfy demand but might be channelled to other (re)use.

The fraction of the returns that can be used to satisfy new orders is denoted by k. To estimate k, we have to study the net demand rate, which is the difference of the demand and return rate. Moreover, (first-time) returns will be available until time $\tau(Q_p) + \lambda$, but useful only until time T. In general, the value of k will depend on

- the dependence on time of the demand rate,
- the length of the delay λ,
- the length of the sales period T,
- and the initial order Q_p,

and will be difficult to uncover. However, under the assumptions that 1) the demand rate is constant (as displayed in the figure), 2) the return rate is constant, and 3) products are only returned once, the figure above illustrates that finding k is not difficult. The upshot is that

$$
k = \begin{cases} \dfrac{\max\{0, T - \lambda\}}{\min\{Q_p/d, T\}} & \text{if } T \le \lambda + \frac{Q_p}{d} \\ 1 & \text{if } T > \lambda + \frac{Q_p}{d}, \end{cases}
$$

where $d = \frac{D}{T}$ is the demand rate.

Calculation of the Optimal Initial Order Quantity Given k

Now assume that the fraction of returns that can (potentially) be used to service demand is a given value k (independent of the initial order). The subsequent analysis will correspond closely to that of the 'news-vendor' problem. We therefore briefly set forth this analysis first. The following symbols are used.

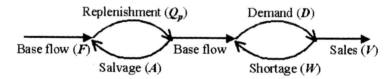

Fig. 8.5. 'News-vendor' problem

symbol	quantity	per unit revenue (symbol)
D	(external) demand	–
Q_p	initial order or replenishment (decision variable)	$-c_p$
W	shortage (or underage)	$-c_b$
A	salvage (or overage)	ω
V	sales or number of units sold (transactions)	v
F	base flow, number of units actually input into the market (sales)	–

As in the previous subsection, demand, as it manifests itself in customer orders and denoted by D, is a stochastic variable. Sales are customer orders that have been filled. The objective is to set Q_p, W, A, and V consistently, such that expected revenue is maximized. Consistency between the variables implies that the flow diagram in Figure 8.5 must be valid.

The solution for Q_p is Q_p^* where Q_p^* is determined from

$$\text{Prob}(D \leq Q_p^*) = \frac{v - c_p + c_b}{v - \omega + c_b}$$

(see e.g. Silver et al., 1998, p. 387). Of course, the unit costs coefficients appearing in the optimal value for the initial replenishment are related to the flows as given in the figure through the table that introduces them. Now consider the case that includes returns. Returns are a fixed fraction γ of sales. The determination of the optimal initial order quantity, which maximizes the total system profit, depends on the collection and recovery strategy and on the costs involved in all the activities.

Several collection and recovery strategies can be distinguished. The following paragraphs discuss some of these in detail.

Full Recovery and Unlimited Reuse

Suppose that all returns that are suitable for re-use are recovered (full recovery). The following additional symbols are used to model the situation.

symbol	quantity	per unit revenue (symbol)
U	number of items returned	$-c_r - v$
R	number of returned items used for satisfying demand	–
A'	number of items salvaged after return	ω
Z	censored demand	?
Ω	censored shortage	?

Furthermore, assume that there is no limit[5] on the number of times a unit can be returned and recovered (unlimited reuse). In the analysis, we assume

---

[5] Also assume that the fraction of returns that is actually used for satisfying demand is k irrespective of the number of times items return.

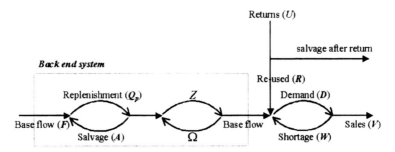

Fig. 8.6. Split model for full recovery and unlimited returns

that demand can be modeled as a continuous variable and can also be filled 'continuously', that is, in non-integer quantities.

We follow the route outlined above, using the split inventory[6] model. However, we have to be a little precise in distinguishing demand from sales, so the splitting requires some extra structure. Also, making returns reuseable requires some processing. Consequently, the part of the model involving returns requires somewhat more care than in the standard case. The ultimate model takes the form as sketched in Figure 8.6.

We first discuss the flows. The auxiliary flows Z and Ω are virtual. We can thus manipulate their values at will as long as consistency with the flow diagram is maintained. Several relationships exist between the recovered flow (R), the censored demand flow (Z), the sales flow (V), the demand flow (D), and the shortage flow (W). For example,

$$U = \gamma V \quad ; F + R + W = D \quad ; \qquad V = D - W;$$
$$R = kU \quad : A' = (1 - k)U = (1 - k)\gamma V$$

Now

$$Z - \Omega = F = D - W - R = V - R = (1 - k\gamma)V = (1 - k\gamma)(D - W) = \frac{1}{\beta}(D - W)$$

where $\beta \equiv \frac{1}{1-k\gamma}$. There is logically no other constraint than $Z - \Omega = F$ that needs to be imposed on the auxiliary flows. However, we choose to put $Z = \frac{D}{\beta}$ and $\Omega = \frac{W}{\beta}$. Second, we turn to costs. A priori the auxiliary flows carried no costs or revenues. We now put a cost of βG per unit on the flow Ω while making the unit costs of W zero at the same time. Then the costs of operating the system remain unaltered. What is the revenue per unit of flow of F that exits the back-end system? As $F = \frac{V}{\beta}$, we can impute a unit costs of β times the revenue of a unit of V while putting the revenue of V to zero without altering the costs. So the question becomes: what is the revenue of

---

[6] The notion of 'inventory' is somewhat spurious in this context, as there are no factual items held at the end of the period.

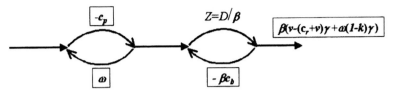

Fig. 8.7. The emulation of the return model: cost and revenue allocation

a unit of V? A unit of flow V implies γ units of returns, $(1 - k)\gamma$ units of salvage after returns (A'), and γk units of recovered items. So a unit of V's revenue is $v - (c_r + v)\gamma + \omega(1 - k)\gamma$.

Now note that the back-end system (see Figure 8.6) is just a standard 'news-vendor' system with the flow F playing the role of sales in the model without returns. So we in fact need to analyze a standard 'news-vendor' model with demand and costs per unit (boxed values) as given in Figure 8.7.

Finding the optimal value for Q_p is now easy. Just apply the unit cost substitution rules $v \leftarrow \beta(v - (c_r + v)\gamma + \omega(1 - k))\gamma$, $c_b \leftarrow \beta c_b$ and $c_p \leftarrow c_p$. Accounting for the factor β in the demand, the standard 'news-vendor' solution now gives the rule: set $Q_p = Q_p^*$ where Q_p^* is determined from

$$\mathrm{Prob}(\frac{D}{\beta} \le Q_p^*) = \frac{\beta(v - (c_r + v)\gamma + \omega(1 - k)\gamma) - c_p + \beta c_b}{\beta(v - (c_r + v)\gamma + \omega(1 - k)\gamma) - \omega + \beta c_b}$$

$$= \frac{(v' + c_b - \omega)\beta + \omega - c_p}{(v' + c_b - \omega)\beta} \equiv \Theta(\beta),$$

where $v' = (1 - \gamma)v + \gamma(\omega - c_r)$.

Full Recovery and One-time Re-use

Under one-time reuse, returns can only be reused once. Once more, we employ the split inventory model (see Figure 8.8).

From one-time recovery, we obtain $U = \gamma F$. From this it follows that $R = kU = k\gamma F$. The flow conditions s(ee Figure 8.8) then further yield

$$Z - \Omega = F = D - W - R = V - R = V - k\gamma F$$

and so $V = (1 + k\gamma)F = \beta' F$, where we define β' as $\beta' = 1 + k\gamma$. We require

$$\beta'(Z - \Omega) = \beta' F = V = D - W.$$

Once more we have some latitude in setting values for Z, the censored demand, and Ω, the censored shortage. We choose to set $Z = \frac{D}{\beta'}$ and $\Omega = \frac{W}{\beta'}$. We now determine costs per unit of Z and Ω which turn the return model into an equivalent standard 'news-vendor' problem. Put

- a cost of $\beta'c_b$ to each unit of Ω (and a cost of zero to W)
- and a revenue of $\beta'v + \beta'\gamma(-c_r - v) + (\beta'\gamma - \beta'k\gamma)\omega = \beta'(v + \gamma(-c_r - v) + (1 - k)\gamma\omega)$ to each unit of F (as each unit of F implies β' units of sales, $\beta'\gamma$ units of returns, $\beta'k\gamma$ units of recovery, and $\beta'\gamma - \beta'k\gamma$ units of salvage after return).

The result for determining the optimal value, Q_p^*, of the initial replenishment can now be written down using the result for the standard 'news-vendor' problem as

$$
\begin{aligned}
\mathrm{Prob}(\frac{D}{\beta'} \leq Q_p^*) &= \frac{\beta'(v + \gamma(-c_r - v) + (1 - k)\gamma\omega) - c_p + \beta'c_b}{\beta'(v + \gamma(-c_r - v) + (1 - k)\gamma\omega) - \omega + \beta'c_b} \\
&= \frac{(v' + c_b - \omega)\beta' + \omega - c_p}{(v' + c_b - \omega)\beta'} = \Theta(\beta')
\end{aligned}
$$

with as before $v' \equiv (1 - \gamma)v + \gamma(\omega - c_r)$. Note that $1 + k\gamma$ is the expansion to first order in $k\gamma$ of $\frac{1}{1-k\gamma}$. The case where items can be reused precisely n times (and where the system is operated with a full-recovery strategy) then allows that one can apply the foregoing end result for obtaining the optimal initial order by substituting $\beta \leftarrow 1 + k\gamma + (k\gamma)^2 + \cdots + (k\gamma)^n$.

Other Cases

Vlachos and Dekker (2000) have studied many more variants of the single decision problem under returns. Additions are fixed costs for recovery operations and the option to recover only a fraction of the returns suitable for recovery. A tree listing the variants analyzed is given in Figure 8.9.

The first branch (B1) of the tree corresponds to the first model examined, while the second model refers to branch B2.1. Branches B2.2 and B2.3 assume that the recovery cost is significant. The difference between them is that in the first one (partial recovery) this cost is paid for every returned item we reuse, while in the second (full recovery) this cost is paid for all returns whether

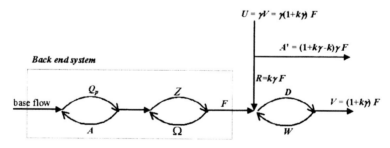

Fig. 8.8. Split model for full recovery and one-time reuse

we reuse them or not. Finally, the recovery of returned products may incur a fixed cost paid per period (e.g., the cost of leasing a packing machine to repack returns). So in partial and full recovery options we have another branch in the tree, which includes (alternatives B2.2.2 and B2.3.2) a fixed recovery cost or not (alternatives B2.2.1 and B2.3.1). The expected profits and the optimality conditions for the other alternative models are presented in Vlachos and Dekker (2000).

A Numerical Example

Figure 8.10 depicts the effect of the return fraction γ and the recovery fraction k to the optimal order quantity and the expected profit for a specific numerical example. The collection and recovery costs are assumed negligible. So, the model allowing re-use only once is used. The other cost parameters are $v = 15$, $c_d = 2$, $c_p = 7$, and $\omega = 5$.

The lines for $\gamma = 0.0$ represent the optimal order quantity and the expected profit of the classical 'news-vendor' problem (without returns). We observe that as the return fraction (γ) or the recovery fraction (k) increase, the effect on Q_p^* and its corresponding expected profit is almost linear. These dependencies also prove the statement that the classical 'news-vendor' optimal quantity is far from optimal when the return rates are high.

Future research directions on single period inventory models include dynamic estimation of expected demand and serviceable returns (quick response) using data from the beginning of the period. This research can be combined with an improved inventory control system for single-period products with returns that includes a second order during the selling period.

In the foregoing models, the ratio between sales (V) and net supply (F) is given by a number β: $V = \beta F$. For the model with unlimited reuse, Mostard and Teunter (2002) go beyond this situation by assuming that each time an

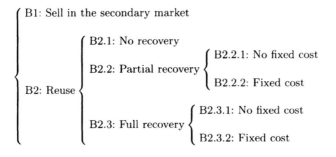

Fig. 8.9. Alternative models for single decision cases with returns

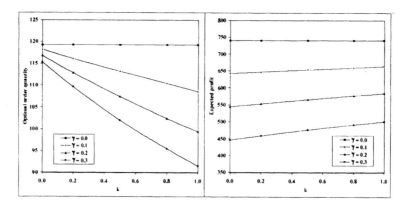

Fig. 8.10. Effect of return fraction γ and recovery fraction k on optimal order quantity Q_p^* and corresponding expected profit

item is sold it has probability of γ of being returned, and that each item returned has a probability of k of being reused. The number of times an item is returned is then geometrically distributed with parameter $1 - k\gamma$: $\mathrm{Prob}(\#\text{ return times} = x) = (1 - k\gamma)(k\gamma)^x$. So the return flow ($R$) can be modeled as $R \cong \sum_{i=1}^{F} \Gamma_i$ where the integer variables $\{\Gamma_i\}$ are independent and each is geometrically distributed with parameter $1 - k\gamma$. Evidently, for any i, $\mathrm{E}(R) = \mathrm{E}(F)\,\mathrm{E}(\Gamma_i) = \mathrm{E}(F)\frac{k\gamma}{1-k\gamma}$ and so from $V = F + R$ one obtains $\mathrm{E}(V) = \frac{\mathrm{E}(F)}{1-k\gamma} = \beta\,\mathrm{E}(F)$ with $\beta \equiv \frac{1}{1-k\gamma}$ as before.

However, knowledge of the value taken by F no longer suffices to determine the values taken by the number of returns (U), the number salvaged after return (A'), the number reused (R), or the number of sales (V). Nonetheless, one readily verifies that each unit of F creates an *expected* revenue of $\beta(v + \gamma(-c_r - v) + (1 - k)\gamma w)$.

Now define[7] the censored demand (Z) and the censored shortage (Ω) through the equations

$$Z + \sum_{i=1}^{Z} \Gamma_i = D \quad \text{and} \quad \Omega + \sum_{i=1}^{\Omega} \Gamma_i = W.$$

Note that $Z \geq \Omega$ (as $D \geq W$). For model consistency, one requires $Z - \Omega = F = D - W - R$, i.e., $Z - \Omega = D - W - R$. However,

$$Z - \Omega + R + W - D = D - \sum_{i=1}^{Z} \Gamma_i - W + \sum_{i=1}^{\Omega} \Gamma_i + R + W - D$$

---

[7] The equations may not result in integer values for Z and Ω. In that case, some rounding is necessary. F.e. $Z \equiv \min\{z|z+\sum_{i=1}^{z} \Gamma_i \geq D\}$ which has the advantage that Z is a stopping time for (Γ_i).

$$= R - \sum_{i=\Omega+1}^{Z} \Gamma_i = 0.$$

So Z and Ω are defined consistently. For any i, $\mathrm{E}(\Omega) + \mathrm{E}(\Omega)\mathrm{E}(\Gamma_i) = \mathrm{E}(W)$ and from this it follows $\mathrm{E}(\Omega) = (1 - k\gamma)\mathrm{E}(W)$ as $\mathrm{E}(\Gamma_i) = \frac{k\gamma}{1-k\gamma}$. Now attribute to each unit of flow of Ω a cost of $\frac{c_b}{1-k\gamma}$ (and put the unit costs of W to zero). Then, as costs were linear in W, the average costs analysis remains the same.

Under all the above unit costs alterations, for convenience summarized in the following list,

- $v \leftarrow \beta(v + \gamma(-c_r - v) + (1 - k)\gamma w)$
- $c_b \leftarrow \frac{c_b}{1-k\gamma} = \beta c_b,$

the back-end system now is a model driven by the censored demand Z. Temporarily discounting for the dependence of Z on Q_p, this leads to the following condition that determines the optimal initial replenishment, Q_p^*,

$$\mathrm{Prob}(Z \leq Q_p^*) = \frac{\beta(v + \gamma(-c_r - v) + (1 - k)\gamma w) - c_p + \beta c_b}{\beta(v + \gamma(-c_r - v) + (1 - k)\gamma w) - w + \beta c_b} = \Theta(\beta),$$

which was previously obtained by Mostard and Teunter (2002).

8.2.4 Multi-period Model

Consider a single inventory facility that carries a single product. Time is segmented into periods. Demand for the product is random with demand in different periods identically and independently distributed (i.i.d.). The facility receives product returns, as well. These product returns in different periods are i.i.d. and independent of demand also. Demand that cannot be met is backlogged. Upon return, products are available for servicing demand after L_r periods. The average of returns per period is smaller than the average demand per period. The difference is furnished by a supplier external to the system considered. To this end, the inventory facility places replenishment orders at the supplier. These orders are delivered with a lead time of L_p periods. Replenishment orders can be placed at the beginning of each period. The decision maker has to determine whether and how much is reordered. Disposal is disallowed. Per unit of product per unit of time, the cost for backlogging demand is c_b and per unit of product per unit of time the inventory carrying cost is h_s. The cost per replenishment order is K_p. As the reader will note, apart from the returns, the model described is the classical inventory model under backlogging.

The analysis of this model lends itself almost immediately as an application of the split inventory model. The one thing that needs some careful consideration is the issue of lead times.

The State Space

In classical stochastic inventory theory, where unmet demand is backlogged, the model just described is best studied on the state space defined by the inventory position. Here, the use of the inventory position avoids the need for dealing explicitly with lead time (lead time is absorbed in the cost function). However, the presence of returns slightly complicates matters, as we now have to deal with two (possibly different) lead times. The state of the system summarizes all the information relevant to the decision at hand. The sequence of events in a period is: 1) arrival of products (returns + ordered) in the serviceable inventory, 2) ordering, and 3) meeting demand and returns. Demand that cannot be met is backlogged. Disposal is disallowed. The state of the system at time period τ is given by a triple $(I_n(\tau), Y(\tau), Z(\tau))$ where

$I_n(\tau)$ is a scalar representing the net stock at the end of Period τ,

$Y(\tau)$ is an L_p-tuple $(Y_{L_p-1}(\tau), Y_{L_p-2}(\tau), \cdots, Y_0(\tau))$ where $Y_j(\tau)$ is the amount reordered, in period $\tau + j - L_p$, due to arrive in period $\tau + j$, and

$Z(\tau)$ is an L_r-tuple $(Z_{L_r-1}(\tau), Z_{L_r-2}(\tau), \cdots, Z_0(\tau))$ where $Z_j(\tau)$ is the amount of returned items becoming available for serving demand in period $\tau + j$.

Introduce the inventory position $I_-(\tau)$ at the beginning of period τ before any events as

$$I_-(\tau) = I_n(\tau - 1) + \sum_{j=0}^{L_p-1} Y_j(\tau) + \sum_{j=0}^{L_r-1} Z_j(\tau) .$$

After product arrivals and reordering, the inventory position is $I_+(\tau) = I_-(\tau) + Y_{L_p}(\tau)$ where $Y_{L_p}(\tau)$ is the replenishment order placed in period τ. The dynamics are given by the equation

$$I_-(\tau + 1) = I_+(\tau) - D_{-1}(\tau + 1) + Z_{L_r}(\tau) ,$$

where $D_j(\tau)$ is the demand during Period $\tau + j$. The net stock I_n can be computed from the inventory position as

$$I_n(\tau + L_p - 1) = I_-(\tau) - \sum_{j=0}^{L_p-1} D_0(\tau + j) + U_{L_r,L_p}(\tau) \qquad (\triangle)$$

with

$$U_{L_r,L_p}(\tau) = \sum_{j=0}^{L_p-1} Z_{L_r}(\tau + j) - \sum_{j=0}^{L_r-1} Z_{L_r}(\tau + L_p - j - 1).$$

Now assume $L_r \leq L_p$. Then, clearly, $U_{L_r,L_p}(\tau) = \sum_{j=0}^{L_p-L_r-1} Z_{L_r}(\tau + j)$. Equally important, in terms of distributions, we have for the dynamics

$$I_-(\tau + 1) = I_-(\tau) + Y_{L_p}(\tau) - \tilde{D},$$

where $Y_{L_p}(\tau)$ is the amount ordered, and $\tilde{D}$ is distributed as $D - Z$ where demands are distributed as D and returns as Z. That is, I is driven, apart from the quantity reordered, by a stream of independent stochastic variables with distribution $\tilde{D}$, the outcomes of which are unknown at Period τ. Note that $\tilde{D}$ can be interpreted as a signed demand, i.e., as a demand that assumes both positive and negative values. The structure of the dynamics of the inventory position in the basic model, therefore, is the same as that in the standard model except that the demand now is signed. Furthermore, for purpose of performance calculation, we can use $(\triangle)$. Indeed, note that the quantity $\sum_{j=0}^{L_p-1} D_0(\tau + j) - U_{L_r,L_p}(\tau)$ is independent of $I_-(\tau)$ (still for $L_r \leq L_p$).

The conclusion is that we can consider the system as being a zero-lead-time system subject to signed demand. To this system, we can apply the full strength of the split-inventory technique. When giving explicit results, the following assumes that $L_r = L_p = 0$. However, from the remarks above it will be clear that the analysis can be carried through analogously for any $L_r \leq L_p$. Below, results are given for the case of optimizing the expected cost per period.

Analysis and Results

The analysis and results are discussed in detail by Fleischmann and Kuik (2003). One peculiarity is that for the equilibrium process of (V, W) the two component processes V and W become independent in the long run: $(W|V = v) = W$ where (V, W) is the equilibrium process. For the stationary case, the cost function for inventory related costs (see Subsection 8.3) takes the simple form

$$H(v) = \mathrm{E}_W(G(v + W))$$

for the inventory in the back-end facility. Somewhat more detailed, the following conclusions hold.

1. An (s, S) inventory policy is optimal for the basic model.
2. The V and W equilibrium processes are independent. The process V can be described through the inventory position process corresponding to an s, S policy for a facility subject to nonnegative demand distributed i.i.d. in periods with distribution

$$(\text{Probability demand} = k) = \begin{cases} \displaystyle\sum_{\ell \geq 0} \pi_\ell \sum_{m \geq 0} \tilde{d}_{\ell-m} & \text{for } k = 0 \\[2ex] \displaystyle\sum_{\ell \geq 0} \pi_\ell \tilde{d}_{k+\ell} & \text{for } k \geq 1 . \end{cases}$$

Here, $\boldsymbol{\pi} = (\pi)_{\ell=0,1,\dots}$ is the invariant distribution for W and $\tilde{d}_i = \mathrm{Prob}(\tilde{D} = i)$. Note that as W is independent of (s, S) so is $\boldsymbol{\pi}$.

3. The costs of the basic model coincide with the costs of the (s, S) process inventory position process described by V with state-dependent period costs given by H. Note that $H(i)$ is the expected cost of a cycle on the half line, incurred by a random walk described by the W-process, starting and ending in the position 0 and with costs in position ℓ given as $G(i+\ell)$.

The three statements together imply that the optimization of the basic model can, at least in principle, be carried out through the processes V and W as an optimization problem for a classic inventory model. This conclusion continues to hold for the case $L_p \geq L_r \geq 0$.

All of the above results immediately carry over to the case of a model in continuous time under continuous review with the obvious modifications (see Fleischmann et al., 2002). The discrete time case is analyzed in more detail in Fleischmann and Kuik (2003).

Remark. The split inventory introduces a front-end facility and a back-end facility. The analysis just reported will carry through in case the back-end system itself is an inventory system consisting of multiple facilities. Thus the analysis, in principle, carries through in case of an assembly system with autonomous returns only at the end-item level (see DeCroix and Zipkin, 2002a).

Heuristics

Several heuristics have been developed for the multi-period model when fixed production costs are disregarded. Simpson (1970) considers a discrete time, periodic review model with a review cycle of m periods. During each period τ there is stochastic demand $D(\tau)$ with mean μ_D and variance σ_D^2 and stochastic recovery output $R(\tau)$ with mean μ_R and variance σ_R^2. Every m periods, a production order is placed with *stochastic* lead time L_p such that the inventory position is raised to S_p. Demand that cannot be fulfilled immediately is backordered. Define the random variable $Y = \sum_{i=\tau+1}^{\tau+m+L_p} D(i) - R(i)$, i.e. Y is the net demand during a review cycle plus a (random) replenishment lead time. The mean and variance of Y are $\mu_Y = (m + \mu_{L_p})(\mu_D - \mu_R)$ and $\sigma_Y^2 = (m+\mu_{L_p})(\sigma_D^2+\sigma_R^2)+(\mu_D-\mu_R)^2\sigma_{L_p}^2$. Under a service objective, i.e. the fraction of demand *not* backordered should be larger than k, it can be shown that the optimal value of S_p is the solution to

$$\int_{S_p}^{0} (y - S_p)g(y)\mathrm{d}(y) = (1 - k)m\mu_D. \tag{8.3}$$

where $g(.)$ is the density of net demand Y. If service is enforced through a penalty cost (c_b per unit backordered) than the optimal S_p is the solution to

$$\int_{S_p}^{\infty} g(y)\mathrm{d}(y) = \frac{mh_s}{c_b}$$

with h_s the serviceable holding cost per unit on stock per period.

A continuous time variant of Simpson's model was recently put forward by Mahadevan et al. (2002). Here, the production lead time, L_p, is a fixed constant and all returned products are initially stocked. As soon as a review epoch occurs, all returns in stock are recovered (fixed lead time L_r) and transferred to serviceable inventory (see Figure 8.11).

One of the heuristics developed for this model approximates the stockout probability $T_p = m + L_p$ time units from the current review epoch, τ and the stockout probability at $\tau + T_r$, which is the time at which the last recovery batch (if any) arrives in the interval $[\tau, \tau + T_p)$. The number of recovery batches, N, that arrive during time T_p equals $\lceil L_p/m \rceil$ if recovery batches always arrive *before* manufacturing batches in a review cycle or $\lceil L_p/m \rceil$-1 if recovery batches always arrive *after* manufacturing batches. Hence, $T_r = (N - 1)m + L_r$. Assuming that demands and returns follow independent Poisson processes, the net demand during time T_p, Y_p, has mean $\mu_{Y_p} = dT_p - uNm$ and variance $\sigma_{Y_p}^2 = dT_p + uNm$. Similarly, the net demand during time T_r, Y_r, has mean $\mu_{Y_r} = dT_r - u(N-1)m$ and variance $\sigma_{Y_r}^2 = dT_r + u(N-1)m$. The optimal value of S_p then follows from

$$\int_{S_p}^{\infty} g_p(y) + g_r(y)\mathrm{d}(y) = \frac{mh_s}{c_b}.$$

where $g_p(.)$ and $g_r(.)$ are the densities of Y_p and Y_r respectively. In this particular push policy, the replenishment order is split into a production portion

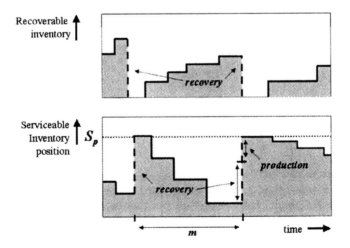

Fig. 8.11. A periodic review with order-up-to level S_p.

and a recovery portion. An interesting side effect is that it pays off to have different lead times for production and recovery so that the incoming serviceable products are more spread out over time. Therefore, increasing one of the lead times may decrease optimal costs.

The only difference between the policies of Simpson ('continuous' recovery) and Mahadevan et al. (periodic recovery) is the way that product returns are handled. Simpson's policy benefits customer service at the price of higher serviceable holding costs, while the policy of Mahadevan et al. delays recovery expenses at the cost of customer service. If recovery is expensive, the latter may be more attractive.

Muckstadt and Isaac (1981) adapted the standard (s, Q) policy to account for a situation with fixed costs for production, K_p. As soon as the inventory position of serviceable items (net serviceable inventory plus everything on order plus all product returns in the recovery facility) drops to s, a production batch of size Q is ordered and will be delivered after a fixed lead time L_p. The recovery facility operates under a one-for-one push policy, and recovery lead times are stochastic. Since the demand and return processes are Poisson streams, the inventory position can be modelled as a Markov chain. The solution procedure is based on the approximation that the net inventory follows a normal distribution and on the assumption that the output of the recovery shop is a Poisson process. The advantage of this procedure is that it results in simple expressions from which one can deduct the optimal values of s and Q very easily. Its disadvantage is that it may not be a very accurate procedure, especially for high return rates (see Van der Laan et al. 1996a). A more accurate procedure was developed by van der Laan et al. (1996a). Instead of approximating the distribution of *net inventory*, the authors approximate the *net demand demand* during time τ, Y_τ, using a Brownian motion with drift and variance equal to $\mu_{Y_\tau} = (d - u)\tau$ and $\sigma^2_{Y_\tau} = (d + u)\tau$ respectively. The optimal reorder quantity, Q_p^*, can be approximated by

$$Q_p^* = \sqrt{\frac{2K_p(d - u)}{h_s}}$$

and the optimal reorder point, s_p^* is the solution to

$$\int_0^{L_p} \int_{s_p}^\infty g(y; \tau)\mathrm{d}y\mathrm{d}\tau = \left(\frac{d - u}{Q_p^*}\right)\left(\frac{h_s}{c_b + h_s}\right) \tag{8.4}$$

(adapted from van der Laan et al., 1996a) where $g(y; \tau)$ is the density of Y_τ. Here it is assumed that the *output* of the recovery process is a Poisson stream. The backorder penalty c_b is per product per time unit. If the backorder penalty is just per product, Equation (8.4) reduces to

$$\int_{s_p}^\infty g(y; L_p)\mathrm{d}y = \left(\frac{d - u}{Q_p^*}\right)\left(\frac{h_s}{c_b}\right)$$

Under a service objective (at least a fraction k of total demand should be fulfilled immediately), we deduct a similar expression as Equation (8.3) so that s_p^* approximately is the solution to

$$\int_{s_p}^{\infty} (y - s_p)g(y; L_p)\mathrm{d}y = (1 - k)Q_p^* .$$

One of the results of this section is that the standard (s, Q) policy is optimal as long as the recovery process is autonomous and its lead time is shorter than the production lead time. In the original formulation by Muckstadt and Isaac, the recovery facility is modelled as a queuing system with (in)finite capacity and stochastic recovery times. The total lead time, i.e. waiting plus processing time, can be well above L_p especially if the system load is high, implying that the (s, Q) policy combined with autonomous recovery is not necessarily optimal. However, a main drawback of the above heuristic approaches is that they are hard to adapt to managed recovery. The issue of managed recovery is studied in the next section.

8.3 Managed Recovery (or Pull Strategies)

In contrast with situations of direct reuse as discussed in the previous section, Section 8.2, value-added recovery involves more elaborate (re)processing. In this case, the throughput times of returned items can be substantial and varying. Moreover, variation and uncertainty of the quality of returns makes it more difficult to predict reprocessing needs and add to the variability of lead times. Typical products that are associated with value-added recovery are products that have been used extensively and are returned/collected in order to be restored to perfect working condition. Due to value-added and/or fixed costs involved with the reprocessing, extra stocking points are needed for managing the flows of returned products.

We begin the discussion on managed recovery by discussing some of the (limited) results on the structure of optimal policies.

8.3.1 Optimality Results for Linear-cost Models

The first results regarding optimal periodic review policies for recovery systems with two stocking points can be found in Simpson (1978), where a model with no fixed costs for production, recovery, and disposal, and zero lead times is considered. It is proven that the optimal periodic policy is determined by three parameters: a produce-up-to level S_p, a recover-up-to level S_r and a dispose-down-to level U. Inderfurth (1997) has shown how the model has to be adapted in cases of positive and equal lead times. Up to now, the optimal policy for the general case of different lead times is not known. For the special case that $L_p = L_r + 1$, the optimal policy turns out to be rather simple and has been given by Inderfurth (1997). In the following, we will describe the model for the situation $L_r = L_p = L$ in more detail.

It is assumed that the inventories are reviewed periodically and that in each period τ it has to be decided how much to produce $(p(\tau))$, how much to recover $(r(\tau))$ and how much to dispose of $(w(\tau))$. Demands in period τ $(d(\tau))$ as well as returns $(u(\tau))$ are random variables, and the corresponding probability density functions are denoted with $\varphi_{\tau,d}$ for the demands and $\varphi_{\tau,u}$ for the returns.

The recovery system can then be described by two state variables, the physical stock of recoverables I_u and the inventory position of serviceables I_s, which is defined as the stock-on-hand of serviceables plus all outstanding orders minus backorders. The inventory balance equation for the recoverables is given as

$$I_u(\tau + 1) = I_u(\tau) - w(\tau) - r(\tau) + u(\tau), \tag{8.5}$$

while the inventory balance equation for the serviceables is given as

$$I_s(\tau + 1) = I_s(\tau) + r(\tau) + p(\tau) - d(\tau). \tag{8.6}$$

In Simpson, the optimal policy is determined with respect to the average total relevant cost over a finite planning horizon T. Thereby, production, recovery, and disposal costs are assumed to be proportional to the number of items, and holding and penalty costs are charged to the net inventory at the end of each period. The cost parameters are denoted as follows.

h_s: unit holding cost for serviceable items
h_u: unit holding cost for recoverable items
c_b: unit backorder cost
c_p: unit production cost
c_r: unit recovery cost
c_w: unit disposal cost

Then the average cost for backorders and keeping serviceables in stock in period τ is given as a function of the inventory position Y_s after the reorder decisions $(Y_s(\tau) = I_s(\tau) + p(\tau) + r(\tau))$ as follows:

$$L_\tau(Y_s) := h_s \int_0^{Y_s} (Y_s - z)\varphi_{\tau,L,D}(z) \, dz + c_b \int_{Y_s}^\infty (z - Y_s)\varphi_{\tau,L,D}(z) \, dz, \tag{8.7}$$

where $\varphi_{\tau,L,D}$ denotes the density function of the cumulative demands in the periods $\tau - L, \tau - L + 1, \ldots, \tau$. Further, we have to include the average cost for stock keeping of recoverable items in period τ which can be computed as a function of the inventory position Y_u of the recoverable inventory after the reorder decisions $(Y_u(\tau) := I_u(\tau) - w(\tau) - r(\tau))$:

$$C_r(Y_u) = h_u \int_{-\infty}^{\infty} (Y_u + z)\varphi_{\tau,u}(z)dz = h_u Y_u + h_u E[u(\tau)]. \qquad (8.8)$$

Since the term $h_u E[u(\tau)]$ in (8.8) has no influence on the optimization, we can neglect it in the sequel. In order to be able to formulate the problem as a stochastic dynamic programming problem, we introduce the average relevant cost f_n for n remaining periods until the end of the planning horizon. This function depends on two variables: the inventory position of the serviceable inventory I_s and the recoverable inventory I_u.

The functional equation of dynamic programming is obtained as follows

$$f_0(I_s, I_u) \equiv 0 \qquad (8.9)$$

and for $n \geq 1$ as

$$f_n(I_s, I_u) = \min_{p,r,w \geq 0} \Big\{ c_p p + c_w w + c_r r + L_{T-n}(I_s + p + r)$$
$$+ h_u(I_u - w - r) + H_n(I_s + p + r, I_u - w - r) \Big\} \qquad (8.10)$$

where

$$H_n(a, b) := \int_{-\infty}^{+\infty} \int_{-\infty}^{+\infty} f_{n-1}(a - z, b + y)\varphi_{n,d}(z)\, \varphi_{n,u}(y)\, dz\, dy \qquad (8.11)$$

with $\varphi_{n,d}(z)$ denoting the density of $d(T-n)$ and $\varphi_{n,u}(y)$ denoting the density of $u(T-n)$. Note that (8.9) and (8.10) also hold for non-identically distributed correlated demands and returns.

An analysis of these equations leads to the structure of the optimal policy (see for details Simpson, 1978). This so-called (S_p, S_r, U) policy (with $S_p \leq S_r \leq U$) is a straightforward extension of the simple (S_p, U) policy which is optimal if stocking of returns is disallowed. The optimal decisions in period τ in the case of two stocking points are determined by the three time-dependent parameters $S_p(\tau)$, $S_r(\tau)$, and $U(\tau)$ as follows.

Items are only disposed of if there are too many in the system, but you can never dispose of items already produced or remanufactured. This leads to:

$$w^*(\tau) = \Big(\min\{I_s(\tau) + I_u(\tau) - U(\tau), I_u(\tau)\} \Big)^+, \qquad (8.12)$$

where $(x)^+$ denotes the $\max\{0, x\}$. Production is used if the total number of items in the system is less than $S_p(\tau)$.

$$p^*(\tau) = \Big(S_p(\tau) - (I_s(\tau) + I_u(\tau))\Big)^+ \qquad (8.13)$$

Since there is only a limited number of recoverables available, the service-able inventory position cannot always be increased to the recover-up-to level $S_r(\tau)$ using recovery, leading to the following recovery decisions:

$$r^*(\tau) = \min\Big\{I_u(\tau), (S_r(\tau) - I_s(\tau))^+\Big\}. \qquad (8.14)$$

For the application of such a control policy in practice, the policy parameters have to be computed.

Omitting the time dependence from the notation, the policy parameters S_p, S_r, and $U \geq 0$ can be determined by solving the following equations.

$$L'(S_p) + \partial_1 H(S_p, 0) = -c_p \qquad (8.15)$$
$$L'(S_r) + \partial_1 H(S_r, U - S_r) = c_w - c_r \qquad (8.16)$$
$$\partial_2 H(S_r, U - S_r) = c_w - h_u \qquad (8.17)$$

where $\partial_1 H$ and $\partial_2 H$ denote the partial derivative of H with respect to the first and second argument respectively. If there exists no solution of (8.17), (8.16) with $U - S_r \geq 0$, one needs to solve (8.16) for S_r with $U = S_r$.

8.3.2 An Exact Modelling Approach

Using the modelling approach of the previous section, it is hard to find, for each period, the optimal policy parameters in the presence of fixed production and recovery costs. This makes it difficult to generalize those results to situations in which batching is necessary. A continuous time, continuous review setting enables one to formulate various inventory control strategies that extend those considered in Section 8.3.1 and that can be optimized and analyzed making use of the theory of Markov Chains. Additionally, this setting facilitates the modelling of stochastic lead times.

Van der Laan et al. (1999b) developed an *exact* procedure that enables one to study a variety of push and pull policies under fairly general conditions, such as stochastic lead times and Markovian return and demand flows. The modelling framework for a system with two stocking points, one for recoverable inventory and one for serviceable inventory, is characterized as follows.

- The *demand and return processes* are stochastic and may be modelled by any Markovian arrival process. The two may even be dependent, but this requires extra state variables to model the number of products in the market (see, for example, Bayindir et al., 2003; Nakashima et al., 2002; Yuan and Cheung, 1998) and/or the time that they have spent there. Although it is common practice to assume simple (compound) Poisson arrivals, an alternative could be to use Coxian-2 arrival processes. These enable one to do a three-moment fit of an arbitrary arrival process, so that a better description of reality can be achieved.

- The *production process* has unlimited capacity. The production costs consist of a variable cost per item and a fixed cost per order. The production lead time L_p is a *discrete* random variable, bounded by some $L_p^{max} < \infty$.
- In principle, the *recovery process* can be any queuing system with Markovian transitions. Alternatively, the recovery process can be modelled as a 'black box' with lead time L_r, a *discrete* random variable bounded by some $L_r^{max} < \infty$. The recovery costs consist of a variable cost per item and a fixed cost per batch.
- If disposals are allowed, the *disposal process* depends on the control policy employed. Next to a variable component, disposal costs may also include a fixed cost per batch.
- The *inventory position* may be defined in various ways (see the discussion in section 8.3). The only restriction is that its transitions are Markovian. For instance, the inventory position may be the net inventory plus all outstanding production orders plus some subset of recoverable products that are currently in the system. As an example, product returns may enter inventory position upon arrival (the arrival process is Markovian) or as soon as some control policy triggers them to be released to the recovery facility. Necessarily, such policies only work on Markovian processes such as the inventory position itself or the stock of recoverables. As we have seen, the optimal policy structure is only known for some very special cases, so in general we have to rely on heuristic policies.
- Although the framework does not pose any restriction on the holding cost parameters, it is reasonable to assume that the recoverable holding cost h_u is smaller than the serviceable holding cost h_s. Moreover, to come to a meaningful performance measure, its numeric values should have a direct relation with the variable costs of production, recovery, and disposal. There is quite some controversy, though, with respect to the correct valuation of holding cost parameters in a reverse logistics setting, but this discussion is left to Chapter 11.
- *Customer service* is modelled in terms of backorder costs, either per product or per product per time unit.
- All system parameters are *stationary*, i.e. do not change over time.

The calculation of the average on-hand serviceable inventory and the average backorder position is difficult since the transitions of the net inventory, $I_n(\tau)$, are usually not Markovian. However, depending on the assumptions with respect to the recovery lead time, we can deduct a handy relation between the net inventory and inventory position $I_s(\tau)$ from which we can calculate the long-run distribution of $I_n(\tau)$. Define

- $W(\tau)$ as the number of recoverables that are included in the inventory position at time τ, but that have not yet been recovered,

  $O(\tau_1 - \tau_2)$ as the output of the recovery process in the interval $(\tau_1, \tau_2]$,
- $D(\tau_1 - \tau_2)$ as the demand in the interval $(\tau_1, \tau_2]$,

- $R(\tau_1, \tau_2)$ as the number of recoverable products that enter the inventory position in the interval $(\tau_1, \tau_2]$ and subsequently enter serviceable inventory at or before time τ_2, and
- $P(\tau_1, \tau_2)$ as the number of products that are ordered at the production facility in the interval $(\tau_1, \tau_2]$ and subsequently enter serviceable inventory at or before time τ_2.

Then we have the following cases.

Case 1: Markovian recovery lead times

$$I_n(\tau) = I_s(\tau - L_p^{max}) - W(\tau) + O(\tau - L_p^{max}, \tau) - D(\tau - L_p^{max}, \tau) \quad (8.18)$$

Case 2a: Discrete recovery lead times bounded by $L_r^{max} \leq L_p^{max}$.

$$I_n(\tau) = I_s(\tau - L_p^{max}) + R(\tau - L_p^{max}, \tau) - D(\tau - L_p^{max}, \tau) \quad (8.19)$$

Case 2b: Discrete recovery lead times bounded by $L_r^{max} \geq L_p^{max}$.

$$I_n(\tau) = I_s(\tau - L_r^{max}) + P(\tau - L_r^{max}, \tau) - D(\tau - L_r^{max}, \tau) \quad (8.20)$$

Relation (8.18) is a generalization of the relation given in Muckstadt and Isaac (1981), whereas Relations (8.19) and (8.20) are taken from van der Laan et al. (1999b). The long-run distribution of $I_n(\tau)$ can be (numerically) found by analyzing long-run and transient behavior of an appropriate Markov chain. For details, we refer to van der Laan (1997). The long-run distribution of net inventory suffices to calculate the long-run expectation of on-hand inventory and the backorder position. All other relevant entities can be obtained from the Markov Chain analysis.

Although the above framework is very general in theory, it suffers from the curse of dimensionality. The state space grows exponentially with the production and recovery lead time and the capacity of the stocking points. Therefore, optimization may be very time consuming and running time grows exponentially in the number of decision variables. The reader should keep in mind, however, that this framework is meant to assess the performance of a wide variety of recovery policies in an *exact* way rather than using approximations. It is not meant as a fast optimization algorithm nor as an efficient numerical recipe. In the case of very large state spaces, simulation, although less accurate, may be an alternative.

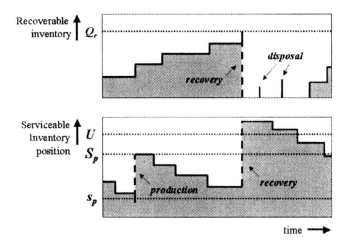

Fig. 8.12. The (s_p, S_p, Q_r, U) PUSH disposal policy.

Push and Pull Policies

The heuristic control policies that have appeared in the literature are all natural extensions of the classical (s, S) policies and may be classified as either push or pull: while production orders are controlled by an (s_p, S_p) policy, recovery batches are either *pushed* through the recovery facility or *pulled* only when they are really needed.

Figures 8.12 and 8.13 give a graphic representation of some of the variants (for details see van der Laan and Salomon, 1997) that can be analyzed with the framework outlined above. With the (s_p, S_p, Q_r, U) PUSH disposal policy, remanufacturing starts whenever Q_r recoverable products are in stock. A production order is placed to increase the serviceable inventory position to S_p as soon as the serviceable inventory position drops to or below the level s_p. Products are disposed of upon arrival as soon as the inventory position exceeds the level U. With the (s_p, S_p, s_r, S_r, U) PULL disposal policy, recovery starts as soon as the serviceable inventory position is at or below s_r, and sufficient recoverable inventory exists to increase the serviceable inventory position to S_r. A production order is placed to increase the serviceable inventory position to S_p as soon as the serviceable inventory position drops to or below the level s_p. Products are disposed of upon arrival as soon as the inventory position exceeds the level U. Note that s_p should never exceed s_r, since otherwise the recovery option would be redundant.

In the above examples, the inventory position is defined as the net serviceable inventory plus all outstanding recovery and production orders. For that case, van der laan et al. (1999a) show that an increase in the recovery lead time or an increase in production lead time variability may lead to lower optimal

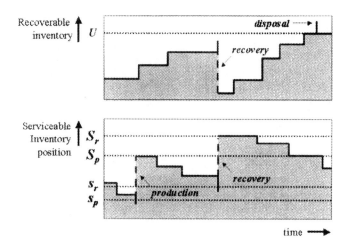

Fig. 8.13. The (s_p, S_p, s_r, S_r, U) PULL disposal policy

costs. It appears that this is due to the sub-optimality of the control policies. Inderfurth and Van der Laan (2001) show that the PUSH-disposal policy is easily improved upon by adjusting the recovery lead time in an appropriate way or by using different information sets for the production, recovery, and disposal decisions. The sub-optimality of the PULL-disposal policy is mainly due to the restriction on the recovery order level ($s_r \geq s_p$). If the recovery lead time is much smaller than the production lead time, s_r will overprotect for recovery lead times. Increasing the recovery lead time then leads to better policy performance. The above effects are further illustrated in Teunter et al. (2002) for push and pull policies without the disposal option.

If the recovery lead time is *smaller* than the production lead time, one could base the production decision on the sum of the serviceable inventory position and the recoverable stock, while the recovery decision could be based on the serviceable inventory position only. If the recovery lead time is longer, one should base both the production and recovery decisions on the serviceable inventory position only. Such policies are easily modelled within the above framework and therefore can be analyzed analytically.

Actually, only orders with a certain remaining service time should be included in the serviceable inventory position (see also the discussion in Section 8.3.3). Such an inventory position, however, is not a Markov process, so our framework cannot be used. In the case that $L_r < L_p$, Teunter et al. (2002) investigate a policy in which the production decision is based on all the inventory in the system, while the recovery decision is based on the serviceable inventory position (serviceables on hand, plus outstanding orders), which includes only those production orders with remaining lead time smaller than L_r (see Figure 8.14). It is shown by simulation that this policy outperforms simple push and pull policies that are based on just one inventory position.

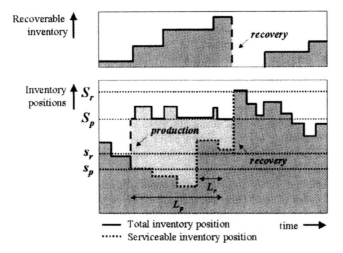

Fig. 8.14. A schematic representation of the double-PULL policy

The drawback of this policy is that it cannot be modelled as a Markov Chain and is therefore difficult to analyze exactly.

8.3.3 Heuristics

In the case of models with managed recovery, the difference with traditional inventory theory goes beyond signed demand. Inventory control now explicitly needs to deal with recovery flows. The source of returned items is unreliable and limited. This makes its control particularly difficult. In traditional inventory theory, one also has to deal with unreliability since supplying inventory facilities might be out of stock. Also in that case, unless special structures in costs and the inventory network exist, one has to resort to rules of thumb or heuristics to find reasonably good solutions. Frequently one finds heuristics and approximations based on the notion of effective echelon inventory position.

Let us first introduce the relevant definitions and concepts. We will do this with an assembly network structure in mind (or special cases thereof, such as series systems or a single location). The installation net stock of a facility is defined as its on-hand minus its backlog. The installation inventory position of a facility is its net stock plus the number of products on order (with its supplier). The nominal echelon stock of a facility is its installation echelon stock plus the installation echelon stock position of all its downstream facilities. In multi-echelon systems, the nominal echelon stock position has no simple relation to the net stock (shifted by a lead time and up to demand during a period of lead-time length), yet in other models this is one of the primary reasons for introducing the concept of inventory position. To restore the simple relationship, traditional inventory management introduces the concept of effective echelon inventory position (see Chen and Zheng, 1994); Van Houtum et al., 1996); and Diks et al., 1996). The effective echelon inventory position of an installation is defined as all stock *in transit* to the installation plus its on-hand stock plus all stock in transit to or on-hand at downstream installations minus the backlog at its end-stock points. Inventory theory with recoveries follows this line of modelling by extending the concept of effective echelon stock policies by including the stock locations for returns.

In the following, we discuss heuristics based on estimating implied costs after a first period, thus essentially reducing the multi-period problem to a 'news-vendor' type problem.

As mentioned in Section 8.3.1, the optimal periodic policy is only known for special cases of lead time pairs, i.e. for the equal lead time case. For the general different lead time case, the optimal policy is expected to be very complex because the dimension of the state space necessary to describe the recovery system accurately is quite large. Therefore, for this situation several heuristics are developed. But also for the equal lead time case it is reasonable to use heuristics since the exact computation of the optimal policy parameters using Equations (8.15), (8.16), and (8.17) can be very time consuming.

Equal Lead Times

Kiesmüller and Scherer (2002) provide two computational approximation schemes to determine nearly optimal policy parameters S_p, S_r, and U. One is based on an approximation of the value-function in the dynamic programming problem (8.10) and leads to excellent results. They obtain in their numerical study an average relative deviation of the average cost of 0.1% and a maximum relative deviation of less than 2%. The other approximation uses a common decomposition technique which is based on a deterministic model and safety stocks. This approach leads to the shortest computation time but also to less accurate policy parameters. The same numerical examples lead to an average relative deviation of the average cost of about 3% and a maximum relative deviation of 10%. Especially in the case of large return rates and large standard deviations for the demands and returns, the first approximation outperforms the latter substantially.

Kiesmüller and Scherer also illustrate that the computation of the policy parameter U can be numerically ill-conditioned when U has hardly any influence on the costs. This is always the case when disposal of items does not play a significant role in the optimal policy. A detailed simulation study by Teunter and Vlachos (2002) shows that in stationary models this is almost always the case. Therefore, an (S_p, S_r) policy, where S_p denotes the produce-up-to-level and S_r the recover-up-to-level is a near-optimal policy for a stationary situation.

For such an (S_p, S_r) policy, simple 'news-vendor' type formulas for the computation of the parameters can be found in Kiesmüller and Minner (2002).

In order to derive these formulas, overage and underage costs are estimated depending on whether it is decided upon the production or the recovery quantity.

First we show how a formula for the recover-up-to-level S_r can be derived. Any underage of a single unit will result in a backorder penalty c_b, assuming that the backorder will only last for a single period. In addition, another recoverable unit could have been remanufactured in the past without affecting serviceable holding costs. Therefore, the cost for each unit short is given by $c_s = c_b + h_u$. The overage cost for the recovery of one unit too many equals the difference of serviceable and recoverable holding cost $c_o = h_s - h_u$. Using c_s and c_o, S_r is determined from the marginal 'news-vendor' approach such that the probability that cumulative demand within the next $L + 1$ periods (the sum of the lead time and the review period) does not exceed S_r is equal to the fraction of underage and overage plus underage cost per unit and unit of time

$$F_{L+1}(S_r) = \frac{c_b + h_u}{c_b + h_s} \qquad (8.21)$$

with F_{L+1} denoting the cumulative distribution function of the demands in $L + 1$ periods.

For the produce-up-to-level, the underage cost simply equals the unit backorder cost per unit of time ($c_s = c_b$). Producing one unit too much leads to serviceable holding costs h_s and might also influence future recovery decisions. Then, the recovery of a unit has to be postponed for some time. This time i can be approximated with a random variable which is distributed according to a geometric distribution with probability $F_{d-u}(0)$. Thereby, $F_{d-u}(0)$ denotes the probability that the net demands $d - u$ are smaller than zero. Therefore, the overage costs can be estimated as follows:

$$c_o = h_s + h_u \sum_{i=1}^{\infty} i \cdot F_{d-u}(0)^i \cdot (1 - F_{d-u}(0)) = h_s + h_u \frac{F_{d-u}(0)}{1 - F_{d-u}(0)}. \qquad (8.22)$$

This leads to the following equation for the determination of the produce-up-to-level:

$$F_{L+1}(S_p) = \frac{c_b}{c_b + h_s + h_u \frac{F_{d-u}(0)}{1 - F_{d-u}(0)}}. \qquad (8.23)$$

The accuracy of these formulas has been tested in a detailed simulation study (see Kiesmüller and Minner, 2002) and reveals excellent results.

Different Lead Times

Now we drop the assumption of equal lead times. Although many authors do not distinguish between the situation with recovery lead time being longer than the manufacturing lead time and the reverse situation, we believe this to be necessary because the control problems are different. In the following, we will provide an approach for the control of a hybrid stochastic production/recovery system with different lead times which can also be found in Kiesmüller (2002). Since we assume a linear cost model and stationary demands and returns, it seems to be reasonable to extend the (S_p, S_r) policy mentioned above to the situation with different lead times. In the following, we assume that the processes for demands and returns are identically distributed in each period.

Long Recovery Lead Time

In case of a long recovery lead time, the system is quite easy to control. For moderate return rates, it is reasonable to push all returns into the recovery process while the faster production supply mode takes care of the items that remain to be produced. Only in the case of large return rates and large lead time differences may some problems occur, if there are low demands in subsequent periods. Then it may happen that more items than required are in the recovery process and there is no more chance for adaptation using the faster supply mode.

In order to use an (S_p, S_r) policy to control the system, we have to know which information should be used for the decisions. Some authors (for example, Gotzel and Inderfurth, 2002, or Inderfurth and van der Laan, 2001) suggest using one inventory position for both the recovery and the production decision. In the context of dual supplier models, this approach would be called a single index policy. Another possibility is to aggregate information in two different variables, which we also call inventory positions (dual index policies). In Kiesmüller (2002), it is shown that the dual index policies outperform the single index policies for recovery systems, especially for large lead time differences. The approach is described in the following.

For the recovery decision, the inventory position $I_m(\tau)$ in period τ includes the current serviceable net-stock plus all outstanding production and recovery orders, including the production order placed at time τ

$$I_m(\tau) := I_n(\tau) + \sum_{i=0}^{L_p} p(\tau - i) + \sum_{i=1}^{L_r} r(\tau - i) \,. \qquad (8.24)$$

The information which is used for the production decision is aggregated in the second inventory position I_s. Since the production decision in period τ influences the stock-on-hand in period $\tau + L_p$, we assume that it is only necessary to consider the outstanding production and recovery orders which arrive in the periods $\tau, \tau + 1, \ldots, \tau + L_p$. This leads to the following definition of I_s:

$$I_s(\tau) := I_n(\tau) + \sum_{i=1}^{L_p} p(\tau - i) + \sum_{i=0}^{L_p} r(\tau - (L_r - L_p + i)) . \qquad (8.25)$$

Based on these inventory positions the following decisions rules are obtained:

$$p(\tau) = (S_p - I_s(\tau))^+ \qquad (8.26)$$

and

$$r(\tau) = \min\{I_u(\tau), (S_r - I_m(\tau))^+\}. \qquad (8.27)$$

In Kiesmüller (2002) it is shown by simulation that such an (S_p, S_r) policy outperforms a similar policy where the decisions are only based on one inventory position because less safety stock is needed. With increasing lead time differences, the cost improvements are increasing and they are much larger than can be obtained with the policy improvement procedure proposed in Inderfurth and van der Laan (2001).

For the policy given by (8.24), (8.25), (8.26), and (8.27), simple formulas exist for the computation of nearly optimal policy parameters (see Kiesmüller and Minner, 2002). Estimating overage and underage costs and using 'newsvendor' type formulas leads to the following two equations for the determination of near optimal policy parameters:

$$F_{L_r+1}(S_r) = \frac{c_b + h_u}{c_b + h_s} \qquad (8.28)$$

and

$$F_{L_p+1}(S_p) = \frac{c_b}{c_b + h_s + h_u \frac{F_{d-u}(0)}{1 - F_{d-u}(0)}} . \qquad (8.29)$$

Long Production Lead Time

In the case of a long production lead time compared to the recovery lead time, the control situation is much more difficult. Due to the longer production lead time, we have to include information with respect to future incoming returns in the production decision. Thus, in this case, the control problem is more complicated.

As an heuristic, we again suggest using an (S_p, S_r) policy based on two inventory positions. For the definition, we use the same principle as above: for the decision with the longer lead time include all outstanding orders in the inventory position and for the decision with the shorter lead time include only the orders which will arrive until the new released order comes in. Therefore, for the production decision the following inventory position is used:

$$I_s(\tau) := I_n(\tau) + I_u(\tau) + \sum_{i=1}^{L_p} p(\tau - i) + \sum_{i=1}^{L_r} r(\tau - i) . \qquad (8.30)$$

For the recovery decision, we use

$$I_m(\tau) := I_n(\tau) + \sum_{i=0}^{L_r} p(\tau - (L_p - L_r + i)) + \sum_{i=1}^{L_r} r(\tau - i) . \tag{8.31}$$

Using (8.30) and (8.31) for the decisions (8.26) and (8.27) leads again to a much better cost performance compared to a policy with one inventory position, although both policies lead to nearly the same system-wide stock-on-hand. The reason for the cost reduction is the partition of the system-wide stock-on-hand in the two stocking points. The policy presented above keeps returned items in the recoverable inventory as long as possible while in the other case most of the items are pushed in the serviceable inventory (see for details Kiesmüller, 2002). Further, the lead-time paradox, decreasing average cost with increasing recovery lead time, cannot be observed for the policy defined by (8.26), (8.27), (8.30), and (8.31).

A nearly optimal recover-up-to-level can be obtained by

$$F_{L_r+1}(S_r) = \frac{c_b + h_u}{c_b + h_s} \tag{8.32}$$

and a-produce-up-to-level by

$$F_\Delta(S_p) = \frac{c_b}{c_b + h_u \frac{1}{1 - F_{d-u}(0)} + h_s F_{d-u}(0)} . \tag{8.33}$$

Thereby, F_Δ denotes the cumulative distribution function of $\sum_{i=0}^{L_p} d(\tau + i) - \sum_{i=0}^{L_p - L_r - 1} u(\tau + i)$.

Extensions

As illustrated above, information plays an important role when deciding about production and recovery. Many problems arise when there is not much information available about returns. On the other hand, additional information can be quite valuable. For the situation when product returns are dependent on the demands, which holds, for example, in the case of rented or leased products, a discussion can be found in Kiesmüller and van der Laan (2001). There it is assumed that the number of returns in a period τ depends on the number of demands in a previous period $\tau - \ell$, where ℓ is some fixed number of periods. If the probability that an item can be recovered is assumed to be known, then the number of returns can be estimated using this probability and the information about the known demands. Using this estimation for system control, it is shown that costs can be reduced compared to the situation where the dependency is ignored. Further it is illustrated that the variance of the inventory processes is reduced. Toktay et al. (2000) come to a similar conclusion. They take a queueing model approach and explicitly model the dependence relation between the demands and returns to investigate the value of return information (see Chapter 3 in this book for further details).

In the models discussed up to now, it is assumed that there is only one option to recover the returned products. But in many situations an old product can be reused in different ways, each yielding different costs and profits. A model with multiple recovery options, one disposal option, but no additional production facility is investigated in Inderfurth et al. (2001). Here the problem is to allocate the limited amount of recoverable products to the different recovery options. The structure of the optimal policy is extremely complicated, due to the inherent allocation problem in the case of scarce recoverables. But under a linear allocation rule, a fairly simple near-optimal policy exists which is characterized by a single dispose-down-to-level and a specific recover-up-to-level for each reuse option.

The research presented in this section is dealing with remanufactured products which are assumed to be as good as new. In Inderfurth (2002), a model is introduced which assumes that remanufactured products differ significantly from new ones and that higher-value new products are offered to the customer if there is a stock-out of remanufactured products. For a single-product, single-period problem with stochastic demands and returns and positive lead times, the optimal policy is determined. It is given by produce-up-to and recover-up-to order functions (of the on-hand serviceable inventories) for manufacturing and remanufacturing.

Other Approaches

A special case of an (S_p, S_r) policy is examined in Tagaras and Vlachos (2001) and Vlachos and Tagaras (2001). These publications refer to an inventory system with two replenishment modes (corresponding to production and recovery), one of which may act as an emergency supply channel which can deliver on short notice. An approximate cost model is provided which can be easily optimized with respect to the decision parameters. This model is used as the basis for an heuristic algorithm, which leads to solutions that are very close to the exact optimal solutions determined through simulation.

8.4 Discussion and Outlook

Inventory control for product recovery is very much different from traditional inventory control due to the highly variable and uncertain nature of the extra resource: product returns. The development of specialized inventory models is essential to analyze and understand the complicated dynamics of stochastic inventory control for product recovery. So far this chapter mainly dealt with *modeling* aspects of stochastic inventory control for product recovery management. Below we list some of the managerial implications of the modeling efforts.

Autonomous versus managed recovery

Specialized techniques should simultaneously determine trigger levels and quantities for production, recovery, and disposal operations given the characteristics of its specific environment, such as procurement lead time, the form of the product life cycle, and seasonal influences. In this context, we distinguished between autonomous recovery and managed recovery. Autonomous recovery (or push recovery) mainly relates to those situations in which minor operations against limited costs suffice for successful recovery and reuse. Managed recovery (or pull recovery) is appropriate if recovery operations are more costly and/or setup costs are present so that it is better to stock recoverables until they are really needed.

Split inventory analysis

In the case of autonomous recovery with only one stocking point (for serviceables) present, the inventory process can be cleverly split into two components: the back system and the front system. The front system depends on the demand and the return process, but is independent of the production orders. The front end can be interpreted as a modified demand process that acts on the back system. In this way, the inventory system can be viewed as a standard single-source system with demand process that takes on positive as well as negative values. In Subsections 8.2.3 and 8.2.4, we showed how this modelling framework applies to both single-period and multi-period decision problems. The managerial implication for the mail-order company of Case A is that in principle it could use standard techniques to calculate the initial order sizes. Finding the correct cost parameters, however, requires a thorough understanding of the underlying model, so this still could be quite problematic.

Naive netting versus sophisticated netting

The 'sophisticated netting' of the demand process that is used in the split inventory technique is very much unlike 'naive netting' through which a part of the expected demand rate is cancelled with the expected return rate. The resulting demand process is then input for traditional inventory control models. This is a very simple approach that unfortunately renders a very poor performance, unless return rates are very low. The variance that is introduced by the return process is completely ignored, while the variance of the demand process is moved away from its real value. It is therefore not recommended to use naive netting in practice unless return rates are very small.

Lead-time effects

In the case that managed recovery is more appropriate, things are more complicated, since we have to take into account non-zero recovery lead times and/or fixed setup costs for the recovery process. Only for very specific situations (no setup costs, fixed production lead time equals fixed recovery lead time) do we know the optimal structure of the inventory policy. In other cases, we have to rely on heuristics. Without fixed setup costs and neglecting disposals one can, as an approximation, reduce the multi-period decision problem to a single-period decision problem that is basically the standard 'news-vendor' problem. Crucial for a good policy is that the recovery decision and the production decision be based on different inventory positions that include different information sets.

Inventory systems that do include fixed recovery costs have mainly been modelled in continuous time rather than discrete time. An exact modeling framework, using the theory of Markov chains, for analyzing a wide variety of (heuristic) push and pull inventory policies was presented in Section 8.3.2. It can be shown that the cost performance of pull policies dominate the performance of push policies for most relevant values of the cost parameters. In practice, a push policy could still be preferred though, since it is easier to implement and its performance could be reasonable in the case of long and/or highly variable recovery lead times. Disposal policies are of importance only when return rates are close to or exceeding demand rates or products are very slow-moving.

Value of information

A crucial assumption of most available models to ensure tractability is that product returns are independent of product demands. In reality, though, there is always some dependency relation between product demands and returns. Knowledge about this dependency relation enables more sophisticated forecasts with respect to timing and quantity of product returns. If good forecasts are available these could be incorporated in the inventory policy and well improve system performance. The issue of forecasting and its impact on inventory management is studied in Chapter 3 of this book.

From a modelling perspective, it is sometimes convenient to assume that the time in market is negative exponentially distributed. This way it is sufficient to keep track of the *total* number of products in the market, rather than all individual products. In de Brito and Dekker (2003), it is shown by company data analysis that this assumption is not always according to real behavior. More research is needed to assess the impact of wrongly assuming exponential lags.

From the massive growth of the literature on inventory control for product recovery we may conclude that at least the scientific community believes that this field is worth studying. Due to the complexity of these systems, however, attention has been limited to single-product, single-component models, while a remanufacturing company like Volkswagen (dis)assembles and recovers thousands of components. At the same time, there seems to be a lack of communication between academics and practitioners, considering the very limited amount of empirical studies that have been conducted up to now.

More case studies will undoubtedly help in bridging the gap between theory and practice.

Managing Dynamic Product Recovery: An Optimal Control Perspective

Gudrun P. Kiesmüller[1], Stefan Minner[2], and Rainer Kleber[2]

[1] Faculty of Technology Management, Eindhoven University of Technology, P.O. Box 513, 5600 MB Eindhoven, The Netherlands
`G.P.Kiesmueller@tm.tue.nl`

[2] Faculty of Economics and Management, Otto-von-Guericke University Magdeburg, P.O.Box 4120, 39016 Magdeburg, Germany,
`minner@ww.uni-magdeburg.de`, `rainer.kleber@ww.uni-magdeburg.de`

9.1 Introduction

In the previous chapters, the impact of the presence of product returns on batching and safety inventory considerations has been addressed under stationary conditions. In this chapter, we will focus on a different aspect, namely the impact of dynamic phenomena. Whereas the insights provided by Chapters 7, 8, and 10 mainly help to improve operational decisions, the benefit of the dynamic optimal control framework lies in the assistance of operational as well as strategic/tactical decisions like competitive strategy, selection of recovery processes, investments in remanufacturing technology, product life cycle decisions, production planning under seasonal demand, etc. As a consequence of reducing model complexity, we will restrict the analysis to deterministic but time varying environments. These dynamic conditions are present in most practical applications, and in the following we will briefly illustrate this aspect using DaimlerChrysler's engine recovery operations (see Driesch et al., 1997).

DaimlerChrysler runs several operations for recovering parts from used cars, e.g. remanufacturing of used engines for Mercedes Benz cars at the plant Berlin-Marienfelde (MTC). Annually, about 12,000 engines from 28 classes and 800 different model variants are remanufactured. An ABC-classification revealed that 60% of the returns are contributed by 3 classes. Orders for remanufactured engines are placed by the main customer of the MTC, the Global Logistics Center (GLC), located in Germersheim, Germany. There, car dealers and licensed garages can exchange dysfunctional engines with remanufactured ones. This exchange induces most of the orders, while others have been sent to external customers or are sold via the internet. Besides the exchange process, a substantial part of the returns originate from collecting used engines from all over the world.

Dynamic issues, i.e. time dependent demands and returns, have to be considered for two reasons. First, demands for an engine class follow the shape of a product life cycle, starting with a phase of increasing sales, followed by the maturing phase and finally declining sales towards the end of a product's life cycle. Returns follow demands in a similar pattern, delayed by the usual lifetime of an engine and reduced by the fact that not all engines are returned. In the growing phase, demand for remanufactured engines is significantly higher than available returns and all returned cores are remanufactured. Later in the maturing phase, demand decreases and returns can exceed remanufacturing orders. Due to corporate policy, no reusable cores are disposed of and therefore have to be kept in inventory. This divides the product life cycle into two main phases, the first with insufficient cores and another one with excessive cores.

Another source of dynamics is seasonal effects in demands and returns, since customer behavior of buying cars (and therefore returning old cars) and having cars inspected in a garage heavily depend on the time of the year. As a consequence, there might be time intervals in which demands exceed returns, followed by time intervals where returns surpass demand. This issue is evaded by the stock-keeping policy, i.e. high inventory levels prevent stock-outs of cores.

Besides the phenomena observed in the engine remanufacturing case, several other real-life phenomena can require a dynamic modeling of product recovery decisions. Regarding the valuation of decisions, cost parameters will, in general, be time-varying. Exogeneous reasons are fluctuating commodity prices (e.g. metals), the expectation of increasing waste disposal costs, and the impact of inflation. The cost structure might also change endogenously as a result of decisions, e.g. exploiting learning curve effects in manufacturing and remanufacturing. Demands and returns depend on the life cycle of the product and therefore there is an interaction with market diffusion and pricing strategies (Simon, 1989). Further, competitive reaction and changing customer behavior (in general, depending on the life-cycle phase) have to be regarded.

As illustrated above, life cycles of products and seasonal effects can have a significant impact on returns as well as on demands. Therefore, it can be necessary to include the dynamic behavior of returns and demands into a quantitative model. There are few research results available that deal with this issue in recoverables inventory control. If dynamic aspects are considered, as done in dynamic lot-sizing (see Chapter 7) and stochastic dynamic inventory control (see Chapter 8), the analysis is mainly carried out numerically. Often restrictive assumptions have to be made, and, further, a discrete time structure has to be imposed that, to some extent, will influence the accuracy of the results. Since there is no obvious choice for such a natural time period, i.e. how to divide the planning horizon into discrete time buckets, especially for strategic decision problems with longer planning horizons of several years, a continuous model appears to be more appropriate. Further, instead of having to forecast parameters for each time bucket, continuous time parameters (described by functions of time) can be estimated from descriptive models, like diffusion models and learning curves, as an alternative to using approaches from continuous time econometrics (see Gandolfo, 1993, for the latter).

Many results are available for traditional continuous time production and inventory models (see Bensoussan et al., 1983, for an overview). For models with a linear cost structure, as assumed for our recovery model, the optimal solution for a model with a single supply mode is to deplete initial inventories and then to manufacture for demand. A build-up of inventories only occurs if manufacturing costs are quadratic (or, more generally, convex), when it is optimal to smooth production and therefore to build up inventories when demand is low and to deplete these inventories when demand is high.

In the following, we first present a simple prototype of a dynamic recovery model (Section 9.2.1) and briefly discuss the solution methodology (Section 9.2.2). The results for the basic model are presented in Section 9.2.3 and are illustrated with a numerical example in 9.2.4. Then, in Section 9.3, several extensions that have been carried out are presented and the main findings discussed. Finally, in Section 9.4, we illustrate important streams of future research in reverse logistics where substantial strategic decision support can be provided by a continuous time-optimal control framework.

9.2 A Basic Model of Dynamic Product Recovery

9.2.1 Model Formulation

In this section, we analyze a variant of the single-product, single-stage product recovery system as it has been introduced in the previous chapters. In contrast to the periodic approach used in Section 8.3.1, the model to be presented incorporates dynamics in continuous time. The system faces deterministic demand and return rates within a finite planning horizon= $[0, T]$. Both are given by non-negative, continuously differentiable functions of time $d(\tau)$ and $u_r(\tau)$, respectively. As an example, consider the following functions for demand and returns over a planning horizon of $T = 4\pi$:

$$d(\tau) = 1 + 0.5\sin(\tau) \quad \text{and} \quad u_r(\tau) = \begin{cases} 0 & \text{if } t < \pi \\ 0.7d(\tau - \pi) & \text{otherwise} \end{cases} . \quad (9.1)$$

The demand function consists of a time-independent base level of 1, which is superimposed by some seasonal influence introduced by a cyclical part $0.5\sin(\tau)$. A fraction of 70% of earlier demand becomes available after staying with the customer for a period of π. This scenario is depicted in Figure 9.1. Demands have to be satisfied instantaneously, i.e. backordering is not permitted. The state of the system at time τ is described by the serviceables (net) inventory $I_n(\tau)$ and the recoverables inventory $I_u(\tau)$. Initial values for the

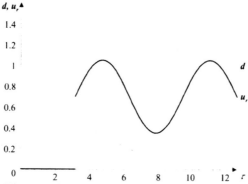

Fig. 9.1. Demands and returns in the example

state variables are given by $I_n(0) = 0$ and $I_u(0) = 0$. Decisions can be implemented at each instant of time, i.e. are given by rates, leading to a continuous and instantaneous change of the system's state because we assume zero lead times. Serviceables inventory can be replenished by producing new items with rate $p(\tau)$ or by remanufacturing returns $r(\tau)$. Thus, the corresponding state transition equation is represented by a differential equation where the rate of change in the serviceables inventory ($\dot{I}_n = \frac{\partial I_n}{\partial \tau}$) equals the sum of production and remanufacturing minus demand rate at time τ

$$\dot{I}_n(\tau) = p(\tau) + r(\tau) - d(\tau). \tag{9.2}$$

In addition to remanufacturing or keeping returns in the recoverables inventory, these can also be disposed of with rate $w(\tau)$. Therefore, the state transition equation for the recoverables inventory is given by

$$\dot{I}_u(\tau) = u_r(\tau) - r(\tau) - w(\tau), \tag{9.3}$$

i.e. the rate of change in the recoverables inventory $\dot{I}_u(\tau)$ is given by the return rate $u_r(\tau)$ minus the sum of remanufacturing $r(\tau)$ and disposal rates $w(\tau)$. Since we do not allow for backlogging, the serviceables inventory has to be non-negative:

$$I_n(\tau) \geq 0. \tag{9.4}$$

The availability of recoverable items limits both remanufacturing and disposal decisions, being expressed by non-negativity of the recoverables inventory:

$$I_u(\tau) \geq 0. \tag{9.5}$$

The state of the system at the end of the planning horizon is given by $I_n(T) = 0$ and $I_u(T) = 0$ because we assume all obligations to end at the planning horizon T. Moreover, production, remanufacturing, and disposal rates (the control variables) are non-negative:

$$p(\tau) \geq 0,\ r(\tau) \geq 0,\ w(\tau) \geq 0. \tag{9.6}$$

The objective is to satisfy customer demands over the planning horizon T at minimal total undiscounted costs. We assume a linear, time-independent cost structure. In order to guarantee a meaningful solution, we assume that the sum of unit production cost c_p and disposal cost c_w must exceed unit remanufacturing cost c_r:

$$c_p + c_w > c_r. \tag{9.7}$$

Otherwise, disposing of a returned item and producing a new one would always be favorable to remanufacturing it. In this case, the model would reduce to a pure production-inventory model with linear costs for which the optimal solution is to synchronize production with demand and to dispose of all returns.

Serviceables and recoverables inventories are subject to inventory holding costs h_s and h_u respectively, each per unit and unit of time. Furthermore, we assume that the holding cost rate for serviceables is larger than the holding cost rate for recoverables items (i.e. $h_s > h_u > 0$) because, in general, more capital is tied up in serviceables. Moreover, if recoverable holding costs are larger than serviceable holding costs, then it is obvious that recoverables inventory is never built up and returns are disposed of and remanufactured at once. Additionally, it is reasonable that out-of-pocket costs are less for recoverables than for serviceables. In the introductory case, for instance, recoverables are stored outside in open boxes whereas serviceables are treated more carefully.

The following linear optimal control model with two state and three control variables has to be solved subject to the state equations, two pure state constraints, and initial and terminal conditions for the state variables, as well as non-negativity conditions of state and control variables:

$$\min_{p(\tau),r(\tau),w(\tau)} \int_0^T \left(c_p p(\tau) + c_r r(\tau) + c_w w(\tau) + h_s I_n(\tau) + h_u I_u(\tau) \right) d\tau$$

$$\begin{aligned}
\text{s.t.} \quad & \dot{I}_n(\tau) = p(\tau) + r(\tau) - d(\tau) && \forall \tau \in [0,T], \\
& \dot{I}_u(\tau) = u_r(\tau) - r(\tau) - w(\tau) && \forall \tau \in [0,T], \\
& I_n(0) = 0,\ I_u(0) = 0,\ I_n(T) = 0,\ I_u(T) = 0, && (9.8) \\
& I_n(\tau) \geq 0,\ I_u(\tau) \geq 0 && \forall \tau \in [0,T], \\
& p(\tau) \geq 0,\ r(\tau) \geq 0,\ w(\tau) \geq 0 && \forall \tau \in [0,T].
\end{aligned}$$

9.2.2 Methodology

Optimal control theory has been developed as an extension of the calculus of variations. Starting with the pioneering work by Pontryagin et al. (1962), it has reached a wide range of applications in economics and management and technical sciences (see Bensoussan, Hurst, and Näslund, 1974, Seierstad and Sydsæter, 1987, Kamien and Schwartz, 1991, or Sethi and Thompson, 2000). It can be compared with dynamic programming methods developed by Bellman (1957) at about the same time but, according to Feichtinger and Hartl (1986), a main advantage of optimal control is the possibility of gaining insights into the general structure of solutions for the entire problem class.

Although there are extensions to solve discrete time problems, optimal control literature mainly deals with continuous time systems. A system in this sense is characterized by one or more state variables which are changed by external influences or by choosing control variables in an appropriate way in order to maximize (or minimize) a given objective function. Thereby both state as well as control variables can be subject to constraints which have to be considered in determining the best solution, given by optimal trajectories (functions of time) of the state and control variables.

Pontryagin's Maximum Principle provides a set of necessary conditions for an optimal solution. For a linear problem (as introduced in the previous section), these are also sufficient if certain regularity conditions hold, which is the case for our model (see Minner and Kleber, 2001). Using the Maximum Principle, the dynamic problem is decomposed into an infinite sequence of interrelated static problems, one for each time instant τ. These are connected by introducing co-state (also called adjoint) variables, being interpreted as the shadow price of changing the system state.

The static objective function, called Hamiltonian, is constructed in a way that it measures the total effects of the decisions at any time τ on the objective. These have both a direct and an indirect effect. The direct effect is given in the objective of problem (9.8) by the (static) cost rate at time τ. Since decisions at this time also have an influence on the future by changing the system's state, e.g. by decreasing or increasing the inventory level, as an indirect effect the value of the state changes, being measured by its rate of change times the corresponding shadow price (given by the co-state).

As in dynamic programming, where an optimal decision is taken at each stage (instant of time), assuming that up to that point all decisions have been taken optimally and the same will hold for future decisions, the Hamiltonian has to be maximized at each instant of time, subject to relevant constraints on control and state variables. This is applied by using standard methods of nonlinear programming. Further necessary conditions include the rate of change of the co-state variables. For a continuously differentiable objective function, a system of differential equations, possibly including the states, for the co-state variables has to be solved.

The complicating issue when applying Pontryagin's Maximum Principle is that an analysis of the necessary optimality conditions in general does not yield a closed expression for the optimal control functions. As a consequence, these conditions can only be used as a vehicle for stating properties of an optimal solution. The approach we are using in the next sections is the following. First, we distinguish different cases with respect to the serviceables and recoverables inventory. Within these intervals where inventory levels are positive or zero, and given that we do not switch to a different case, we can state closed form expressions for the control functions. Since remaining in one case in general will not be optimal for the whole planning horizon, we check under which circumstances switching to a different case does not violate the necessary conditions. Together with using other properties that characterize the intervals between the so-called transition (switching) points, we are able to derive a solution algorithm that numerically determines the optimal solution.

9.2.3 Optimal Recovery Policy

In a dynamic situation when the return rate is larger than the demand rate, it is evident that it is profitable, to some extent, to collect returns for satisfying future demands. There exists a trade-off between the remanufacturing cost advantage and the holding cost for recoverables which has to be balanced. Therefore, it is an important issue to determine the optimal starting point for keeping excessive recoverables for later use.

In the following, we will not explicitly show how Pontryagin's Maximum Principle can be used to determine these optimal starting points and the optimal policy structure because this involves a lot of technical and mathematical work. We will only present the important results. For proofs and a more detailed discussion of the solution algorithm, we refer to Minner and Kleber (2001).

Optimal Policy Structure

Since there is no capacity restriction for remanufacturing and production, and due to a holding cost advantage for recoverables, it is always superior to hold an item in recoverables instead of in servicable inventory, and to postpone remanufacturing and production until an item is demanded. This leads to $I_n(\tau) = 0$ for all $\tau \in [0, T]$ in the optimal solution.

Therefore, in an optimal solution, only the following system states are possible.

- Case 1: $I_n(\tau) = 0$ and $I_u(\tau) > 0$
- Case 2: $I_n(\tau) = 0$ and $I_u(\tau) = 0$

Analyzing the optimality conditions given by Pontryagin's Maximum Principle yields the optimal policy structure denoted with $(p^*(\tau), r^*(\tau), w^*(\tau))$, for a given system state (and assuming that we do not leave the case).

- Case 1:
  $p^*(\tau) = 0$, $r^*(\tau) = d(\tau)$ and $w^*(\tau) = 0$
- Case 2:
  $p^*(\tau) = \max\{d(\tau) - u_r(\tau), 0\}$, $r^*(\tau) = \min\{d(\tau), u_r(\tau)\}$
  and $w^*(\tau) = \max\{u_r(\tau) - d(\tau), 0\}$

Note that if units are disposed of, this is done instantaneously. This optimal structure is intuitively clear, but the power of Pontryagin's Maximum Principle is that it additionally enables us to determine the optimal time points for entering and leaving the cases mentioned above.

Optimal Transition Points

The time points where transitions between cases occur are called transition points. Since it is not possible to give an explicit analytical expression for the optimal transition points that holds for any demand/returns situation, we rather provide conditions and properties to be used in an algorithm that determines the optimal solution. Given the optimal policy structure for Case 1 and Case 2, we have to answer the question of in which time intervals are the inventories positive or zero. We recognize that applying the optimal policy in Case 1 will lead to a transition into Case 2 once the inventory level reaches zero. Case 2 will not be left automatically, however staying in Case 2 will contradict the necessary conditions for an optimal solution at certain points. Candidates where it is necessary to have a positive recoverables inventory are identified by the location property. Using these points in time, a complete interval of Case 1 can be constructed. This construction underlies certain conditions which are described by an integral property and a maximal length property.

Location Property

It is easy to see that time intervals with a positive recoverables inventory are located around intersection points of the demand and return rate. Time intervals with $I_u(\tau) > 0$ are related to the points $\theta_u \in \mathcal{N}$ with

$$\mathcal{N} := \{\theta_u \mid d(\theta_u) = u_r(\theta_u), d(\theta_u - \epsilon) < u_r(\theta_u - \epsilon),$$
$$d(\theta_u + \epsilon) > u_r(\theta_u + \epsilon), 0 \le \theta_u \le T\}. \qquad (9.9)$$

for a sufficient small value of ϵ. Further, we define points $\theta_o \in \mathcal{M}$ with

$$\mathcal{M} := \{\theta_o \mid d(\theta_o) = u_r(\theta_o), d(\theta_o - \epsilon) > u_r(\theta_o - \epsilon),$$
$$d(\theta_o + \epsilon) < u_r(\theta_o + \epsilon), 0 \le \theta_o \le T\}. \qquad (9.10)$$

These two sets identify intersection points where returns intersect demands from above (intervals with excessive returns are followed by excessive demands) and from below (excessive demands are followed by excessive returns).

Integral Property

Integral conditions refer to (non)negativity of the inventories during Case 1 intervals, and zero recoverables inventory at the end of such an interval. Let us denote a time interval with positive recoverables inventory with $[\tau_{s,1}, \tau_{e,1}]$.

Then, cumulative returns have to be larger than cumulative demands during the whole time interval $[\tau_{s,1}, \tau_{e,1}]$, and, at the end of such an interval, the cumulative return rate has to match the cumulative demands. This leads to the following integral condition:

$$\int_{\tau_{s,1}}^{t} (u_r(s) - d(s))\, ds \geq 0 \quad \forall t \in [\tau_{s,1}, \tau_{e,1}] \quad \text{and} \quad \int_{\tau_{s,1}}^{\tau_{e,1}} (u_r(s) - d(s))\, ds = 0.$$
(9.11)

Interval Length Property

The maximum length of a collection interval is determined by a marginal criterion. If an additional return unit is collected and stored at the beginning of the interval $\tau_{s,1}$, this additional unit has to be stored until the end of the interval $\tau_{e,1}$. Therefore, it is only beneficial as long as the associated holding costs $h_u(\tau_{e,1} - \tau_{s,1})$ do not exceed the recovery cost advantage $c_p + c_w - c_r$, giving the following criterion for the optimal interval length $\tau_{e,1}^* - \tau_{s,1}^*$:

$$\tau_{e,1}^* - \tau_{s,1}^* \leq \frac{c_p + c_w - c_r}{h_u}.$$
(9.12)

Note that in the optimal solution the length of a time interval in Case 1 can be smaller than the upper limit as given in (9.12) due to the integral condition. For example, if there are not enough returns available to save further units for future demands, then a Case 1 time interval can be shorter than $\frac{c_p + c_w - c_r}{h_u}$ and such an interval will start or end at an intersection point $\theta_o \in \mathcal{M}$ or at the boundaries of the planning horizon.

Based on the location property, the integral property, and the time length property, a solution algorithm can be constructed to compute the optimal starting and end points of Case 1 intervals. In the first step of the algorithm, intervals with maximal possible length (upper limit in (9.12)) are constructed around the intersection points in $\mathcal{N}$. If the obtained intervals in step 1 are overlapping, or they do not satisfy the integral condition, then their length is reduced in the second part of the algorithm until all necessary conditions are satisfied. Afterwards, the optimal production, remanufacturing, and disposal rates for the whole planning horizon can be determined. For a proof of the above mentioned conditions and the solution algorithm, we refer to Minner and Kleber (2001).

As a by-product of the analysis, the optimal trajectory of the adjoint variable is determined. Let $\lambda(\tau)$ denote the co-state variable associated with the recoverables inventory movement equation. It gives the value (shadow price) of an additionally returned item at τ. A unit that is immediately disposed of has a value of $\lambda = -c_w$ whereas a unit that is immediately used for remanufacturing and satisfaction of demand has a value of $\lambda = c_p - c_r$. Within an interval with positive inventory, the value of a returned unit increases by the holding cost rate h_u, because a later return causes less inventory holding costs and therefore has a higher value. This economic interpretation of the value of a return is also important for accounting issues (see Chapter 11) as it illustrates that there is no unique value of a return in a dynamic environment and that the value rather depends on the timing of product return and usage for recovery.

9.2.4 Numerical Example

In this section, the results of the basic model will be illustrated for the example as given in (9.1). According to the location property, intervals with positive recoverables inventory will be located around time points $\theta_u \in \mathcal{N} = \{5.92, 12.21\}$. When not having maximal length, they will either start or end at time points $\theta_o \in \mathcal{M} = \{3.50, 9.79\}$ or at $\tau = 0$ or $\tau = T$.

Cost parameters are set as follows: $c_p = 2$, $c_r = 1$, $c_w = 1$, and $h_u = 1$. Therefore, the Case 1 interval length will not exceed 2. Using the solution algorithm shown in Minner and Kleber (2001), the following two collection intervals are determined: $[\tau_{s,1}^{*1}, \tau_{e,1}^{*1}] = [4.85, 6.85]$ and $[\tau_{s,1}^{*2}, \tau_{e,1}^{*2}] = [11.82, T]$. The first interval has maximum length, but the second one is shorter because it reaches the planning horizon where no further demand is available. Figure 9.2 shows the optimal Case 1 intervals as well as the optimal development of the co-state λ.

Outside optimal Case 1 intervals, demand is satisfied either from remanufacturing and production if it exceeds the return rate, or from remanufacturing alone if the opposite applies. In the first case, an additional return would have a value of $c_p - c_r = 1$ since it could be used immediately for replacing production. In the second case, the additional return ought to be disposed of, thus having a value of $-c_w = -1$. Inside Case 1 intervals, demand is met by remanufacturing current or previously collected returns. The value of an

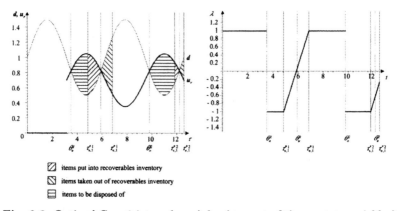

☑ items put into recoverables inventory
☒ items taken out of recoverables inventory
☰ items to be disposed of

Fig. 9.2. Optimal Case 1 intervals and development of the co-state variable λ

additional (marginal) return increases with time since the time it prevails in inventory reduces.

9.3 Model Extensions

9.3.1 Discounting

Especially in long-term planning problems, a discounted cash flow criterion appears to be more appropriate than an average cost criterion because of the more realistic modeling of the time value of payments. In the basic product recovery model, applying a discounted cash flow criterion is straightforward. Let $\alpha > 0$ denote the (continuous time) interest rate. Then, each payment made at time τ has to be discounted by the factor $e^{-\alpha\tau}$ to express the present value of this payment at time zero. In a cash flow approach, h_s and h_u represent out-of-pocket holding costs, whereas in the average cost approach the cost of capital is included as well (see Chapter 11 for a discussion). In order to assure a meaningful solution, we assume that holding an item in recoverables inventory is more expensive than the interest gained from delaying disposal, i.e. $h_u > \alpha c_w$. Otherwise, items are never disposed of but are kept in inventory instead, regardless of whether there is any future need for them.

The total cost in problem (9.8) has to be replaced by

$$\int_0^T e^{-\alpha\tau}\Big(c_p p(\tau) + c_r r(\tau) + c_w w(\tau) + h_s I_n(\tau) + h_u I_u(\tau)\Big)d\tau. \qquad (9.13)$$

In the analysis of necessary conditions, the current value Hamiltonian has to be used. The main difference compared to the basic model is the evolution of the adjoint, i.e. the value of a returned unit. Under a discounted cash flow criterion,

$$\dot{\lambda}(\tau) = \alpha\lambda(\tau) + h_u \qquad (9.14)$$

within intervals with positive recoverables inventory. This first-order differential equation indicates that the value of a later return not only consists of direct savings in inventory holding costs but it additionally postpones payments and thus saves opportunity costs of capital with an amount which equals the interest rate on the current value of returns.

Solving this differential equation only has an impact on the interval length property. Instead of the maximum optimal length presented for the undiscounted model, we find (see Kleber et al. 2002)

$$\tau_{e,1}^* - \tau_{s,1}^* \le \frac{1}{\alpha}\cdot ln\left(\frac{\alpha(c_p - c_r) + h_u}{-\alpha c_w + h_u}\right). \qquad (9.15)$$

9.3.2 Backordering of Demand

In comparison to the basic model, additional cost savings can be realized if it is permitted to delay the fulfillment of demands. This may happen in cases in which demand is satisfied by producing new items when there are many returns in the near future. On the other hand, additional costs are caused by backorders, e.g. price discounts for a customer that is willing to wait. Backlogs are therefore valued by a unit cost rate c_b with $c_b > h_u > 0$. The cost trade-off for accepting backorders is between the remanufacturing cost advantage from satisfying current demands from future returns and the costs of backordering current demands until these returns become available.

Serviceables are differently valued for positive and negative net inventory and the objective function has to be modified accordingly

$$h(I_n) = \begin{cases} h_s I_n & \text{if } I_n > 0 \\ -c_b I_n & \text{if } I_n \leq 0 \end{cases}. \tag{9.16}$$

Since serviceables net inventory may become negative in case backorders occur, condition (9.4) is not relevant. Compared to (9.8), the optimal control problem with backlogging has only a single pure state constraint. Additionally, the total cost function is not continuously differentiable because of the discontinuity of the holding cost function (9.16) at $I_n = 0$. Therefore, an extension of the solution methodology that addresses problems with discontinuous objective functions is required (see Hartl and Sethi, 1984). Then, differential inclusions for the co-state variables are obtained.

Under backordering, we find an additional case with respect to the values of serviceables and recoverables inventories. Since it can never be optimal from an economic point of view to allow backorders when recoverables are held on stock, this additional case and the associated decisions are

- Case 3: $I_n(\tau) < 0$ and $I_u(\tau) = 0$
  $p^*(\tau) = 0$, $r^*(\tau) = u_r(\tau)$ and $w^*(\tau) = 0$.

For a detailed description of the analysis and a solution algorithm, we refer to Kiesmüller et al. (2000).

Location Property

Intervals where backorders are beneficial are located around intersection points where demands intersect returns from above as described by the set $\mathcal{M}$.

Integral Property

Let us denote a time interval with negative serviceable inventory with $[\tau_{s,3}, \tau_{e,3}]$. Then, cumulative returns have to be less than cumulative demands during the whole time interval $[\tau_{s,3}, \tau_{e,3}]$ and, at the end of such an interval, the cumulative return rate has to match the cumulative demands.

$$\int_{\tau_{s,3}}^{t} (u_r(s) - d(s))\, ds \le 0 \quad \forall t \in [\tau_{s,3}, \tau_{e,3}] \quad \text{and} \quad \int_{\tau_{s,3}}^{\tau_{e,3}} (u_r(s) - d(s))\, ds = 0.$$

$$(9.17)$$

Interval Length Property

Allowing backorders is only profitable if backorder costs over the length of the time interval do not exceed the recovery cost advantage. Using this trade-off, we can determine an upper limit for the optimal length $\tau_{e,3}^* - \tau_{s,3}^*$ of a time interval of Case 3:

$$\tau_{e,3}^* - \tau_{s,3}^* \le \frac{c_p + c_w - c_r}{c_b}. \qquad (9.18)$$

Condition (9.18) together with (9.12) from the basic model reveal that the upper limits for the optimal time intervals in Case 1 and Case 3 can be quite large for a small backorder cost rate and/or a small recoverable holding cost rate. In this case there is no time interval with positive length between Case 1 and Case 3 where we have zero inventories. In order to determine the maximum length of such consecutive intervals, the marginal cost trade-off is the following. Assume an interval with backordering (Case 3) is followed by a collection interval (Case 1). At the point in time where Case 3 ends and Case 1 starts ($\tau_{e,3}^* = \tau_{s,1}^*$), the excessive return unit can either be used to satisfy an additional backorder of demands from $\tau_{s,3}^*$ associated with backordering costs $c_b(\tau_{e,3}^* - \tau_{s,3}^*)$ or to satisfy a future demand at $\tau_{e,1}^*$ associated with holding costs $h_u(\tau_{e,1}^* - \tau_{s,1}^*)$. In an optimal solution, one has to be indifferent between these two options. This yields the following relation between the lengths of consecutive intervals.

$$\frac{\tau_{e,3}^* - \tau_{s,3}^*}{\tau_{e,1}^* - \tau_{s,1}^*} = \frac{h_u}{c_b} \qquad (9.19)$$

In the case that a collection interval is immediately followed by a backorder interval, the same cost trade-off holds for the excessive demand unit at $\tau_{e,1}^* = \tau_{s,3}^*$. For a detailed description of the analysis and a solution algorithm, we refer to Kiesmüller et al. (2000).

9.3.3 Lead Times

In this section, we extend the basic model and additionally allow positive lead times, not necessarily equal, for the remanufacturing and the production process. The lead times have an effect on the dynamics of the serviceable inventory. Instead of (9.2) we get the following equation

$$\dot{I}_n(\tau) = p(\tau - L_p) + r(\tau - L_r) - d(\tau), \qquad (9.20)$$

with L_p denoting the production lead time and L_r representing the remanufacturing lead time.

In contrast to stochastic models with different lead times, where it is quite difficult to find the optimal recovery policy (see Chapter 8.3), the optimal recovery policy for a deterministic model as described above can easily be obtained after a straightforward transformation into a model with zero lead times, which can be solved with the methods described above. Depending on the lead-time relation, different transformations of the demand rate, the servicable inventory, and the production rate have to be considered. We have to distinguish between equal lead times, larger production lead time, and larger remanufacuring lead time. The same transformations can be used if additionally backorders are allowed. For a more detailed description of these transformations we refer to Kiesmüller (2003).

9.3.4 Capacity Constraints

Reconsidering the results of the basic model, there exists a number of reasons that require highly flexible processes. First, since the serviceables inventory is not used, production and remanufacturing rates are synchronized with demand and undergo (in sum) the same variations. Second, in intervals with a positive recoverables inventory, demand is served from remanufacturing alone, as production is zero. Further on, when the recoverables inventory runs empty, the production rate jumps from zero up to the difference of demand and return rates.

Under the presence of capacity constraints, these processes will not be that flexible and therefore smoothing of manufacturing and remanufacturing volumes is required. Gaimon (1988) presents an application of optimal control to a production/inventory model where, in addition to production quantities, optimal prices and capacity expansion are determined in order to maximize total profits. Another option to smooth processes is to use convex cost functions, as applied by Kistner and Dobos (2000) for a product recovery system.

Capacity constraints can be present in different ways. If production and remanufacturing take place in a common facility, a joint capacity usage can be limited. Given total capacity $\bar{a}$ and capacity requirements coefficients a_p and a_r for both processes. $a_p p(\tau) + a_r r(\tau) \leq \bar{a} \ \forall \tau \in [0, T]$ must hold. If the processes take place in different shops, independent capacity constraints, i.e. maximum production $\bar{p}$ or remanufacturing rate $\bar{r}$, are given. A similar approach can be applied if, due to legislative regulations, the disposal rate is limited, as well.

In the following, we show the effects of a single capacity constraint in the production process.

$$p(\tau) \leq \bar{p} \quad \forall \tau \in [0, T] \tag{9.21}$$

The respective optimization problem is given by (9.8) with additional constraint (9.21). Since both sources to satisfy demand are limited, a feasible solution exists if the cumulative demand does not exceed the sum of maximal production and cumulative returns, i.e.

$$\int_0^\theta d(\tau) d\tau \leq \int_0^\theta (\bar{p} + u_r(\tau)) d\tau \quad \forall \theta \in [0, T]. \tag{9.22}$$

Optimal Policy Structure

The analysis of bottleneck situations adds another motive for holding recoverables inventory. If demand exceeds current production and remanufacturing capacity

$$d(\tau) > \bar{p} + u_r(\tau), \tag{9.23}$$

it is necessary to have a positive (serviceables or recoverables) inventory at τ. An optimal solution must therefore answer the question of which inventory to use. In a pure production/inventory model, the solution is to produce in advance to cover later bottleneck situations and to hold items in a serviceables inventory. In a model with recovery, a second option exists which is to save returns for use during such a bottleneck situation. This is especially attractive because remanufacturing is not limited and holding costs are less for recoverables than for serviceables.

The policies for Case 2 as described in Section 9.2.3 still apply. Due to the fact that additional returns would save holding costs when the capacity constraint becomes binding, they can have a higher value (measured by the co-state λ) than $c_p - c_r$. Therefore, a Case 1 situation can show one of the following two subcases:

- Case 1(a): $I_n(\tau) = 0$, $I_u(\tau) > 0$ and $\lambda(\tau) \leq c_p - c_r$
  $p^*(\tau) = 0$, $r^*(\tau) = d(\tau)$ and $w^*(\tau) = 0$
- Case 1(b): $I_n(\tau) = 0$, $I_u(\tau) > 0$ and $\lambda(\tau) > c_p - c_r$
  $p^*(\tau) = \min\{d(\tau), \bar{p}\}$, $r^*(\tau) = \max\{d(\tau) - \bar{p}, 0\}$ and $w^*(\tau) = 0$

In Case 1(b), production is used instead of remanufacturing in order to save more returns than in Case 1(a) for a later use. The recoverables stock therefore only decreases if (9.23) holds.

In contrast to the basic model, it can be optimal to have a positive serviceables stock. This is only possible connected with a positive recoverables stock, and represents a further inventory case, denoted by Case 4. The corresponding optimal decisions are

- Case 4: $I_n(\tau) > 0$ and $I_u(\tau) > 0$
  $p^*(\tau) = \bar{p}$, $r^*(\tau) = 0$ and $w^*(\tau) = 0$

In a Case 4 situation, demand is satisfied by producing at the upper limit and thereby enlarging the serviceables stock if $d(\tau) < \bar{p}$ and reducing it if $d(\tau) > \bar{p}$.

The general solution is characterized by a combination of the two motivations for stock keeping. This complicates the solution process because in intervals with many returns followed by a bottleneck situation (9.23), stock keeping can also be motivated by using the recovery cost advantage. In the following, since we concentrate on the effects of capacity constraints, we restrict to situations where the recovery cost advantage plays no role as motivation for stock keeping. This is especially the case if returns never exceed demand within the planning horizon, i.e. $u_r(\tau) < d(\tau) \ \forall \tau \in [0,T]$, because independently of the capacity constraint all returns will be used for serving demand. This limitation has two direct implications. First, Case 1(a) intervals, where the cost advantage of remanufacturing motivates stock keeping, do not appear in an optimal solution. And second, production capacity always becomes binding in intervals where (9.23) holds. In the following, these will be called bottleneck intervals.

Optimal Case Transitions

In an optimal solution, Case 2 cannot be followed by Case 4 or vice versa, which means that before starting an interval where both inventories are positive (Case 4), there must be an interval where only returns are collected (Case 1(b)). Since in a Case 4 interval returns are saved but not used, there can only be a transition back to Case 1(b). Therefore, the following sequences can occur in an optimal solution.

Case 1(b) → Case 2 → Case 1(b) and Case 1(b) → Case 4 → Case 1(b)

Location property

Transitions from Case 1(b) to Case 2 only take place at the end of bottleneck intervals, being characterized by intersection points $\theta_r \in \mathcal{R}$ of demand rate with the sum of maximal production and return rate, for which it also must hold that after θ_r current production and remanufacturing capacity is sufficient to satisfy demand, i.e.

$$\mathcal{R} := \{\theta_r \mid d(\theta_r) = \bar{p} + u_r(\theta_r), d(\theta_r - \epsilon) > \bar{p} + u_r(\theta_r - \epsilon),$$
$$d(\theta_r + \epsilon) < \bar{p} + u_r(\theta_r + \epsilon), 0 \leq \theta_r \leq T\}. \tag{9.24}$$

The time of transitions from Case 2 to Case 1(b) depends on how many items have to be stored to cover bottleneck intervals and how long it takes to collect them, and can therefore not be given in a similar way. Transitions from Case 1(b) to Case 4 require demand to be less than the capacity constraint and vice versa from Case 4 to Case 1(b). Therefore, Case 4 intervals will reside around time points $\theta_s \in \mathcal{S}$, where it is also optimal to have a positive recoverables inventory, and at which demand increases and equals production capacity.

$$\mathcal{S} := \{\theta_s \mid d(\theta_s) = \bar{p}, d(\theta_s - \epsilon) < \bar{p}, d(\theta_s + \epsilon) > \bar{p}, I_u(\theta_s) > 0, 0 \leq \theta_s \leq T\} \tag{9.25}$$

Integral Property

Since there exists a holding cost advantage, bottlenecks are primarily served using previously collected returns. The size of the bottleneck determines how many returns to save and when to start collecting returns. The period from this point until the end of a bottleneck interval is called the collection interval. Building up a serviceables stock during the collection interval can be used in order to shorten it, because in Case 4 more returns are stored than in Case 1(b). Therefore, for each bottleneck situation given by its end points (collected in set $\mathcal{R}$) there exists an optimal sequence of intervals of Case 1(b) and Case 4, starting and ending with a Case 1(b) interval. Let $\tilde{\tau}_{s,1b}$ denote the time where collecting returns starts first for a given bottleneck and $\tilde{\tau}_{e,1b} \in \mathcal{R}$ the end point of the respective bottleneck interval. Since there may be several points $\theta_s \in S$ within the collection period $[\tilde{\tau}_{s,1b}, \tilde{\tau}_{e,1b}]$, we assume that there are $n \geq 0$ Case 4 intervals $[\tau_{s,4}^i, \tau_{e,4}^i]$, $i = 1, 2, ..., n$. Analogous to (9.11), the following integral condition must hold.

$$\int_{\tau_{s,4}^i}^{t} (\bar{p} - d(\tau))d\tau \geq 0 \quad t \in [\tau_{s,4}^i, \tau_{e,4}^i]$$

$$\text{and} \quad \int_{\tau_{s,4}^i}^{\tau_{e,4}^i} (\bar{p} - d(\tau))d\tau = 0, \quad i = 1, 2, ..., n \tag{9.26}$$

Comparing the optimal policies between a Case 1(b) and a Case 4 interval, in the latter case additional returns can be stored with a quantity of $\int_{\tau_{s,4}^i}^{\tau_{e,4}^i}(\max\{\bar{p} - d(\tau), 0\})d\tau$. Using this result, integral conditions regarding the recoverables inventory can be given also. Skipping the non-negativity condition, the inventory is emptied at the end of a bottleneck situation if

$$\int_{\tilde{\tau}_{s,1b}}^{\tilde{\tau}_{e,1b}} (r(\tau) - \max\{d(\tau) - \bar{p}, 0\})d\tau + \sum_{i=1}^{n} \int_{\tau_{s,4}^i}^{\tau_{e,4}^i} (\max\{\bar{p} - d(\tau), 0\})d\tau = 0. \tag{9.27}$$

Interval Length Property

Increasing the length of a Case 4 interval leads to a decreasing collection period. But this is only profitable as long as the induced serviceable holding costs lead to a higher reduction of recoverables holding costs. Using this trade-off, a balancing equation which is valid except for situations described later can be given

$$h_s(\tau_{e,4}^i - \tau_{s,4}^i) = h_u(\tau_{e,4}^i - \tilde{\tau}_{s,1b}), \quad i = 1, 2, ..., n. \tag{9.28}$$

The reasoning behind marginal criterion (9.28) is as follows. Assume there are two possibilities to satisfy the last demand unit of a bottleneck interval. This can be done either by reducing $\tilde{\tau}_{s,1b}$ (increasing the collection period) or by decreasing $\tau_{s,4}^i$ (increasing the length of the Case 4 interval). Independent of this (marginal) decision, the same amount of returns will be on stock at $\tau_{e,4}^i$, i.e. at the end of the Case 4 interval. Therefore, only holding costs up to this time need to be compared, and in an optimal solution, one has to be indifferent between both options. As an interesting implication of (9.28), Case 4 interval lengths will increase during a collection period.

There exist two types of situations where (9.28) does not hold. First, the collection interval starts at $\tilde{\tau}_{s,1b} = 0$, or at another bottleneck situation, i.e. $\tilde{\tau}_{s,1b} = \theta_r \in R$. Second, a Case 4 interval may not be extended because if it is integral property (9.26) will be violated. A detailed consideration of such situations is presented in Kleber (2002).

Optimal Recovery Policy

Adding a capacity constraint to the basic model leads to an additional motivation for keeping stock, namely to prevent against bottleneck situations, which makes the solution process more complicated. In contrast to the results presented in Section 9.2.3, in general both inventories will be used. Differences in holding cost rates lead to a preference towards the usage of the recoverables inventory, but using the serviceables inventory reduces costs as it shortens the collection interval. The optimal policy balances this trade-off by using the developed interval length property.

Effects of a Remanufacturing Constraint

Now we will shortly sketch the implications of a capacity restriction for the remanufacturing process in the absence of other constraints. Since production is not restricted, there is no bottleneck situation comparable to (9.23), because demand can always be satisfied by producing new items. But, having a positive serviceables inventory becomes preferable in situations where the remanufacturing constraint is active. Remanufacturing in advance and putting items in the serviceables inventory for use during an interval where the constraint is binding saves production costs. But these savings have to be balanced with the induced serviceables holding costs. Therefore, a similar marginal criterion as the interval length property (9.12) can be given.

9.3.5 Multiple Reuse Options

Until now we have assumed that there is only one possible way to remanufacture returned products. But in practice it can also happen that different remanufacturing options are possible, for example different quality classes or different products etc., related with different costs and profits. Then the question arises regarding for which option the returned products should be used. Inderfurth et al. (2001) have investigated a stochastic remanufacturing system with multiple reuse options and no production facility. Then, backorders can occur and it has to be determined how to allocate returns to satisfy the different demand options. Hartl et al. (1992) discuss a deterministic model with multiple reuse options for a given amount of waste from ore production. In this section, we will extend the basic model and include m different remanufacturing options for a single return stream. This leads to m remanufacturing and production facilities. The m different demand streams are replenished from the corresponding inventory for the serviceable items. We get the following optimization problem:

$$\min_{p_i(\tau), r_i(\tau), w(\tau)} \int_0^T c_w w(\tau) + h_u I_u(\tau) + \sum_{i=1}^m \left(c_i^p p_i(\tau) + c_i^r r_i(\tau) + h_i^s I_i^n(\tau) \right) d\tau,$$

$$(9.29)$$

subject to the state dynamics

$$\dot{I}_i^n(\tau) = p_i(\tau) + r_i(\tau) - d_i(\tau), \quad i = 1, 2, \ldots, m \qquad (9.30)$$

and

$$\dot{I}_u(\tau) = u_r(\tau) - \sum_{i=1}^m r_i(\tau) - w(\tau), \qquad (9.31)$$

the initial and terminal conditions

$$I_u(0) = I_u(T) = 0 \quad \text{and} \quad I_i^n(0) = I_i^n(T) = 0, \quad i = 1, 2, \ldots, m, \qquad (9.32)$$

and the non-negativity Constraints

$$I_u(\tau) \geq 0 \quad \text{and} \quad I_i^n(\tau) \geq 0, \quad i = 1, 2, ..., m \text{ and } \forall \tau \in [0, T] \qquad (9.33)$$

$$w(\tau) \geq 0, \; p_i(\tau) \geq 0, \; r_i(\tau) \geq 0, \quad i = 1, 2, ..., m \text{ and } \forall \tau \in [0, T] . (9.34)$$

Besides the condition

$$c_i^p + c_w > c_i^r, \quad i = 1, 2, \ldots, m, \qquad (9.35)$$

we additionally assume, without loss of generality, that the different products are numbered such that

$$c_i^p - c_i^r > c_{i+1}^p - c_{i+1}^r, \quad i = 1, 2, \ldots, m - 1, \qquad (9.36)$$

which implies that all remanufacturing options are numbered in decreasing order of the cost advantage of remanufacturing over production.

Besides the question of whether excess returns should be stored for future recovery or disposed of, we have to answer the question regarding for which remanufacturing option the returns should be used. Again, Pontryagin's Maximum Principle can be used to determine the optimal control policy $(p_1^*(\tau), p_2^*(\tau), \ldots, p_m^*(\tau), r_1^*(\tau), r_2^*(\tau), \ldots, r_m^*(\tau), w^*(\tau))$. The solution technique is the same as in Section 9.2.3. First, the optimal policy structure is determined for a given system state, and then the transition points are considered.

Optimal Policy Structure

In order to determine an optimal policy structure in the basic model, we have defined different cases based on the state variables. For the model with multiple remanufacturing options, these cases are further divided into subcases.

If the recoverables inventory is equal to zero (Case 2) then $m + 1$ subcases (denoted with Case(2)(k)) are defined depending on the relation between the demand and the return rates.

- If it holds $u_r(\tau) \leq d_1(\tau)$ and $I_u(\tau) = 0$, then the system is in Case(2)(1).
- If it holds $\sum_{i=1}^{k-1} d_i(\tau) < u_r(\tau) \leq \sum_{i=1}^{k} d_i(\tau)$ and $I_u(\tau) = 0$, then the system is in Case(2)(k), $(k = 2, \ldots, m)$.
- If it holds $\sum_{i=1}^{m} d_i(\tau) < u_r(\tau)$ and $I_u(\tau) = 0$, then the system is in Case(2)(m+1).

For each subcase, the optimal policy structure is as follows.

- Case(2)(1): $r_1^* = u_r, p_1^* = d_1 - r_1^*$ and $r_i^* = 0, p_i^* = d_i, (i = 2, \ldots, m)$, $w^* = 0$
- Case(2)(k): $r_i^* = d_i, p_i^* = 0, (i = 1, \ldots, k - 1), r_k^* = u_r - \sum_{i=1}^{k-1} d_i,$ $p_k^* = d_k - r_k^*, r_i^* = 0, p_i^* = d_i, (i = k + 1, \ldots, m), w^* = 0$
- Case(2)(m+1): $r_i^* = d_i, p_i^* = 0, (i = 1, \ldots, m), w^* = u_r - \sum_{i=1}^{m} d_i$

This means that in an optimal policy, given that recoverables inventory will be empty, as many demands as possible are satisfied from remanufactured parts in the order of decreasing recovery cost advantage.

The subcases for the situation with positive recoverables inventory are not defined using the demand and return rates, but the definition of the m subcases of Case 1 is based on the co-state variable λ (the value of the returns) which is given as a solution of the following differential equation (together with the choice of an appropriate starting value).

$$\dot{\lambda}(\tau) = h_u \tag{9.37}$$

Based on the value of λ, the following m subcases of Case 1 are defined as follows.

- If it holds $c_k^p - c_k^r > \lambda(\tau) > c_{k+1}^p - c_{k+1}^r$ and $I_u(\tau) > 0$, then the system is in Case(1)(k) $(k = 1, \ldots, m - 1)$.
- If it holds $c_m^p - c_m^r > \lambda(\tau) > -c_w$ and $I_u(\tau) > 0$, then the system is in Case(1)(m).

For each subcase with $I_u(\tau) > 0$, the optimal policy structure is given as

- Case(1)(k): $r_i^* = d_i, p_i^* = 0, (i = 1, \ldots, k),$ $r_i^* = 0, p_i^* = d_i, (i = k + 1, \ldots, m), w^* = 0$
- Case(1)(m): $r_i^* = d_i, p_i^* = 0, (i = 1, \ldots, m), w^* = 0$

Only those demand classes with a larger recovery cost advantage compared to the value of returns are satisfied from remanufactured parts. During an interval with positive inventories, the value of returns increases with rate h_u and therefore remanufacturing for lower advantage classes is terminated in favor of saving returns for demand classes with a higher cost advantage.

Optimal Transition Points

We still have to answer when it is optimal to start keeping returns in stock. Therefore, we have to determine the optimal transition points for the transition from Case 2 into Case 1. It can be shown that most of the transitions between subcases of the different cases cannot be optimal. Only transitions from Case (2)(k) to Case (1)(k-1) and from Case (2)(k) to Case (1)(k) can occur in an optimal solution. Depending on the continuity property of the adjoint variable λ at the transition point, the first or the latter transition takes place.

Again, it is not possible to give analytical expressions for these time points. But similar to the basic model, location, integral, and time length properties related to the subcases can be obtained for an optimal recovery policy. For a detailed discussion of these properties, and for a solution algorithm, we refer to Kleber et al. (2002).

Comparison with the Basic Model

Comparing the optimal recovery policy for the basic model with the optimal policy for a situation with multiple remanufacturing options, we can observe a different behavior of the system. While the optimal recovery policy for the basic model never allows remanufacturing to stop if recoverables are available, the optimal recovery policy for multiple reuse options terminates a remanufacturing option to save returns for future remanufacturing of another option with a higher recovery advantage.

9.3.6 Returns Acquisition

Contrary to the aforementioned models, we now assume that the return stream can be actively influenced by acquisition management. Demands are still given as an exogeneous variable. The importance of such an active balancing of returns and demands has been pointed out in Guide (2000) and Guide and van Wassenhove (2001) (see also Chapter 3 for a discussion of this issue). The main instrument for acquisition is pricing, i.e. offering a buy-back price to customers that currently use the product. However, one can also consider advertising expenditures to inform customers about return opportunities, or a mix of both instruments. We assume that the amount of returned products depends on the expenditure at each point of time, i.e. the offered buy-back price. The relationship between the return quantity and the offered buy-back price is expressed by a return response function. Further, we assume a non-negative buy-back price $b(\tau) \geq 0$. As an extension, one could allow for a disposal fee which implies that the buy-back price becomes unrestricted. For the sake of the simplicity of the following presentation, we use a linear response, i.e. there is an autonomous return stream $m_1(\tau)$ if no active acquisition is promoted. By increasing the buy-back price $b(\tau)$ by a unit, additional $m_2(\tau)$ return units can be acquired. However, the analysis may be carried out in the same manner for more elaborate response functions (see e.g. Simon, 1989, for an overview on different mathematical representations of price response functions).

$$u_r(\tau) = m_1(\tau) + m_2(\tau)b(\tau) \quad \forall \tau \in [0, T] \tag{9.38}$$

Since returns are no longer exogeneous, the basic model presented in Section 9.2 has to be modified. There exist additional costs $b(\tau)u_r(\tau)$ for the acquisition of returns, i.e. we now have a quadratic cost expression for recoverables acquisition. Further, the state transition equation for the recoverables inventory has to be modified accordingly.

$$\dot{I}_u(\tau) = m_1(\tau) + m_2(\tau)b(\tau) - r(\tau) - w(\tau) \quad \forall \tau \in [0, T] \tag{9.39}$$

Then, the dynamic optimization problem for a joint determination of the optimal remanufacturing and buy-back pricing policy is given by

$$\min_{p(\tau), r(\tau), w(\tau), b(\tau)} \int_0^T \Big(c_p p(\tau) + c_r r(\tau) + c_w w(\tau) + b(\tau) u_r(\tau) + h_u I_u(\tau) \Big) d\tau$$

$$\begin{aligned}
s.t. \quad & \dot{I}_u(\tau) = m_1(\tau) + m_2(\tau)b(\tau) - r(\tau) - w(\tau) && \forall \tau \in [0, T] \quad (9.40) \\
& I_u(\tau) \geq 0, \ I_u(0) = 0, \ I_u(T) = 0 \\
& p(\tau) \geq 0, r(\tau) \geq 0, w(\tau) \geq 0, b(\tau) \geq 0 && \forall \tau \in [0, T].
\end{aligned}$$

As a result of the analysis of the necessary and sufficient optimality conditions of Pontryagin's Maximum Principle (for a detailed description, see Minner and Kiesmüller (2002)), the optimal buy-back price and the resulting return quantity are given by

$$b^*(\tau) = \begin{cases} \frac{\lambda^*(\tau)}{2} - \frac{m_1(\tau)}{2m_2(\tau)} & \text{if } \lambda^*(\tau) > \frac{m_1(\tau)}{m_2(\tau)} \\ 0 & \text{if } \lambda^*(\tau) \leq \frac{m_1(\tau)}{m_2(\tau)} \end{cases} \tag{9.41}$$

$$u_r^*(\tau) = \begin{cases} \frac{m_1(\tau)}{2} + \frac{\lambda^*(\tau)m_2(\tau)}{2} & \text{if } \lambda^*(\tau) > \frac{m_1(\tau)}{m_2(\tau)} \\ m_1(\tau) & \text{if } \lambda^*(\tau) \leq \frac{m_1(\tau)}{m_2(\tau)} \end{cases} \tag{9.42}$$

Depending on the value of the co-state variable, the optimal buy-back strategy is to accept autonomous returns only, i.e. not to pursue an active returns acquisition management, if the value is sufficiently small or to offer a buy-back which is not larger than a half of the inventory-cost-related value of the returned item.

As shown for the basic model, we find two cases with respect to recoverables, zero and positive inventory.

- Case 1: $I_u(\tau) > 0$
  $b^*(\tau) = \max\{0, \frac{\lambda^*(\tau)}{2} - \frac{m_1(\tau)}{2m_2(\tau)}\}, \ u_r(\tau) = \frac{m_1(\tau)+m_2(\tau)\lambda^*(\tau)}{2}, \ p^*(\tau) = 0,$
  $r^*(\tau) = d(\tau), \ w^*(\tau) = 0$
- Case 2: $I_u(\tau) = 0$

(a) Synchronization $\frac{m_1(\tau)-c_w m_2(\tau)}{2} \leq d(\tau) < \frac{m_1(\tau)+m_2(\tau)(c_p-c_r)}{2}$

$b^*(\tau) = \max\{0, \frac{d(\tau)-m_1(\tau)}{m_2(\tau)}\}$, $u_r(\tau) = d(\tau)$, $p^*(\tau) = 0$, $r^*(\tau) = d(\tau)$,

$w^*(\tau) = 0$

(b) Manufacturing cut-off $d(\tau) \geq \frac{m_1(\tau)+m_2(\tau)(c_p-c_r)}{2}$

$b^*(\tau) = \max\{0, \frac{m_2(\tau)(c_p-c_r)-m_1(\tau)}{2m_2(\tau)}\}$, $u_r(\tau) = \frac{m_1(\tau)+(c_p-c_r)m_2(\tau)}{2}$,

$p^*(\tau) = d(\tau) - u_r(\tau)$, $r^*(\tau) = u_r(\tau)$, $w^*(\tau) = 0$

(c) No acquisition and disposal $d(\tau) < \frac{m_1(\tau)-c_w m_2(\tau)}{2}$

$b^* = 0$, $u_r = m_1(\tau)$, $p^*(\tau) = 0$, $r^*(\tau) = \min\{m_1(\tau, d(\tau)\}$,

$w^*(\tau) = \max\{0, u_r(\tau) - d(\tau)\}$

In Case 1 with $I_u > 0$, we find that the value of the adjoint variable increases with rate $\dot{\lambda} = h_u$ and therefore the buy-back price increases within such a collection interval. Items being returned later and more timely for remanufacturing use are rewarded at a higher price. The exact value of the adjoint value, however, depends on the exact starting and ending points of such a Case 1 interval, which will be discussed in the following.

In Case 2 with $I_u = 0$, we now have three different subcases. In subcase (a), as many returns as are needed to satisfy demands from remanufactured parts are acquired. If demands are large, this option (because of the quadratic cost function) becomes too expensive and there is a maximum acquisition quantity of $\frac{m_1(\tau)+m_2(\tau)(c_p-c_r)}{2}$. Excessive demands are satisfied from manufactured parts. If demands are quite low, there is no need to buy back returns because a higher profit can be achieved from just accepting autonomous returns and producing for the remaining requirements.

An optimal sequence of these intervals, respective of the transition points, are characterized in the following.

Location Property

(1) Intersection points between demands and autonomous returns m_1
(2) Points where synchronization yields a larger slope in λ than inventory holding (h_u) would.
The first motive for inventories is the one known from the basic model. The second motive applies to Case 2a, if further synchronization would yield a higher increase in the adjoint than the holding cost rate. This means that it is cheaper to acquire an additional unit and to store it instead of adjusting the buy-back price.

Integral Property

The integral property is the same shown in the basic model except that exogeneous return rates now have to be replaced by the returns acquired depending on the chosen buy-back price.

Interval Length Property

Maximum length depends on the policies (subcases of Case 2) when entering and when leaving Case 1. If we dispose excessive returns before entering the collection interval (Case 2c), and if we have a subcase with manufacturing after recoverables are depleted (Case 2a), the maximum length is the same shown for the basic model. In all the other cases, the maximum length results from the difference of the adjoint value after leaving that interval minus the value before entering the interval.

Minner and Kiesmüller (2002) provide a more detailed discussion of these aspects and an algorithm for the construction of and transitions between such intervals. Here, we assume that only returns can be influenced by a pricing policy. A further balancing of demands and returns can be achieved if demands can also be influenced by an appropriate sales pricing policy.

9.4 Further Research Issues

9.4.1 Stochastic Dynamic Models

Uncertainty might occur with respect to several model parameters, especially demands and returns. Whereas uncertainty about cost parameters can be incorporated into the model by using an expected value framework and therefore transforming the stochastic problem into a deterministic one, stochastics in demands and returns have an impact on the movement of the state variables (see Chapter 8 for a general overview on stochastic inventory models). The usual differential equations have to be replaced by Itô's stochastic differential equations:

$$dI_n(\tau) = p(\tau) + r(\tau) - d(\tau) + \sigma_d(\tau)d\epsilon_d(\tau) \tag{9.43}$$

$$dI_u(\tau) = -r(\tau) - w(\tau) + u_r(\tau) + \sigma_u(\tau)d\epsilon_u(\tau) , \tag{9.44}$$

where $d(\tau)$ and $u_r(\tau)$ represent the expected demand and return quantities whereas $d\epsilon_d(\tau)$ and $d\epsilon_u(\tau)$ are the increments of a Wiener process (Brownian motion), i.e. $\epsilon(\tau_1) - \epsilon(\tau_0)$ is an independently, normally distributed random variable with mean zero and variance $\tau_1 - \tau_0$.

The resulting optimization problem can be solved by a generalized Hamilton-Jacobi-Bellman dynamic programming approach or by applying the stochastic maximum principle. An overview on modeling and solution techniques in stochastic optimal control is provided by Neck (1984) (see also Yong and Zhou, 1999). With zero lead times and an uncapacitated manufacturing supply mode, there is still no incentive to have positive serviceables inventory. However, keeping recoverables (instead of disposal) becomes more attractive.

9.4.2 Competition

In a continuous time framework with competition, differential games have to be analyzed. This might be used in order to investigate price competition (to influence demands and returns) and aspects of market entry. The latter is of special interest in product recovery because most original equipment manufacturers (OEM) have the desire to avoid competition from firms that collect used products and remanufacture them (see Chapter 12 for an overview on coordination and competition issues). A popular case example is the market of toner cartridges (Narisetti, 1998), because products being remanufactured and sold (at a lower price than new products) by remanufacturing firms will reduce the market share and profits of the OEM.

In a differential game, each player solves an optimal control problem. The incorporation of the player's interaction can be modelled in several ways. One aspect is the incorporation of available information about states and decisions. The second aspect concerns the concept of equilibrium (see Dockner et al. 2000, for an overview). Open-loop strategies only depend on the initial state and the time index, but not on intermediate decisions. This implies that both players initially commit to their optimal strategies. A justification might be that the ability to make future changes is limited (e.g. after making an investment in remanufacturing technology) or that future states and competitor actions cannot be observed. Closed-loop strategies depend on time and the current state. They allow for feedback to changes in the environment and the competitor's action. With respect to the concept of equilibrium, Nash- and Stackelberg-equilibria are distinguished. In a Nash-equilibrium, all players simultaneously decide about their strategies (e.g. sales and return acquisition prices) and the equilibrium is characterized by the property that it is not optimal for a single player to deviate from an equilibrium point. In contrast to this concept, in a Stackelberg game there exists a leader and a follower and decisions are made sequentially. The follower optimizes the strategy under a given decision of the leader. The leader, however, anticipates the optimal reaction of the follower in the strategy optimization, e.g. an original equipment manufacturer of toner cartridges who anticipates (and eventually tries to avoid) a market entry of a competing remanufacturer.

9.4.3 Learning in Product Recovery

The learning curve effect is another phenomenon that can be found quite often in practical applications (for a literature review, see Kantor and Zangwill, 1991). It shows a (potential) relationship between cumulative output and variable production costs. In its classical form, the learning curve predicts a reduction of unit production costs by a constant percentage each time the cumulative output doubles. Since this only represents the results of a system's inherent learning process, efforts have been taken to explain the underlying causes, resulting in refinements of the functional forms (Adler and Clark,

1991, investigate the influence of forced engineering changes and training on productivity improvement).

Cost reductions due to learning are not restricted to production alone. It can also make sense for remanufacturing processes, if there are repeated operations. In our introductory case, this especially applies for "new" types of engines, for which workers have to develop special skills (e.g. how to dismantle them). As an implication, remanufacturing can be profitable in the long run, even if there is no immediate cost advantage, because subsequent unit remanufacturing costs are lowered. Here, the advantages of an optimal control framework can be used because, as explained in Section 9.2.2, the indirect effect of current decisions on future costs can be valued, as well.

The following changes to the basic model are necessary in order to account for learning. An additional state variable has to be introduced for each learning process (remanufacturing $R(\tau)$ and/or production $P(\tau)$) representing the cumulative output of the process, which is added up by a respective state equation, e.g. $\dot{R}(\tau) = r(\tau)$, $\dot{P}(\tau) = p(\tau)$. The corresponding co-state variable values the impact of the current decision on future remanufacturing or production costs. The direct effect is rated by a function measuring the learning curve effect realized so far, which replaces unit variable costs in the objective function. Therefore, recovery unit costs are given as a decreasing function of cumulative output $c_r = f(R)$, e.g. by using the mixed exponential learning approach (Kantor and Zangwill, 1991)

$$c_r(R) = c_r^0 \cdot e^{-bR}, \tag{9.45}$$

with b as learning parameter and c_r^0 denoting initial unit remanufacturing costs.

9.5 Conclusions

In this chapter, we have introduced continuous dynamic models for recovery systems and provided a solution methodology to determine optimal recovery policies. All the models use simple differential equations to describe the evolution of the system. The advantage of these simple models is that optimal policies can be calculated and insights into the optimal systems behavior can be obtained. The presented models and the streams for future research cover several important problems in operations strategy of reverse logistics systems. The basic model provides a framework to decide about the strategic collection and remanufacturing of excessive returns instead of disposing of and land-filling used products. The main cost trade-off is between recovery cost advantages and inventory holding costs until returns are reused. As a byproduct of the analysis, we find a dynamic valuation of product returns being important for accounting issues. Extensions to the basic model illustrate the smoothing effect that remanufacturing has in capacitated manufacturing systems. Moving from aspects of manufacturing to customer related issues, the approach helps to decide how returns may be used in the most profitable way if there are several different options (associated with different cost structures) for recovery. Instead of simply taking back used products from customers, further improvements can be achieved from an active returns acquisition management by buying back products from customers. Then, our approach enables the simultaneous choice of an optimal recovery and buy-back pricing policy. The framework of optimal control provides an additional tool to analyze further important strategic decision problems, e.g. how learning and reductions in remanufacturing unit costs impact the recovery policy, when and how much to invest in remanufacturing technology, and how competitive reaction influences decisions.

Production Planning for Product Recovery Management

Karl Inderfurth[1], Simme Douwe P. Flapper[2], A.J.D.(Fred) Lambert[2],
Costas P. Pappis[3], and Theodore G. Voutsinas[3]

[1] Faculty of Economics and Management, Otto-von-Guericke University
   Magdeburg, P.O.Box 4120, 39016 Magdeburg, Germany,
   inderfurth@ww.uni-magdeburg.de
[2] Faculty of Technology Management, Eindhoven University of Technology,
   P.O. Box 513, 5600 MB Eindhoven, The Netherlands,
   s.d.p.flapper@tm.tue.nl, a.j.d.lambert@tm.tue.nl
[3] Department of Industrial Management and Technology, University of Piraeus, 80
   Karaoli & Dimitriou Str., 18534 Piraeus, Greece,
   pappis@unipi.gr, vutsinas@unipi.gr

10.1 Introduction to Production Planning in Product Recovery Systems

Production Planning and Control (PPC) in product recovery systems faces complications due to several characteristics which typically require tools different from, and in addition to, those known from traditional forward production and logistics systems (e.g., see Guide, 2000, and Inderfurth and Teunter, 2002). Many of these characteristics are due to additional specific operations necessary for disassembling, reprocessing, and rearranging recoverable products. Furthermore, in product recovery management, considerable sources of uncertainty have to be taken into consideration concerning the arrival of recoverables as well as the outcome of disassembly and reprocessing activities. The specific environment that poses challenges for PPC under product recovery will be demonstrated by two case examples, one of them concerning remanufacturing of used products in the field of discrete products' manufacture, and the other considering rework of by-products in the field of process industries.

Case Example A: Remanufacturing of Copiers at Xerox
(see Thierry, 1997)

The Xerox Company in Europe has already a quite long tradition of incorporating product recovery in its copier manufacturing branch. Used copiers are taken back from the costumers, and, after passing some inspection, they are shipped to a central facility where both manufacturing and remanufacturing of copiers take place. In the remanufacturing process, cores are disassembled and components and parts are tested and refurbished so that they meet the quality standards of virgin products. If reprocessing is out of the question for either technical or economical reasons, used items are replaced by new ones. Reassembly of remanufactured products and assembly of new ones are performed on the same assembly line. Under these circumstances, production planning incorporates a coordination of production and remanufacturing orders of respective products to secure timely fulfillment of customer demands at minimum cost. In the recovery context, it has to be planned in which sequence and to which extent used products should be disassembled before the resulting parts are assigned to different dispositions like refurbishment, recycling, and disposal. Furthermore, shop floor control of reprocessing orders of refurbishable parts and modules is a major planning issue.

Case Example B: Rework of By-products at Schering
(see Teunter et al., 2000)

Schering AG is a major German company engaged in the pharmaceutical industry. As a main component of its pharmaceuticals, it produces active ingredients out of basic raw materials in complex, multi-stage batch production processes. These chemical processes are such that valuable by-products accrue which, for environmental and economic reasons to a large extent, are undergoing specific processing steps and performing some kind of rework operations to regain valuable substances. Reworked by-products can be used as input to regular production steps so that we face material flows with cyclical structure. Both rework and regular production processes use the same operators and multi-purpose facilities. Due to high setup costs, production at many stages is performed in large lots of single batches. The major production planning issue in this context is to coordinate production and rework batches over a planning horizon of multiple weeks in such a way that the medium-term master production schedule of active ingredients is met and materials flow over all production stages is guaranteed to be at minimum cost. Hereby, the cyclical flow structure caused by the occurrence of reworkable by-products within the production process has to be taken into account with respect to quantity and time. Additional planning complexity can arise for reworkable substances which have a limited shelf life or are losing value due to deterioration.

The different types of product recovery systems and their major subsystems as reported in Case Examples *A* and *B* are visualized in Figure 10.1. Remanufacturing performs a loose closed-loop system where serviceable products, after being used by customers, re-enter the production system as far as they are returned and re-collected. These external product returns usually enter a complex disassembly subsystem. In contrast, rework is imbedded in a strict closed-loop system with internally generated returns and normally without a

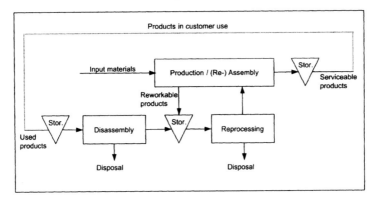

Fig. 10.1. Scheme of product recovery systems

disassembly part. Figure 10.1 also reveals how storage processes at different stages of the recovery system are interwoven with disassembly, reprocessing, and production/assembly activities. Production planning problems now have to be solved with respect to various subsystems of a complete product recovery system.

A basic issue in product recovery systems like that mentioned in Case Example *A* is the planning and scheduling of disassembly operations. This task incorporates establishing the best disassembly sequence for each core and determining the optimal disassembly depth together with deciding upon the best way to recover and use disassembled components and parts. An additional aspect in this context is the task of determining the minimum number of cores for disassembly which is necessary to fulfill given requirements for components and parts of a complex product. Section 10.2 gives insight into models which support respective disassembly and recovery planning. In a remanufacturing environment there are often shop floor control problems which are of extraordinary complexity due to the serious uncertainties found for respective operations. Specific control rules have been developed and tested for coordinated scheduling of disassembly, reprocessing, and reassembly orders. Section 10.3 is dedicated to quantitative research in this area. In a hybrid manufacturing/remanufacturing system as described in Case Example *A*, specific coordination problems have to be solved, which basically refer to materials planning as both remanufactured items and originally produced ones can be used as alternative sources to fulfill material requirements. Also, capacity planning can be affected and must be coordinated if common resources are used for manufacturing and remanufacturing operations. These situations require integrated planning approaches ranging from application of large-scale LP models to classical MRP procedures. Respective approaches are presented in Section 10.4. Section 10.5 describes specific PPC issues for product recovery environments where rework plays a major role (as described in Case Example B). Here we mostly face materials and capacity coordination problems as mentioned above. However, the immediate connection between production and rework (due to production processes being the direct generator of recoverable items) creates a need for an even stronger coordination. Also, lot sizing and scheduling have to be integrated in a closer way in a rework environment. Additionally, specific aspects such as limited storeability and deterioration of recoverables, which is often the case in process industries, has to be taken into consideration. Finally, in Section 10.6, some general conclusions are drawn.

10.2 Disassembly and Recovery Planning

Recovery of complex mechanical and electronic products has to be considered within the framework of their complete life cycle (Lambert, 2001a). Complex products usually consist of many components and materials, some of them with a considerable environmental impact, such as the presence of heavy metals in batteries and of CFCs in cooling systems and insulating foam (see Lambert and Stoop, 2001, for the case of refrigerators).

Disassembly planning is involved with the search for the appropriate design and organization of the disassembly process. It includes two basic functions: 1) selection of appropriate tasks and facilities to perform these tasks, and 2) generating a minimal set of ordering constraints on these tasks (Fox and Kempf, 1985). Our research has been focused on the second topic, which is a higher-level approach. It defines tasks as the detachment of definite components. A disassembly task might be complex, but all of this complexity is aggregated in one single figure, which represents the cost of the task.

Disassembly planning is indispensible, particularly if the process is automated. Adequate planning is important in adapting the process for unpredictable changes, which are due to, for example, quality loss of the product or its modules and components. It also has a major impact on scheduling, as proper planning generates the optimum amount of definite components to meet the demand without the performance of superfluous tasks and with the minimum production of redundant components. The generation of ordering constraints is within the field of disassembly sequencing, which has been developing since the 80s (see Lambert, 2002b, for a recent review of the literature in this particular field). Disassembly sequencing has been influenced by two basically different approaches: the mechanical and the hierarchical. The mechanical approach was originally developed for and aimed at the assembly of mechanical products. A typical example of such a product is a gearbox. This approach is briefly explained in Subsection 10.2.1. It is a general method that can be used for every type of disassemblable complex product. The hierarchical approach has been developed with a tree structure, such as in MRP algorithms, as a starting point. It has a simpler structure, which considerably reduces the size of the model. However, it is only applicable if the product has a multi-level hierarchical structure. This approach is useful if families of products are considered, whose members share multiple components. It is frequently applied in the description of the structure of electronic appliances such as computers, monitors, printers, and mobile phones (see Krikke et al., 1998, 1999). The theory that has been developed for the mechanical approach can be adapted to the hierarchical approach, which shows a much simpler structure. In this case, it can be applied to scheduling problems such as those that appear in remanufacturing theory. We apply the theory to recovery planning in Subsection 10.2.2.

A typical aspect of end-of-life disassembly is uncertainty. The quality of the different modules of a product can be unknown initially. Dependent on test results, the disassembly depth (i.e. the extent to which the disassembly process is carried through) and the choice of processing methods should be adapted. This is briefly described in Subsection 10.2.3.

10.2.1 Disassembly Sequencing

Disassembly sequencing is part of disassembly planning. It is applied in three distinct cases.

1. As a tool for assembly studies
2. In maintenance and repair problems, in which both disassembly and reassembly is required
3. In end-of-life disassembly problems

Disassembly study has emerged from assembly studies in which assembly is considered as reverse disassembly, which is true under certain conditions, such as the absence of forces and the assumption that components are rigid bodies. This is complete disassembly, which always proceeds until all components are detached. Usually, disassembly is not performed until the final state, which is seperation into all the individual components. The case when the final state is obtained is called complete disassembly. If the final state is not obtained, one deals with incomplete, partial, or selective disassembly. Maintenance and repair require the detachment of a set of predefined components.

End-of-life disassembly is carried out for both economic and environmental reasons. Here, not only the ordering of disassembly steps but also the disassembly depth is a variable, as disassembly is usually not carried through to its full extent. The extent to which the disassembly process is carried through is a decision variable here. The main incentive of end-of-life disassembly is the recovery of reusable components (such as spare parts or in good-as-new products), and recyclable materials. Apart from this, definite materials and components should be separated from the product because these are either hazardous for humans or for the environment (safety or environmental constraints) or they contaminate the potentially recyclable materials (technical or economic constraints).

Disassembly sequencing is the search for the set of all possible disassembly sequences, and the selection of the optimum sequence out of this set, which includes both the disassembly ordering and the disassembly depth. A sequence consists of a list of consecutive disassembly actions. A disassembly action is defined as the separation of a parent assembly into two child subassemblies such that their union is the parent assembly and their intersection is empty. The principal characteristic of a product is its number of components N. The number of a priori possible sequences dramatically increases with N. If $N = 4$, there are 41 sequences. With $N = 6$, there are $3,797$ sequences possible, if $N = 10$, this number increases to $314,726,117$. A recurrence relation that generates these numbers is given in Lambert (1997).

In practice, the number of sequences is far below this worst-case scenario, because disassembly processes are subjected to several types of constraints according to the following classification.

1. Topological constraints, which assure that a subassembly is connected
2. Geometrical constraints, which assure that a subassembly can be created by disassembly
3. Technical constraints, which deal with forces and other real-world aspects of processes

This is illustrated with the example in Figure 10.2, with $N = 6$, which is taken from the founder of disassembly theory, Bourjault (1984).

The connection diagram is an undirected graph, which is derived from the assembly drawing. It depicts the connection relations between the different components. Each of $2^N - 1$ nonempty subsets of the set of components $A, .., F$ is a potential subassembly. The numbers in the connection diagram correspond to the connections between two distinct components. *Topological constraints* imply that the diagram that represents the subassembly has to be connected. Consequently, AC is not a subassembly. This leaves us with 24 of 63 subassemblies.

Geometrical constraints account for obstruction of motion if the components are considered rigid and subjected to neither internal (friction, adhesion) nor external (gravity) forces. Component B, for example, cannot be removed from subassembly ABF if A and F are kept in place. This implies that subassemblies containing both A and F but that don't contain B are

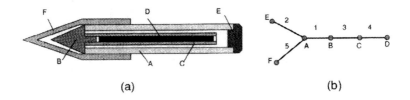

(a)   (b)

Fig. 10.2. Assembly drawing and connection diagram of a simple mechanical product, called Bourjault's ballpoint (see Bourjault, 1984)

geometrically infeasible. This is called a selection rule. Two selection rules are applicable here: AF not B and BE not D. These rule out subassemblies such as AEF and ABE, which can be checked in Figure 10.2. Due to these constraints, the number of subassemblies is further reduced to 18. Geometrical constraints are reversible with respect to assembly and disassembly.

Technical constraints might be different for assembly and disassembly. In this case, the ink D can only be removed from container C if B is removed first. This can be effectuated by an additional constraint: BC not D, which results in a further reduction to 15 feasible subassemblies. Technical constraints depend on the availability of tools and processes, and are also a consequence of physical features, stability of subassemblies, accessibility of components, etc.

The disassembly graph, which is a directed AND/OR graph, can be derived from the list of feasible subassemblies. This list can be generated semi-automatically from a set of precedence relations that in turn can be derived manually or automatically, dependent on the product structure and the availability of geometric information (Lambert, 2001b). In such a representation, nodes correspond with subassemblies and arcs with disassembly actions. These arcs are pointing from the parent subassembly to two different child subassemblies (AND relation) and are called hyperarcs for this reason. The hyperarc 6 in Figure 10.3(b), for example, points from the parent subassembly $ABCDF$ toward the child subassemblies ABF and CD; the hyperarc 4 points from $ABCDF$ to $ABCD$ and F. As the latter is a single component, this branch is not depicted.

The number of sequences is determined starting with the final state and going in the opposite direction as the arcs. AB, for instance, can be separated in one way, or it can be left intact. The latter feature makes the number of possible sequences increase at every node. As CD can be processed in 2 ways, and ABF in 3 ways, the parent $ABCDF$ can be processed along these lines in $2 \cdot 3 = 6$ ways. As $ABCD$ can be processed in 8 ways, $ABCDF$ can be processed in $6 + 8 = 14$ ways. It can also be kept intact, which adds to 15 possible sequences for $ABCDF$ (see Figure 10.3 (a)). If a subassembly consists of an individual component, the corresponding branch of the hyperarc is not depicted. Actions 5, 6, and 8 are classified as *parallel disassembly*.

(a)   (b)

Fig. 10.3. Disassembly graph of Bourjault's ballpoint in AND/OR notation: (a) determining the number of sequences and (b) enumeration of nodes and arcs

Such actions are particularly useful in disassembly line balancing. There are still 31 sequences possible, including incomplete disassembly. This number can be graphically determined by proceeding from the right to the left in Figure 10.3(a). The number underneath each subassembly corresponds with the number of sequences according to which it can be disassembled. CD, for example, can be left intact or can be separated into C and D. This number is transferred to the entering flows. Numbers in branches of AND relations are multiplied, those in OR relations are added. The resulting number is increased by 1 in each node. It is illustrative to compare the resulting number with the worst case maximum, which equals to 3,797 for $N = 6$.

With the disassembly graph as a starting point, the selection of the optimum sequence can be carried out. This procedure is described by Lambert (1999), and is based on a Binary Linear Programming approach that relaxes to a Linear Programming approach if no constraints are imposed on capacity of actions. We will illustrate this with Bourjault's ballpoint as a case. The objective function is the profit. Parameters are the revenues r_i, which are obtained by selling subassemblies or components, and the costs c_j, which are assigned to the different disassembly actions. There are 15 subassemblies and 14 actions here, including the 'zero' action of setting as available the complete product. The structure of the disassembly graph is in the (15 x 14) transition matrix T. The enumeration proceeds according to Figure 10.3(b). The components A through F, which are omitted in this figure, are labeled 10 through 15. The elements T_{ij} in the transition matrix are put to 1 if the definite action j creates the subassembly i, and they are put to -1 if the action j destroys the subassembly i. Other elements are zero. The 8th row of the matrix, corresponding with $ABCD$, can thus be written as:

$$[0 \quad 0 \quad 0 \quad 1 \quad 1 \quad 0 \quad 0 \quad -1 \quad -1 \quad 0 \quad 0 \quad 0 \quad 0 \quad 0] \quad ,$$

with j running from 0 to 13. Other rows are established along similar lines. These elements act as structural parameters. Binary flow variables x_j, corresponding with actions, are calculated. These equal 1 if an action is carried out, and are 0 if an action is not carried out. The problem is stated by the following expressions:

Maximize the profit Π

$$\Pi = \sum_i \sum_j (T_{ij} \cdot r_i - c_j) \cdot x_j$$

subject to the node constraints

$$\sum_j T_{ij} \cdot x_j \geq 0 \qquad \forall i$$

and the initialization constraint

$$x_0 = 1 \quad .$$

As a solution, the components of the x-vector are returned. These correspond to a connected subgraph of the disassembly graph, for example,

$$[1 \quad 1 \quad 0 \quad 0 \quad 0 \quad 1 \quad 0 \quad 0 \quad 0 \quad 1 \quad 0 \quad 1 \quad 1 \quad 0] \ .$$

This corresponds with $x_0 = x_1 = x_5 = x_9 = x_{11} = x_{12} = 1$, which means that those actions in Figure 10.3(b) are performed. The node constraints guarantee that the resulting subgraph is connected. Additional restrictions can be imposed by legislation, for example, which might require that a hazardous substance (the ink D in the example) has to be removed.

The method can be applied to problems with large values of N. Geometrical constraints essentially reduce the size of the problem, which is expressed by the number of nodes and arcs in the graph. Because Binary Linear Programming is applied, one is not forced to exploit the complete search space. The compact notation allows for extensions and modifications to be easily carried through.

10.2.2 Disassembly and Product Remanufacturing

In many cases from actual practice, not a detailed mechanical structure but rather a hierarchically organized modular structure is encountered. This topic has been studied by Lambert and Gupta (2002), who applied it to remanufacturing problems. The main problem to be solved involves meeting the demand on various components in an optimum way by properly disassembling a mix of different product types, which is called a disassembly-to-order problem. A similar problem, but without regard to the product's structure, has been treated by Uzsoy and Venkatchalam (1998). Related research has been carried out by Spengler et al. (1997), who worked on integrated disassembly depth and recycling planning problems, applied to dismantling and recycling decisions for domestic buildings. A comparison between mechanically and hierarchically structured products is extensively discussed by Lambert (2002a). A simple example is presented in Figure 10.4, which depicts an electronic device (a). It is represented by a connection diagram (b) with a two-level tree structure, called a hierarchical tree. In contrast with Figure 10.2(a), connections here are no longer mating relations, but hierarchical relations between roots, and between roots and leaves. The leaves represent useful components and the roots correspond with parts that are obstructing the detachment of the leaves. In contrast with a mechanical example such as Bourjault's ballpoint (Figure 10.2), this type of product is considered a mine of components, which has several different hierarchical levels rather than a coherent structure with equivalent components. It is also observed that mechanical components such as casings and frames are not considered valuable here, but rather act as constraints. In computers, for example, one might be interested in the Printed Circuit Boards (PCBs) only. Consequently, the additional components, such as casings and frames, are only considered insofar as they obstruct the removal

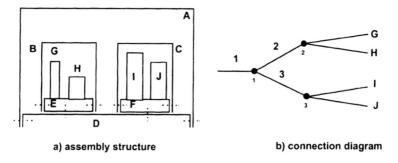

a) assembly structure   **b) connection diagram**

Fig. 10.4. Hierarchical tree structure, example

of the desired components (meaning that they have to be removed prior to
the detachment of the PCBs.

There are two hierarchical levels in the case of Figure 10.4. Prior to attaining the valuable components G through J (which are considered as leaves),
we have to take apart the roots, which are the nodes in the connection graph.
Prior to taking apart root 2 or 3, we have to take apart root 1. Thus, the
hierarchy is expressed by precedence relations. Study of the problem is aimed
at remanufacturing. A basic problem is meeting a given demand for different
types of components at minimum costs. The components might be present in
different product types (commonality), in different quantities (multiplicity),
and at different positions in the product. Indices include h for product type,
i for the component type, and j and k for the root number. We will present
here a particular product type, say $h = 1$.

The hierarchical structure, which is given by the connection diagram, is
condensed in the configuration matrix S, which is the equivalent of the transition matrix in the mechanical approach. Its elements S_{hjk}, with $j < k$, equal
1 if root number j of product type h is connected to the higher-level root with
number k. The condition $j < k$ guarantees that subsequent hierarchical levels
are distinguished. By convention, $S_{h,j=1,k=1}$ is put equal to 1. All the other
elements are zero. In the example, $S_{h=1,j=1,k} = \begin{bmatrix} 1 & 1 & 1 \end{bmatrix}$ and the other rows
have zero elements. The elements Y_{hji} of the yield matrix Y are nonnegative
integers that equal the number of components of type i that are available as a
leaf of root number j in product type h. In the example of Figure 10.4, we have
$Y_{h=1,j=2,i} = \begin{bmatrix} 1 & 1 & 0 & 0 \end{bmatrix}; Y_{h=1,j=3,i} = \begin{bmatrix} 0 & 0 & 1 & 1 \end{bmatrix}$. Costs c_{hj} are assigned
to the taking apart of the roots. Taking apart the basic root 1 also includes
the costs of having available the desired quantity of this product. The demand
for components is expressed by the demand vector with vector-components
d_i, which are nonnegative integers. Additionally, the flow vector with flow
variables z_{hj} as vector-components is defined. The z_{hj} are nonnegative integer variables that reflect how many items of root j of product type h have
to be taken apart. Revenues of the components only depend on the demand,

which is a given parameter. Therefore, we have to minimize the disassembly and waste-processing costs. The waste-disposal costs w_i reflect the costs of disposal of the redundant components that are in the disassembled products but that are not required for meeting the demand. The complete problem can be written in the following way.

Minimize the total costs TC for disassembly and waste processing

$$TC = \sum_h \sum_j c_{hj} \cdot z_{hj} + \sum_i \left(\left(\sum_h \sum_j (Y_{hji} \cdot z_{h,j=1}) - d_i \right) \cdot w_i \right)$$

subjected to the node constraints

$$\sum_k S_{hkj} \cdot z_{hk} \geq z_{hj} \qquad \forall h, j$$

and the demand constraints

$$\sum_h \sum_j Y_{hij} \cdot z_{hj} \geq d_i \qquad \forall i \quad .$$

Solution of this Integer Linear Programming model proceeds straightforwardly. As a matter of fact, this model can be applied to various cases, and many extensions and modifications are possible.

In Lambert and Gupta (2002), it has been applied to a basic example, which has been taken from Veerakamolmal and Gupta (1998). This case referred to 3 different types of computer server, each with a maximum of 5 roots, which were arranged in 3 hierarchical levels, and containing 27 different relevant components. The optimum batch of products to disassemble and the disassembly depth for each product to meet the demand at minimum costs was obtained at negligible CPU time with a standard solver.

This makes the method adequate for adaptive schedulers that can adapt the result to varying demand, test results of different components, etc. It can also be extended to encompass a multi-period environment, thus enabling optimization over a period of time, with a time-dependent demand vector, which is a powerful extension of the simple Dynamic Programming models such as described by Hoshino et al. (1995).

10.2.3 Uncertainty in End-of-life Disassembly

The methods that are described so far refer to essentially deterministic systems. In end-of-life disassembly, however, the quality of components and subassemblies is uncertain and should be established by testing. As we have seen, products are often hierarchically organized. Consequently, the complete product is considered 0th level and its basic modules 1st level. A basic module is composed of 2nd-level modules, etc. A monitor, for example, has a tube, a video unit, and a power supply unit. The power supply unit consists of a transformer, a chassis, a PCB, etc. Test results of the complete product provide information on the possible quality of its 1st-level modules. Testing a 1st-level module reveals information on its 2nd-level modules and so on. Not only quality, but other parameters that influence the value are determined, such as materials composition, mass, etc. This problem has been addressed by Krikke et al. (1998, 1999). These authors define quality classes q_i, which are assigned to modules according to their test results. Index i refers here to the class number. Probabilities $p_k(q_i|q)$ correspond with the probability of finding a module k in a certain class q_i if its parent module is in class q. The better the class q is, the higher the probability that a definite child module is also in a better quality class. If the product's quality class is high, for example $q = 1$, it will be upgraded as a whole; if it is intermediate ($q = 2$), the 1st-level modules will be disassembled because the probability of a good quality of the 2nd-order modules is rather high. If the product falls in a low class ($q = 3$), it will be shredded for materials reuse.

The model thus consists of a hierarchical tree structure, a quality classification scheme for assigning test results to quality classes, and a set of conditional probabilities. Apart from this, disassembly costs of the different modules and revenues of recovery/disposal options are required as parameters.

This results in a Stochastic Dynamic Programming model, which optimizes the overall net profit. From optimization through using a standard backward recursion scheme, we receive optimal conditional decisions for each module telling us how to proceed (i.e., with further disassembly or some kind of processing) if the respective module is in a specific quality class.

In case of complete disassembly (as experienced in the field of engine remanufacturing, for instance), uncertainty with regard to the quality of modules and parts necessitates deciding upon the number of used products to be disassembled in order to fulfill a given product demand, before the number of recoverable parts is known. This results in a complex stochastic yield problem for which promising heuristics have been developed in Inderfurth and Langella (2003).

10.3 Shop Floor Control for Disassembly and Recovery

As mentioned in the introduction to this chapter, the remanufacturing process is characterized by some complicating characteristics that affect the PPC activities. Such characteristics include the uncertainty regarding the recovery rate, the quality of returns and the reprocessing times, the need to balance returns with demands, and the need for disassembly and inspection of the quality of the materials recovered from returned products. These characteristics in many respects specifically complicate scheduling and shop floor control problems. In the following Subsection, we address some of the respective aspects.

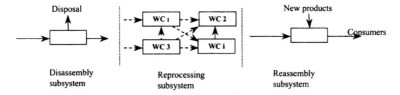

Fig. 10.5. Structure of a remanufacturing facility

10.3.1 Control Rules for Combined Disassembly/Remanufacturing

Specific complexity will be generated if the remanufacturing system performs serial-number-specific reassembly operations. In a serial-number-specific environment, parts must be reassembled to the same unit from which they were recovered (this is common in the case of heavy equipment engines, where customers want to take back the unit they forwarded for remanufacture). In this case, we usually find the need for complete disassembly of used products. Thus, here we face the most comprehensive form of disassembly operations discussed in Section 10.2. A remanufacturing facility usually is divided into three subsystems: disassembly, reprocessing, and reassembly. Figure 10.5, which represents a part of the complete product recovery system in Figure 10.1, visualizes the respective structure.

A unit of a used product enters the disassembly subsystem where it is completely disassembled into n parts. Parts that are in good condition are reconditioned at the Work Centers (WC) of the reprocessing subsystem. Each part may need to be processed by more than one WC (e.g. cleaning and inspection). The reassembly subsystem reassembles remanufactured and new products into units ready for use. The coordination of each of these subsystems usually generates problems of high complexity. Coordinating the overall system is much more complicated but is necessary in order to provide a smooth flow of materials between the subsystems and to achieve the predetermined objectives, such as minimizing make-span or tardiness. Unfortunately, there is no global strategy in choosing the appropriate control rules for the coordination of a remanufacturing system. The decision maker must take into account the complicating characteristics of the remanufacturing system and include them in the mathematical model that will be created to tackle the problem.

Guide et al. (1997a) study several Priority Dispatch Rules (PDR) and Disassembly Release Mechanisms (DRM) in a remanufacturing facility via a simulation model. It is shown that (for the case studied) there are no significant differences between the DRM, thus practicing managers should favor the simplest to implement and manage. It is also shown that simple-due-date based PDR perform well for a variety of performance measures. Guide et al. (1997b) study the impact of product structure complexity on operational decisions such as scheduling. It is shown that in a remanufacturing environment,

with its greater inherent uncertainty, product structure complexity significantly affects the choice of scheduling policies used.

Apart from the complicating characteristics mentioned above, there are
others that are associated with a special category of remanufactured products (e.g. value reduction of used products over time) or a special category of
remanufacturing processes (e.g. environmentally damaging emissions of products in the waiting line for remanufacture).

10.3.2 Control Rules for Product Recovery Under Value Deterioration

A special category of remanufactured products is that of high technology
products. In such a case, a complicating factor for the remanufacturing system
often is the value of the finished products, which deteriorates over time due to
reasons such as technological devaluation, physical degradation, etc. PCs, cars,
etc., are characterized by rapid value decreases over time and their components
can have different deterioration rates.

An example in this area of problems: suppose that a product, composed
of n different components, after reaching the end of its useful life-time, is
disassembled in order to be remanufactured. During the disassembly process,
a certain quantity of each component i $(i = 1, \ldots, n)$ is generated, which must
be remanufactured. To describe a model for optimizing the job sequence in
controlling the product recovery process, we use the following notation.

Y_i	: the quantity of components i produced during disassembly
L_i	: the processing time of one unit of i
J_i	: the job of reprocessing Y_i units of i
K_i	: the value of component i at time zero
$V_i(t)$	: the (deteriorating over time) value of one unit of i after a time span of t
a_i	: the value deterioration rate for component i $(a_i < 0)$

Then J_i requires $Y_i \cdot L_i$ time units to be completed. If we assume that
the value of each component deteriorates polynomially over time with an
individual rate a_i, the value function $V_i(t)$ is of the form

$$V_i(t) = K_i \cdot t^{a_i} \quad .$$

$V_i(t)$ is assumed to be computed at the time the processing of all units has
finished, that is, at time t, which is equal to the sum of the processing times of
all preceding jobs plus $Y_i \cdot L_i$. Assuming that all components will be processed
on a single machine, the problem is to find the sequence of the n jobs (each
job corresponding to the remanufacture of the whole lot of one component)
so that the total value V of the components remanufactured is maximized.

Let $< J_1, J_2, \ldots, J_n >$ be a given sequence. Then the respective value V
of this sequence is

$$V = \sum_{i=1}^{n} Y_i \cdot K_i \cdot \left(\sum_{j=1}^{i} Y_j \cdot L_j \right)^{a_i} .$$

The complexity of the problem of finding the value-maximizing job sequence is conjectured to be NP-hard. In Voutsinas and Pappis (2002), a heuristic algorithm of complexity $O(n^2)$ for solving this problem has been proposed, which is based on a mixed application of different problem-specific sorting policies of jobs. Numerical investigations show that this algorithm generates useful solutions. For computing optimal results, these solutions can be used as initial solutions for an effective Branch and Bound (B&B) method in order to prune branches of the solution tree (as shown in Voutsinas and Pappis, 2000). The B&B and the heuristic method were applied to a case study referring to personal computers (PC) remanufacturing. The study is based on Ferrer (1997), where it is shown that the values of the components of a PC deteriorate as time goes on. For example, the value of the PC component 'CPU' decreases according to $V_{cpu}(t) = K_{cpu} \cdot t^{-0.6}$, where $V_{cpu}(t)$ is the value of the component at time t and K_{cpu} is the value of the component at time 0. Future research in this area (among others) has to investigate the impact of different kinds of deterioration functions on the solution of the scheduling problem.

10.4 Integrated Production and Recovery Planning

Product recovery faces the most challenging planning tasks if it is embedded in a situation where both regular production and product recovery processes take place within the same organization. This is usually the case in production systems with rework, but it is also present if a manufacturer of original products is engaged in remanufacturing products from returned cores (or their major components), thus using an additional source for providing customers with serviceable products. Under such circumstances, remanufacturing (or rework) activities have to be integrated with manufacturing decisions in order to coordinate material requirements and capacity needs over all production, disassembly, and recovery stages in the most economical way.

10.4.1 Optimization Procedures for Integrated Planning

Material and capacity planning in a hybrid manufacturing/remanufacturing environment is concerned with regular production as well as with disassembly and reassembly stages. Thus, specifically in the field of materials coordination, a large number of different options, including the option to dispose of items, has to be integrated into the planning system. Depending on how detailed disassembly and recovery options are taken into account, different planning

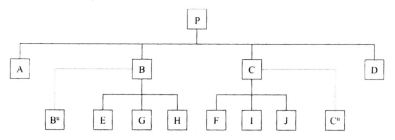

Fig. 10.6. Assembly product structure diagram

approaches can be used. On the one hand, integrated planning can be developed from an extension of the detailed disassembly and remanufacturing models described in Subsection 10.2.2., when manufacturing is also considered and component demands are derived from a Master Production Schedule. On the other hand, integration can be performed by a regular production and materials planning approach which incorporates disassembly, remanufacturing, and disposal on a somewhat aggregate level without focusing on single disassembly operations. The latter approach will be presented in more detail.

We refer to a single-product case and use the example introduced in Figure 10.4. of Section 10.2.2. Instead of the disassembly diagram, we present the respective (extended) assembly product structure diagram which is depicted in Figure 10.6.

In this case, it is assumed that, apart from regular production, the major components B and C can also be generated by remanufacturing used items B^u and C^u (marked by dashed lines in the diagram), which are regained from disassembling used products P^u. Thus we find two parallel procurement sources for items B and C. From the development of a disassembly strategy (as described in Subsection 10.2.1), it may turn out to be most economical to disassemble all accepted cores P^u down to their major components B^u and C^u, which are then stocked and remanufactured according to the respective requirements of B and C. Given this product and process structure, and given known external requirements for the products and known returns of recoverable components, the coordination problem is deciding how many units of each item in the above product structure to manufacture, and how many recoverable components to remanufacture or dispose of, in each period in order to minimize the total relevant cost. This problem can be formulated as a Linear Program (LP) with the following variables.

p_{it} : quantity of item i gained from production in period t

r_{it} : quantity of item i gained from remanufacturing in period t

w_{jt} : disposal quantity of recoverable item j

I_{it}^s : inventory on hand of serviceable item i at the beginning of period t

I_{jt}^u : inventory on hand of recoverable item j at the beginning of period t

As additional notation, we use the following.

d_{it} : (external) requirements of item i during period t

u_{jt} : (external) returns of recoverable items j during period t

a_{ik} : number of item i necessary to produce one unit of item k

b_{jk} : number of item j necessary to remanufacture one unit of item k

SI_i : minimum stock level of item i (serving as safety stock)

DI_j : maximum stock level of item j

c_p^i : unit variable production cost of item i

c_r^i : unit variable remanufacturing cost of item i

c_w^j : unit variable disposal cost of item j

h_s^i : unit holding cost of item i (per unit of time)

h_u^j : unit holding cost of recoverable item j (per unit of time)

The total cost TC to be minimized is

$$TC = \sum_t \sum_i c_p^i \cdot p_{it} + \sum_t \sum_i c_r^i \cdot r_{it} + \sum_t \sum_j c_w^j \cdot w_{jt}$$
$$+ \sum_t \sum_i h_s^i \cdot I_{it}^s + \sum_t \sum_j h_u^j \cdot I_{jt}^u \quad .$$

The connection between periods is given by inventory balance equations:

$$I_{i,t+1}^s = I_{it}^s + p_{it} + r_{it} - \sum_k a_{ik} \cdot p_{kt} - d_{it} \quad \forall i, t$$

$$I_{j,t+1}^u = I_{jt}^u - w_{jt} - \sum_k b_{jk} \cdot r_{kt} + u_{jt} \qquad \forall j, t$$

In order to protect against uncertainties (specifically in requirements and returns), safety stocks for all or certain items are frequently introduced, resulting in additional constraints

$$I_{it}^s \geq SI_i \qquad \forall i, t.$$

To avoid overstocking recoverable items or in case of limited inventory capacities for returns, additional maximum stock levels can be introduced.

$$I_{jt}^u \leq DI_j \quad \forall j, t$$

All respective variables are restricted to be non-negative. This basic LP approach can easily be extended to incorporate multiple final products, multi-period processing time, capacity constraints for jointly used resources, multi-stage remanufacturing processes, and many more aspects. Models with several of these refinements are published in Clegg et al. (1995) and Uzsoy and

Venkatachalam (1998). A respective model for the case of a production system with rework is found in Teunter et al. (2000). These LP models of integrated planning in a product recovery environment can be considered to be major extensions of respective aggregate planning models in a conventional production context (see e.g. Nahmias, 1997, 139-144), which incorporate the impact of various product recovery options. In the basic model above, for instance, additional variables for remanufacturing, disposal, and stockholding of recoverables are modeled with consequences for costs and restrictions leading to a higher computational effort, especially if variables have to be treated as integers. This is one aspect which may draw attention to more simple methods of integrated planning in hybrid production/recovery systems.

10.4.2 MRP-based Approaches

Most companies are still reserved about applying advanced planning systems to production planning. Thus, LP approaches for materials and capacity planning are not (yet) widely used in practice. Instead, MRP-based procedures are regularly found, often integrated in an MRP-II system for including capacity checks. Thierry (1997) found that MRP-based approaches are also often employed in the product recovery business. Guide (2000) confirmed these findings for the US remanufacturing industry. In order to support disassembly operations, an MRP system must use both assembly and disassembly Bills of Material (BOM) (see Thierry, 1997). Furthermore, for integrating regular production and remanufacturing, the assembly BOM must be extended to include multiple supply sources generated by the additional recovery option. In an early approach, Flapper (1994) proposed a respective BOM scheme which is completed by a priority rule for determining which source of procurement to use first. By exploiting this approach in a slightly extended form, it is possible to incorporate integrated materials planning in an MRP system using the standard level-by-level procedure of exploding the BOM. However, for all items with multiple supply sources (like for components B and C in Figure 10.7), the standard MRP record has to be replaced by an extended one, in which not only planned order releases for production but also planned order releases for remanufacturing and disposal have to be determined period by period.

In order to demonstrate this procedure, we will present a numerical example showing the respective extended MRP record for component B in Figure 10.6 for an horizon of 4 periods. We assume that both production and remanufacturing of B have a lead time of 1 period. Furthermore, we request to hold a safety stock of 5 units for B (i.e., $SI = 5$) and want the stock of recoverables B^u not to exceed a level of 10 units, i.e. we use an inventory of $DI = 10$ as a so-called disposal level. As far as recoverables are available to satisfy net requirements, remanufacturing is always preferred to production. Under these conditions, the extended MRP record looks like that presented in Table 10.1.

Table 10.1. Extended MRP record for component B

	Current	Time period 1	2	3	4
Gross Requirement		**20**	**10**	**5**	**15**
Scheduled Receipts Production		**16**	-	-	-
Scheduled Receipts Remanufacturing		**3**	-	-	-
Projected Serviceables on Hand $[SI = 5]$	**8**	7	5	5	5
Net Requirements		0	8	5	15
Expected Returns Recoverables		**9**	**8**	**5**	**5**
Projected Recoverables on Hand $[DI = 10]$	**5**	9	10	5	
Planned Order Receipts Remanufacturing			5	5	10
Planned Order Receipts Production				0	5
Planned Order Release Disposal		0	2	0	
Planned Order Release Remanufacturing		5	5	10	
Planned Order Release Production		3	0	5	

This MRP record is read like a regular one, but several additional lines (typed in italics) have to be taken into account. These lines provide information on expected returns and inventories of recoverables (B^u) as well as on-receipt and release data for remanufacturing orders for B. Under a disposal option for recoverables, an additional line for planned disposal orders has to be incorporated. All numbers in Table 10.5.1 appearing in bold figures are input data. The other numbers are calculated following standard MRP rules, and note that the above mentioned priority rule for remanufacturing and the disposal rule for excess inventory of recoverables is applied. As shown in Inderfurth and Jensen (1999), a similar scheme can be employed for materials planning in the case of rework.

It has been argued that, due to the specific uncertainties in a product recovery environment, MRP-based approaches may not work sufficiently well under these circumstances (see e.g. Guide, 2000). Therefore, adjusted scheduling systems like proposed in Guide (1996) are advocated instead of MRP concepts. On the other hand, it has been investigated that if in applying MRP, in addition to implementing it in a rolling horizon framework, the usage of safety stocks can be a reasonable measure to offer sufficient protection against the additional uncertainties in product recovery. However, determining the most effective size of safety stocks in a hybrid manufacturing/remanufacturing system is a very challenging problem. Simulation studies give the impression that a rough sizing of safety stocks may not lead to satisfactory results (see Guide, 1997). Numerical investigations by Inderfurth (1998) show that employing a newsboy-type determination of safety stocks in addition to a cost-based fixing of disposal levels can result in a fairly good MRP performance. Furthermore, Gotzel and Inderfurth (2002) present improved safety stock formulas, based on separate safety stocks for production and remanufacturing, which guarantee extremely good MRP-based decisions, even in situations with considerable uncertainties. Chapter 8 presents several rules for safety stock determinization under different lead-time settings which can easily be implemanted in MRP-based approaches. Additionally, it seems that in many situations high uncertainty in remanufacturing lead times can very well be responded to by implementing safety lead times in the MRP approach. These findings are theoretically underpinned by the fact that it can be shown, for a class of problems, that an MRP application with separate safety stocks and a predetermined disposal level results in the same structural decisions rule as we find for the optimal policy developed by a stochastic inventory control approach as described in Chapter 8 of this book (see Inderfurth and Jensen, 1999).

A further possible shortcoming of MRP under conditions of uncertainty is that in a rolling horizon framework it can result in a considerable lack of planning stability. In the product recovery context, recent research has revealed that the additional presence of return uncertainty will not necessarily increase planning nervousness (see Heisig and Fleischmann, 2001, and Heisig, 2002).

10.5 Production Planning in the Case of Rework

Rework is defined as the set of activities to transform products not produced well initially into products that do fulfill certain preset requirements before they are distributed. The reasons why companies are reworking correspond with the reasons for doing product recovery in general: lower costs than producing anew, higher sales prices or lower disposal costs, expensive or limited supply of input materials, and less time required for rework than for producing or purchasing anew. Although rework may concern products that have already been sold to distributors but have not been used (and are still kept in store by the distributors), we shall focus hereafter on reworking products that did not leave the production site.

10.5.1 Lot Sizing in the Case of Rework

Probably the most essential difference between the planning and control of rework when compared with the planning and control of the recovery of products from outside the production plant is that the flows that may be recovered are essentially only generated during production/testing, whereas in the case of other product recovery situations, the products become available more or less continuously. Another essential difference is that the configuration and condition of what may be reworked is often well known, in contradistinction to the configuration and condition of products that are from outside, especially when these products have been used. The above two characteristics seem to make planning and controlling rework easier than product recovery in general.

At first sight, situations with rework seem to have a lot in common with production situations with re-entrant jobs, i.e. situations where products-in-process are visiting one or a number of work stations several times in order to be produced. A big difference between the two situations is the uncertainty with respect to whether rework is required or not. In the case of re-entrant jobs, such an uncertainty does not exist.

From a planning and control point of view, a distinction has to be made between single- and multiple-stage rework situations. In the former situation, all production processes have to be executed before it can be estimated via testing whether or not a product or production batch was produced well. In multiple-stage rework situations, it may be possible to do tests after one or a number of the total set of production processes have been executed and start the rework directly after instead of having to wait until all processes have been gone through. In this way, unnecessary costs may be avoided. A second important subdivision of rework situations from a planning and control point of view concerns whether rework is done with partly or completely the same resources as used for production.

Hereafter, we consider the relatively simple single-stage situation with a production facility dedicated to one product that is also used for rework. As a specific detail of the general product recovery system in Figure 10.1, Figure 10.7 shows this situation, which, among others, effectively occurs in the production of integrated circuits (see, for example, So and Tang, 1995). Among the reasons for using the same resources are high investments in machines and tools, or to realize quality improvements by letting operators rework the products that they did not produce well (see e.g. Robinson et al., 1990).

In the literature, the following two strategies are usually considered for situations with setup times when going from production to rework or vice versa (for an overview of rework strategies in the literature, see Flapper and Jensen, 2002)).

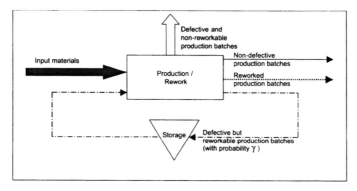

Fig. 10.7. A production situation with shared resources for production and rework

Strategy I : Rework after a certain number Q_p of products/production batches has been finished

Strategy II : Rework once there is a predetermined number Q_r of products/production batches with a certain defect

In the case of discrete products like cars, individual products may turn out to be defective and may be reworked. Therefore, in this case, Strategies 1 and 2 are based on individual products. In the case of batch process industries, products like liquids are produced in tanks. The quantity produced in one tank, usually denoted by batch, is the smallest quantity that can be produced and therefore the contents of one tank (i.e. one batch) is the smallest quantity that may have to be reworked.

For the basic models presenting an application of *Strategy I* and *Strategy II*, we make several assumptions. There are no machine breakdowns. Batches become randomly and mutually independently reworkable defective with a constant probability γ. There occurs only one type of defect, so all defective batches will be reworked. It is assumed that rework is always successful. In order to keep things as simple as possible, we consider a situation where everything that is produced or reworked well can be immediately sold with no difference in sales price. Further, it is assumed that the tests to estimate whether or not a produced batch needs to be reworked take no time and that these tests give always 100% reliable results. Holding costs are charged according to the maximum space required for storing batches to be reworked, and the space has to be reserved for them and can only be used for these batches.

For the above situation, under the given assumptions and *Strategy I*, the objective function to be maximized concerns the average profit Π per hour.

$$\Pi(Q_p) = \frac{(s - c_p) \cdot Q_p - (1 - (1 - \gamma)^{Q_p}) \cdot c_r \cdot Q_r}{L_p \cdot Q_p + (1 - (1 - \gamma)^{Q_p}) \cdot (F_p + F_r) + L_r \cdot Q_r} - h_u \cdot Q_p$$

Hereby, we use the following notation.

Q_p : the number of production batches after which rework is started

Q_r : the average number of production batches that have to be reworked

γ : the probability that a production batch is defective

s : the sales price of a good batch

c_p : the production cost per production batch (including cost for materials)

c_r : the rework cost per production batch

h_u : the storage costs per hour for storing production batches to be reworked

L_p : the time to produce one batch (in hours)

L_r : the time to rework one batch (in hours)

F_p : the fixed time required to setup the production facility for production (in hours)

F_r : the fixed time required to setup the production facility for rework (in hours)

The numerator in profit function $\Pi(Q_p)$ denotes the expected costs of a production-rework cycle, whereas the denominator denotes the expected time of a production-rework cycle. The term $(1-(1-\gamma)^{Q_p})$ denotes the probability that one or more of the Q_p batches that are produced directly after each other turn out to be reworkable defective. Once we know that defective batches have been produced, we know that on average $\gamma \cdot Q_p$ defective batches will have been produced. Therefore: $Q_r = \gamma \cdot Q_p$.

The optimal value for Q_p can be determined via standard methods, using the first- and second-order derivatives of the objective functions with respect to Q_p after which, in case of a non-integral value for Q_p, the value of the objective function for the two closest integer values to the found value have to be estimated and compared.

Under *Strategy II*, when Q_r is the decision variable, the objective function reads:

$$\Pi(Q_r) = \frac{[(s-c_p) \cdot Q_p - c_r \cdot Q_r]}{[L_p \cdot Q_p + F_p + F_r + L_r \cdot Q_r]} - h_u \cdot Q_r, \tag{10.1}$$

where we use the expected value $Q_p = \frac{Q_r}{\gamma}$.

For extensions of the situation considered above, including stochastic times, probabilities, several types of defects, and unreliable test results, see the literature overview given by Flapper and Jensen (1999). Chapter 7 of this book presents lot-sizing models for rework systems in the case of deterministic production yields.

10.5.2 Planning and Control of Rework in the Process Industries

Rework is of specific importance in many process industries. There are a number of issues which make planning and control of rework in the process industry different from rework in other industries. As stressed in Flapper et al. (2002), among the most salient issues from a rework point of view are storage time restrictions. Due to the decreasing quality of work-in-process, the change in quality of defective batches waiting to be reworked has to be taken into account. In this section, we will point out how the latter influences the planning and control of rework for the relatively simple situation considered in Subsection 10.5.1.

We consider a situation with a production tank dedicated to one product, where this tank is used both for production and rework. In order to be able to rework all defective production batches, there should be space to store them. There are Q_p tanks only available for storing the production batches of the product that have to be reworked. Note that, as is usual in the process industry, it is not allowed to store more than one batch in one tank due to the high required level of traceability. Investments in storage tanks are taken into account via depreciation and loss of interest. It is assumed that the loading and unloading of tanks do not require time. Moreover, it is assumed that the required time for rework linearly depends on the time span t between the moment that a batch has been produced and the moment that reworking the batch is actually started, so that the rework time L_r becomes a function of t, i.e. $L_r(t)$. In the case of linear dependency, we have

$$L_r(t) = L_r^f + L_r^v \cdot t.$$

with L_r^f as fixed and L_r^v as the variable rework time component.

Further, the same assumptions as in 10.5.1 are made. Hereafter, we only indicate the consequences for the objective function related to *Strategy I* for the special case $Q_p = 2$. The consequences for the objective function related to *Strategy II* are much more complex. For the latter and further details, including the concrete expression for the objective functions, see Flapper and Teunter (2001) and Teunter and Flapper (2003).

When the time required for rework does not depend on the time that a production batch has to wait before it is actually reworked, it is sufficient to know only the number of reworkable defective batches. However, in the case of linearly increasing rework time, it also becomes important to know which production batches are reworkable defective and in which sequence these defective batches are reworked. Note that, from the latter point of view, the situation considered here has a lot in common with the situation with value deterioration that has been discussed in Subsection 10.3.2. If only one of the two batches is reworkable defective, then, assuming that rework is always successful, it is important to know which of the two, because the first production batch has to wait longer than the second. If both batches are defective, it matters which defective batch is reworked first and which second. As is shown in Flapper and Teunter (2001), a Last Come First Served (LCFS) priority rule for reworking is optimal in the environment described above.

It will become clear that the expression for the objective function for the situation with linearly increasing processing times for rework is more complex than for the situation discussed in Subsection 10.5.1. For the specific situation considered, assuming $Q_p = 2$, the rework time term in the profit function in Subsection 10.5.1 (i.e. $L_r \cdot Q_r$) would have to be replaced by the following more complex expression:

$$L_r \cdot Q_r = [\gamma \cdot (1 - \gamma) \cdot (L_r^f + L_r^v \cdot (L_p + F_r))] + [(1 - \gamma) \cdot \gamma \cdot (L_r^f + L_r^v \cdot F_r)] +$$

$$+\gamma^2 \cdot [(L_r^f + L_r^v \cdot F_r) + (L_r^f + L_r^v \cdot (L_p + F_r + L_r^f + L_r^v \cdot F_r))] ,$$

where the first term is related to the situation that only the first production batch turns out to be defective (i.e. $L_r = L_r^f + L_r^v \cdot (L_p + F_r)$), the second term is related to the situation that only the second production batch turns out to be defective (i.e. $L_r = L_r^f + L_r^v \cdot F_r$), and the third term denotes the situation where both production batches turn out to be reworkable defective and an LCFS rule is used (i.e. $L_r = L_r^f + L_r^v \cdot F_r$ for the batch reworked first and $L_r = L_r^f + L_r^v \cdot (L_p + F_r + L_r^f + L_r^v \cdot F_r)$ for the batch reworked second).

Generally, one may expect that the optimal value for Q_p will be less in situations where the rework time increases with the period of time that a production batch has to wait for its rework, because keeping batches to be reworked longer means also longer rework times and thereby less time available for producing new batches.

A lot of rework situations in practice require further research (see Flapper and Jensen, 1999, and Flapper et al, 2002). For instance, hardly any attention has been paid so far to the growing number of situations where different products share resources. Also, with respect to the process industries, in a lot of food and pharmaceutical companies there exist restrictions with respect to the maximum period of time that a product-in-process can be stored in order to allow rework at all. This has not been dealt with so far.

10.6 Conclusions

In this chapter, it has been demonstrated why traditional PPC approaches are not sufficient to overcome the specific planning and scheduling problems that occur in a product recovery environment. This holds for systems with external as well as internal return flows of recoverables, i.e. for situations where products enter a production system externally after being used by customers, and for environments where by-products or defective items are generated internally due to incomplete controllability of production processes. Several complicating characteristics have been described for PPC in systems with product recovery. From a general point of view, the additional processes in these systems like disassembly, reprocessing, and reassembly can be treated as specific production processes. However, these processes incorporate such special tasks and generate such complicated coordination problems that additional or modified PPC tools are needed. These needs are increased by the fact that respective planning and control systems have to be able to cope with serious additional uncertainties which often are typical for product recovery problems.

In recent years, many research contributions helped to gain more insight into the way the above mentioned complicating factors are influencing the results of decision making in product recovery management. Several scientific tools have been developed to support PPC tasks in the many different fields of this area. In this Chapter, we gave many examples of recent research contributions in the whole planning area ranging from disassembly and recovery planning to production control and job shop scheduling on the one side, to integrated materials and capacity planning on the other. These approaches are dedicated to help managers in improving their methods to solve production planning problems when they are dealing with product recovery. However, it is indicated that despite all scientific progress, a lot of respective planning problems still are not yet supported satisfactorily by quantitative tools, thus demanding further research.

Valuation of Inventories in Systems with Product Recovery

Ruud H. Teunter[1] and Erwin A. van der Laan[2]

[1] Rotterdam School of Economics, Erasmus University Rotterdam, P.O. Box 1738, 3000 DR Rotterdam, The Netherlands, `teunter@few.eur.nl`
[2] Rotterdam School of Management / Faculteit Bedrijfskunde, Erasmus University Rotterdam, P.O. Box 1738, 3000 DR Rotterdam, The Netherlands, `elaan@fbk.eur.nl`

11.1 Introduction

Valuation of inventories has different purposes, in particular accounting and decision making, and it is not necessary for a firm to use the same valuation method for both purposes. In fact, it is not uncommon to use accounting books as well as management books. In this chapter, we will only consider inventory values from the perspective of decision making. More specifically, we will analyze the effect of inventory valuation on inventory control decisions (and not the corresponding financial results) for systems with product recovery. Of course, inventory valuation also influences other strategic and operations management decisions concerning product recovery, as is illustrated by the following real-life example.

Case: Product Recovery of Copiers

A copier producer/remanufacturer has a single European (re)manufacturing facility (CF) and a number of National sales/lease Organizations (NO). The NO operate independently, purchasing new and/or remanufactured copiers from the CF and returning used copiers to the CF. The product flows are controlled by the CF using internal purchase prices and return fees. The return fees for used copiers and the purchase prices for remanufactured copiers are based on the valuation by the CF of used and remanufactured copiers. Using different valuation methods over the last few years has lead to some important insights. Higher return fees lead to more returns of newer models. Though this implies more recovery opportunities, there is also the associated risk of NO returning copiers that can still be leased/sold at a reasonable profit. As expected, higher purchase prices for remanufactured copiers lead to less demand from the NO. Lower values for used and remanufactured copiers indicate that remanufacturing is less expensive than production, motivating designers to build 'green' copiers.

But our focus is solely on inventory control decisions. We start by describing the link between inventory values and inventory control decisions. This discussion is not restricted to systems with product recovery. In fact, it is based on literature for systems without product recovery.

Both in the modeling theory and in practice, inventory control decisions are often based on an *average cost (AC)* model of reality. Instead of analyzing the effect of inventory decisions on cash flows, the AC approach transforms cash flows into costs because those are easier to work with. Consider, for instance, the purchase of a product at price c at time t. Assume that this product is kept in stock until a demand occurs at time $T \geq t$. The associated purchase cash flow, discounted at rate α to time T, is $ce^{\alpha(T-t)} \approx c + c\alpha(T - t)$. The AC approach transforms this cash flow into a purchase cost c at time t and a holding cost $h(T - t)$ during period (t, T). where h is the so-called opportunity holding cost rate. Clearly, for this example, setting $h = c\alpha$ approximately transforms the *discounted cash flow (DCF)* into costs.

This simple example illustrates that the inventory holding cost rates are crucial in that transformation process. The AC approach adds an opportunity (non-cash flow) holding cost rate to the 'true' out-of-pocket (cash flow) holding cost rate. The opportunity holding cost rate is generally calculated by multiplying the value of a product with the discount/interest rate (see also the above example). See, for instance, Naddor (1966), Silver *et al.* (1998), and Tersine (1988). So, in AC approaches, product values influence inventory control decisions via holding cost rates.

For many models with forward logistics only, it has been shown that the approximate AC approach with holding cost rates calculated as described above leads to (nearly) the same inventory control decisions as the exact DCF approach (see Klein Haneveld and Teunter, 1998, and Corbey *et al.*,1999). But does the same hold for models with product recovery? And if so, how do we determine the 'right' values for recoverable and recovered products? These questions were avoided in Chapters 7 to 10, where many AC models of inventory systems with product recovery were analyzed. There, the holding cost rates were simply assumed to be given, without referring to the underlying cash flows. In this chapter, we will deal with the above questions. To keep the discussion and analysis transparent, we do not consider out-of-pocket holding costs, so costs related to inventory investment are assumed to be 'opportunity costs' only. We remark, however, that out-of-pocket holding costs can easily be included in an AC approach by adding the out-of-pocket holding cost rate to the opportunity holding cost rate.

The remainder of this chapter, is organized as follows. In Section 11.2, we review those studies that compare the AC model and the DCF model analytically. All consider the simple inventory system that is depicted graphically in Figure 11.1. (Some studies include the 'dotted' disposal option, but others do not.)

The results of these studies show that there indeed exist values (and corresponding holding cost rates) that approximately transform the AC into the DCF. Some of those values are surprising, in the sense that they would not result from the traditional 'cost price reasoning'. In Section 11.3, we discuss a simulation study (of a model based on the inventory systems depicted in Figure 11.1) with similar findings, and offer intuitive explanations. Moreover, based on those explanations, we discuss the conditions under which we expect certain values to approximately transform AC into DCF. In Section 11.3, this discussion is restricted to the inventory system in Figure 11.1, but in the following section we also consider more complex inventory systems with (dis)assembly of (returned) products. In Section 11.5, we discuss the implications of these findings for systems without recovery but with multiple sources for obtaining serviceable products. We end with conclusions and directions for future research in Section 11.6.

Throughout this chapter, we use the notations listed in Table 11.1.

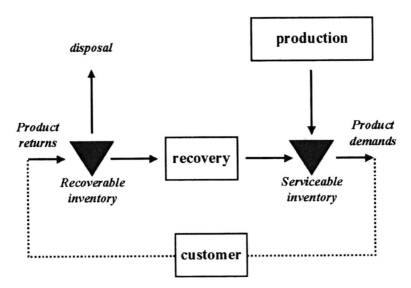

Fig. 11.1. A simple inventory system with product recovery

11.2 Analytical Comparisons of Average Cost and Discounted Cash Flow

11.2.1 Discounted Cash Flow, Net Present Value, and Annuity Stream

Consider a series of deterministic cash flows $C_1, C_2, ..., C_N$ occurring at times $T_1, T_2, ..., T_N$. The total discounted cash flow, i.e. the *net present value (NPV)*, of this series can be calculated as

$$\sum_{n=1}^{N} C_n e^{-\alpha T_n}.$$

For stochastic systems, let us use the NPV as the expectation of the total discounted cash flow over an infinite horizon. As an example, consider a cash flow C that occurs at stochastic times $T_1, T_2, ...$ with independent inter-occurrence times, $S_1 = T_1 - 0$ and $S_n := T_n - T_{n-1}$ for $n > 1$, with probability density function

$$f_{S_n}(t) = \begin{cases} z_1(t). & n = 1 \\ z(t). & n > 1 . \end{cases}$$

Note that inter-occurrence times are identically distributed from time T_1, i.e.. the cash flow is cyclic from time T_1. But the cash flow is not necessarily cyclic from time 0, and therefore the first inter-occurrence is treated separately. Such cash flow series are basic elements that return in many of the subsequent models. The NPV of this series of cash flows, discounted at rate α, equals

Table 11.1. Notations

d	demand rate
u	return rate
α	discount rate
c_p	production cost (per product)
c_r	remanufacturing cost (per product)
c_w	disposal cost (per product)
K_p	set-up cost for production
K_r	set-up cost for remanufacturing
K_w	set-up cost for disposal
AC	average cost
NPV	net present value
AS	annuity stream αNPV
Q_p	production order quantity
Q_r	remanufacturing order quantity
h_s	holding cost rate for serviceable products
h_r	holding cost rate for remanufacturable returned products
h_w	holding cost rate for disposable returned products

$$NPV = \mathrm{E}\left\{\sum_{n=1}^{\infty} C e^{-\alpha T_n}\right\} = C \sum_{n=1}^{\infty} \int_0^{\infty} z_1 * z^{*(n-1)}(t)\, e^{-\alpha t}\, \mathrm{d}t$$

$$= C\tilde{z}_1(\alpha) \sum_{n=1}^{\infty} \tilde{z}(\alpha)^{(n-1)} = \frac{C\tilde{z}_1(\alpha)}{1 - \tilde{z}(\alpha)},$$

where the asterisk denotes convolution and $\tilde{z}(s) = \int_0^{\infty} z(t)e^{-st}\mathrm{d}t$ denotes the Laplace transform of time function $z(t)$, $t \geq 0$, for complex s.

If a stream of cash flows has net present value X, then a continuous cash flow of αX has the same net present value, i.e. X. Therefore, we define the so-called *Annuity Stream (AS)* as

$$AS = \alpha NPV.$$

The annuity stream is useful since it can be compared directly with average costs.

11.2.2 A Stochastic Inventory Model with Production and Instantaneous Remanufacturing

For a first analysis, we consider the inventory system depicted in Figure 11.1. To keep the analysis tractable, we assume that the demand process and the return process are *independent* Poisson processes with rates d and $u < d$ respectively. We further assume that the lead times for both manufacturing and remanufacturing are zero, and that backorders are not allowed.

There is a setup cost K_p per production order, a variable production cost c_p per product, and a variable remanufacturing cost c_r per product. Here $c_r < c_p$, otherwise remanufacturing would never be preferred over production.

Inventory is controlled by a continuous review PUSH strategy, which is defined as follows (see Section 8.3.2 of Chapter 8 for a detailed discussion). Remanufacturing is instantaneous, i.e. starts as soon as a product is returned. So, there is no disposal nor stocking of remanufacturables. Production occurs in batches of fixed size Q_p, and starts whenever the inventory drops below zero. Note that this is the optimal strategy structure under the above assumptions and cost structure (see Fleischmann and Kuik, 2003). The (serviceable) inventory level at time zero equals I_0. Under the NPV criterion all cash in- and out-flows are discounted with opportunity cost rate α.

For simplicity, we do not consider cash flows related to sales and acquisition of product returns, but they could easily be included in the analysis.

Define $\{R_n | n \geq 1\}$ and $\{P_n | n \geq 1\}$ as the occurrence times of remanufacturing orders and production orders, respectively. Note that the stream of remanufacturing orders is a Poisson process, since the return process is a Poisson process also, and all returns are instantaneously remanufactured. Since $\sum_{n=1}^{\infty} \mathrm{E}\left(e^{-\alpha R_n}\right) = u/\alpha$, the annuity stream as a function of order size Q_p reads

$$AS(Q_p) = \alpha \sum_{n=1}^{\infty} \mathrm{E}\left(c_r e^{-\alpha R_n} + (K_p + c_p Q_p)e^{-\alpha P_n}\right)$$

$$= c_r u + \alpha(K_p + c_p Q_p)\frac{\tilde{f}_{I_0}(\alpha)}{1 - \tilde{f}_{Q_p-1}(\alpha)}, \tag{11.1}$$

where $f_i(t)$ is the probability density function with regard to the first-occurrence time of production orders, given that the process starts with inventory level i. These functions cannot be calculated directly, so instead we develop the following procedure.

Suppose that at time zero the inventory level is $i \geq 0$. Either the next occurrence is a demand at time t. with probability density $g(t) = de^{-(d+u)t}$. which moves the inventory down to $i - 1$. or the next occurrence is a return. with probability density $h(t) = ue^{-(d+u)t}$. which moves the inventory level up to $i + 1$. Thus we have

$$f_i(t) = \begin{cases} g(t) + h * f_1(t), & i = 0, \\ g * f_{i-1}(t) + h * f_{i+1}(t), & i > 0. \end{cases} \tag{11.2}$$

Taking Laplace transforms and evaluating at α, (11.2) becomes

$$\tilde{f}_i(\alpha) = \begin{cases} \tilde{g}(\alpha) + \tilde{h}(\alpha)\tilde{f}_1(\alpha), & i = 0, \\ \tilde{g}(\alpha)\tilde{f}_{i-1}(\alpha) + \tilde{h}(\alpha)\tilde{f}_{i+1}(\alpha), & i > 0, \end{cases} \tag{11.3}$$

where $\tilde{g}(\alpha) = \frac{d}{d+u+\alpha}$ and $\tilde{h}(\alpha) = \frac{u}{d+u+\alpha}$ are the Laplace transforms of $g(t)$ and $h(t)$ respectively, and $\tilde{f}_i(\alpha)$ denotes the discounted first occurrence time of a production order. given that the inventory level starts at state i. Solving equations (11.3) for $\tilde{f}_i(\alpha)$ gives

$$\tilde{f}_i(\alpha) = \left(\frac{1 - \sqrt{1 - 4\tilde{g}(\alpha)\tilde{h}(\alpha)}}{2\tilde{h}(\alpha)}\right)^{i+1}, \quad i \geq 0. \tag{11.4}$$

Combining (11.4) and (11.1) it can be shown (see Van der Laan, 2003) that the linearization of $AS(Q_p)$ is written as

$$\overline{AS}(Q_p) = c_r u + \left[c_p(d - u) + (K_p + a)\left(\frac{d - u}{Q_p}\right)\right.$$

$$\left. + \alpha c_p\left(\frac{Q_p - 1}{2} + \frac{u}{d - u}\right) + b\right]\tilde{f}_{I_0}(\alpha). \tag{11.5}$$

with $a = \alpha K_p\left(\frac{d+u}{2(d-u)^2}\right)$ and $b = \alpha(c_p + K_p/2)$. We may interpret a as the (relevant) opportunity cost per batch of a production order.
Note that $\tilde{f}_{I_0}(\alpha)$ is a constant, so it is left out of the linearization. The traditional average cost approach computes the average cost function as

$$AC(Q_p) = c_r u + c_p(d - u) + K_p \left(\frac{d - u}{Q_p}\right) + h_s \left(\frac{Q_p - 1}{2} + \frac{u}{d - u}\right). \quad (11.6)$$

The first three terms correspond to the average variable production cost, the average variable remanufacturing cost, and the average ordering cost for production, respectively. The last term is the average serviceable inventory (see e.g. Muckstadt and Isaac, 1981) times the serviceable holding cost parameter h_s. From the traditional average cost point of view, it is not immediately clear what the value of h_s should be. The interpretation that opportunity costs of holding inventories are proportional to the average inventory investment suggests that h_s should be set to $\alpha \left(\frac{d-u}{d}c_p + \frac{u}{d}c_r\right)$, since the inventory of serviceable products is a mixture of produced (fraction $\frac{d-u}{d}$) and remanufactured (fraction $\frac{u}{d}$) products with *different* marginal costs c_p and c_r, respectively.

However, a comparison of the (approximately) optimal order quantities

$$Q_p^{\overline{AS}} = \sqrt{\frac{2(K_p + a)(d - u)}{\alpha c_p}}$$

and

$$Q_p^{AC} = \sqrt{\frac{2K_p(d - u)}{h_s}},$$

which can easily be obtained by putting the derivative of (11.5) and (11.6), respectively, with respect to Q_p equal to zero and solving for Q_p, shows that choosing $h_s = \alpha c_p$ (independent of c_r!) results in similar optimal ordering quantities for both approaches. Note that for moderate values of u, the influence of a is limited. Van der Laan (2003) shows that the linearization is very accurate for moderate return probabilities (< 0.8). Using the linearization for optimization gives very good performance even for return rates that are close to the demand rate.

We end this section with a simple example that illustrates the importance of setting the right holding cost rate for the AC approach. *The yearly demand rate is $d = 100$ and the yearly return rate is $u = 80$. Production is much more expensive than remanufacturing ($c_p = 5$, $c_r = 1$). The set-up cost for production is $K_p = 10$. The yearly discount rate is 20% ($\alpha = 0.20$). If we set $h_s = \alpha c_p = 1$, then $Q_p^{AC} = 20$, which is close to $Q_p^{\overline{AS}} = 20.2$. But if we set $h_s = \alpha \left(\frac{d-u}{d}c_p + \frac{u}{d}c_r\right) = 1.8$, then the resulting order quantity $Q_p^{AC} = 33.3$ is far from optimal.*

11.2.3 A Deterministic System with Production, Remanufacturing, and Disposal

In this section, we would like to illustrate the complications that arise if we allow for batch remanufacturing and disposal. Unfortunately, the stochastic approach of the previous section is too difficult to apply in that situation, so instead we take a deterministic approach. We remark that it is common (in forward logistics systems) to calculate order quantities using a deterministic model. The resulting order quantities are generally near-optimal for the corresponding stochastic model also. The model that we will consider is similar to those discussed in Chapter 7. The main difference is that our model includes a set-up cost for disposal, and therefore we analyze strategies with batch disposal.

Consider the model of the previous section, but instead of remanufacturing all product returns upon arrival, we collect returns during a time interval T, i.e., we collect uT products. At the end of that interval, we remanufacture $Q_r = rT$ and dispose of the rest (batch disposal because there is a set-up cost for disposal). Here, $r \leq u$ may be interpreted as the recovery rate. So, apart from the decision variable Q_p, we also have the recovery rate r as a

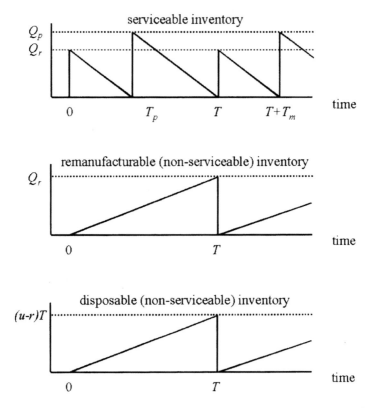

Fig. 11.2. The inventory processes for the system with production, remanufacturing, and disposal

decision variable. There are fixed costs K_r and K_w associated with remanufacturing and disposal respectively. The unit 'cost' related to disposal, c_w can be either positive (for instance, if products contain hazardous materials which need to be processed in an environmentally friendly manner) or negative (for instance, if product returns have a positive salvage value and can be sold to a third party). For ease of explanation, we assume here that the inventory system is controlled by repeatedly producing one production batch of size Q_p, succeeded by one remanufacturing batch of size Q_r. However, the analysis is easily extended to arbitrary sequences of production and remanufacturing batches. Figure 11.2 shows the inventory processes involved. Figure 11.3 shows the associated cash flows. See also Chapter 7, where this type of lot-sizing strategy for a deterministic system is discussed. We make a distinction between disposable inventory and remanufacturable inventory, because the associated cash flows differ. By the same argument, we do not distinguish between produced and remanufactured products because they represent the same value with respect to sales price.

Note that the decision variables, Q_p, Q_r, and r, are mutually dependent through the relation $Q_r = \frac{r}{d-r}Q_p$. We assume that at time 0 we start with zero inventory of both serviceables and remanufacturables. Thus, to start up the system and to guarantee a monotonous ordering strategy at the same time, we have to start with a *production* batch of size Q_r. The first *regular* production batch of size Q_p then occurs at time $T_p = Q_r/d$ and the first *remanufacturing* batch occurs at time $T = (Q_p + Q_r)/d$. Continuing this way, production batches and remanufacturing batches occur every T time units. We remark that it would not be 'fair' to compare strategies with different initial inventories, and hence the start-up batch is needed. Lead times are assumed to be zero.

In general, if T denotes the cycling time of a discrete cash flow C, with first occurrence time T_1, then the annuity stream is given by

$$AS = \alpha C \sum_{n=0}^{\infty} e^{-\alpha(T_1 + nT)} = \frac{\alpha C e^{-\alpha T_1}}{1 - e^{-\alpha T}}, \qquad (11.7)$$

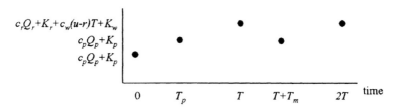

Fig. 11.3. Relevant cash flows for the system with production, remanufacturing, and disposal

and its linearization in α (using a Tayler series approximation) by

$$\overline{AS} = \frac{C}{T} + \alpha C \left(\frac{1}{2} - \frac{T_1}{T} \right).$$

Using those results and choosing C and T_1 according to the cash flows in Figure 11.3, we find that for $0 \leq r \leq u < d$, the total annuity stream is given by

$$AS = \alpha \left(K_p + Q_r c_p + \frac{(K_p + Q_p c_p)\mathrm{e}^{-\alpha T_p}}{1 - \mathrm{e}^{-\alpha T}} \right.$$
$$\left. + \frac{(K_r + K_w + Q_r c_r + (u - r)Tc_w)\mathrm{e}^{-\alpha T}}{1 - \mathrm{e}^{-\alpha T}} \right).$$

which can be approximated by the function

$$\overline{AS} = c_p(d - r) + c_r r + c_w(u - r)$$
$$+ (d - r)K_p/Q_p + r(K_r + K_w)/Q_r + \alpha K_p - \alpha(K_r + K_w)/2$$
$$+ \alpha c_p Q_r + \alpha(K_p + c_p Q_p)\left(\frac{1}{2} - \frac{T_p}{T}\right) - \alpha c_r Q_r/2 - \alpha c_w(u - r)T/2$$
$$= c_p(d - r) + c_r r + c_w(u - r)$$
$$+ (d - r)K_p/Q_p + r(K_r + K_w)/Q_r + \alpha K_p - \alpha(K_r + K_w)/2$$
$$+ \alpha(K_p + c_p Q_p)\left(\frac{1}{2} - \frac{r}{d}\right) + \alpha(c_p - c_r/2)Q_r - \alpha c_w\left(\frac{u}{r} - 1\right)Q_r/2.$$

Next, we apply the traditional AC approach. Clearly, the numbers of produced, recovered, and disposed items per time unit are $d - r$, r, and $u - r$, respectively. The fractions of demand satisfied by production and recovery are $(d - r)/d$ and r/d, respectively. Combining this with Figure 11.2 and using $Q_r = rT$ gives a total average cost of

$$AC = c_p(d - r) + c_r r + c_w(u - r)$$
$$+ (d - r)K_p/Q_p + r(K_r + K_w)/Q_r$$
$$+ h_s\left[(1 - \frac{r}{d})Q_p/2 + \left(\frac{r}{d}\right)Q_r/2\right] + h_r Q_r/2 + h_w\left(\frac{u}{r} - 1\right)Q_r/2,$$

where h_s, h_r, and h_w are the holding cost rates for serviceables, remanufacturables, and disposables, respectively. Using the Relation $Q_r = \frac{r}{d-r}Q_p$, it is easily verified that we can transform AC into $\overline{AS}$ (up to a constant) by using the following transformation of c_p, h_s, h_r, and h_w.

$$c_p \;\to\; c_p + \alpha K_p/d$$

$$h_s \;\to\; \alpha c_p$$

$$h_r \;\to\; \alpha(c_p - c_r) \tag{11.8}$$

$$h_w \;\to\; -\alpha c_w$$

This transformation is unique with respect to h_s, h_r, and h_w. Note the value of h_s is consistent with the results of the previous section. Furthermore, in the special case that nothing is disposed of, i.e. $r = u$, the value of h_r has been validated for several deterministic models (see van der Laan and Teunter, 2002) and stochastic models, as well (Teunter, 2002, Teunter et al., 2000, van der Laan, 2003). The next section attempts to give an intuitive explanation of the above results.

11.3 Intuitive Explanations

In the previous section, the average cost per time unit and the annuity stream of the total discounted cost (net present value) were compared analytically for some simple models. A set of holding cost rates for the average cost expression was sought that approximately transforms it into the annuity stream of the discounted cost expression, or, for short, that approximately transforms the average cost into the discounted cost. For ease of notation, we will refer to those rates as *transformation rates* in what follows. The corresponding values, calculated by first subtracting the out-of-pocket holding cost rate and then dividing by the discount rate, will be referred to as the *transformation values*.

Recall that all analytical studies were based on the inventory system as depicted in Figure 11.1. Some studies considered strategies that remanufacture all returned products, but others considered strategies that also use the disposal option. For all models that were considered, it turned out that the transformation rates are as follows.

$$h_s \;=\; \alpha c_p$$

$$h_r \;=\; \alpha(c_p - c_r) \tag{11.9}$$

$$h_w \;=\; -\alpha c_w$$

Note that there are different holding cost rates for remanufacturable returned products and disposable returned products, though these products are identical. Also note that all serviceable items, whether produced or remanufactured, have the same holding cost rate. These two aspects of the transformation rates are not in agreement with the traditional 'cost price reasoning' usually applied to find the holding cost rates for average cost inventory systems without product recovery. That reasoning would lead to identical holding cost rates for disposables and remanufacturables (since their collection cost is identical), but different holding cost rates for produced and remanufactured products (since different costs are involved). From a 'cash flow point of view', however, the observed two aspects of the transformation rates do make sense. Disposables and remanufacturables have different future cash flows and hence tie up different amounts of capital, but produced and remanufactured products do not. Below, we will continue this reasoning to intuitively derive values and the corresponding holding cost rates.

First consider production and remanufacturing of serviceable products. The capital tied up in a produced product is c_p, and so its value is c_p. A remanufactured product has the same quality as a produced product and is used to satisfy the same demands, so its value is also c_p. The cost of remanufacturing a product is c_r, after which a remanufactured product with value c_p is obtained. So the value of a remanufacturable item is $c_p - c_r$. This reasoning leads to the following holding cost rates: $h_r = \alpha(c_p - c_r)$, $h_s = \alpha c_p$. See also Teunter, et al. (2000) (they use different notations). Note that these are the transformation rates in (11.9).

Next consider disposal. The disposal cost is c_w. So the capital tied up in a disposable item is $-c_w$, and hence its value is $-c_w$. We remark that c_w can be negative and hence the value positive if, for instance, disposal means the lucrative selling of remanufacturable products to a broker. If c_w is positive, however, then the value of a remanufacturable product is negative. Indeed, in such a case a remanufacturable product is a liability rather than an asset. This reasoning leads to $h_w = \alpha(-c_w)$. See also Teunter et al. (2000). Again, this is the transformation rate in (11.9).

The analytical results of the previous section and the above reasoning provide some confidence that the holding cost rates in (11.9) roughly transform an average cost model into a discounted cost model (for the system depicted in Figure 11.1). The simulation results of Teunter (2002) provide some additional evidence. He assumes that both demand and return are stochastic (driven by Poisson processes). He analyzes EOQ-strategies with fixed order quantities for production and remanufacturing (no disposal). These strategies imply that a new batch should be ordered if the stock drops below 0 (zero lead times); a remanufacturing batch is ordered if enough remanufacturable products are available and a production batch is ordered otherwise. For all the examples that he considers, the average cost model with holding cost rates as in (11.9) and the discounted cost model lead to approximately the same (simulated) optimal order quantities.

But the analytical results and their intuitive explanation also show the limitations of the average cost approach. The transformation rates (11.9) can only be applied if returned products are marked as either disposable (i.e. to be disposed of) or remanufacturable (i.e. to be remanufactured) directly upon arrival. In situations with stochastic demand and return processes, however, it seems better to apply inventory strategies that base remanufacturing and disposal decisions on the serviceable inventory level. See e.g. van der Laan et al. (1999). In such situations where a distinction between disposables and remanufacturables can not be made, the same holding cost rate has to be used for both. It seems reasonable to then use some mixture of $\alpha(c_p - c_r)$ and $-\alpha c_w$, but that will in general lead to sub-optimal remanufacturing decisions as well as sub-optimal disposal decisions (see also van der Laan and Teunter, 2002).

In the next section, we try to generalize our findings for more complex inventory systems with disassembly of returned products.

11.4 Systems with Disassembly

So far, we have restricted ourselves to the inventory system depicted in Figure 11.1. In this section, we consider more complex systems with disassembly of returned products. Most of the material presented in this section is based on Teunter (2001).

Our discussion will be for inventory systems with product (dis)assembly in general, but we will illustrate it using the example of a product that consists of three components (no subdivision of components into parts, etc.). The product structure and corresponding production costs are depicted in Figure 11.4. For ease of notation, we will refer to products as well as components, parts, etc., as assemblies.

Based on the results/arguments of the previous sections, we expect that the average cost approach is not capable of (roughly) including capital costs if different recovery/disposal options are used for one or more assemblies. Assume, for instance, that the product in Figure 11.4 is disposed of in some cases but disassembled into its three components (which are then recovered) in other cases. The decision to dispose or disassemble could depend on the

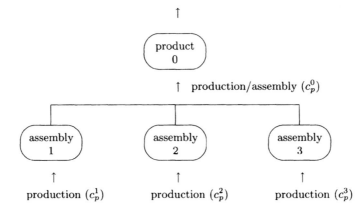

Fig. 11.4. Example of a product structure and corresponding production costs

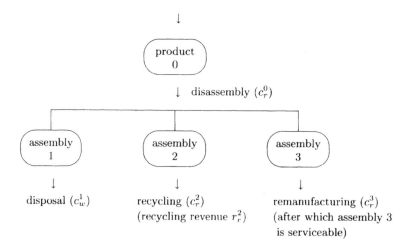

Fig. 11.5. A recovery strategy for the product of Figure 11.4 and the associated recovery/disposal costs. For assembly 2, the recycling revenues, i.e. the value of the obtained materials, is also given.

stocks of remanufacturable and produced/remanufactured serviceable components. Since there are different costs and revenues associated with the disposal option and the disassembly option, there will not be a single set of values (and corresponding holding cost rates) that approximately transforms the average cost into the discounted cost.

But we do expect that a set of transformation values exists if the recovery strategy is fixed. By recovery strategy we mean a description of the (partial) disassembly scheme and of the recovery/disposal operations for all disassembled assemblies. An example of a fixed recovery strategy for the product of Figure 11.4 and the associated costs/revenues are given in Figure 11.5. This is just one possible recovery strategy and not necessarily the optimal one. We refer readers who are interested in the determination of the optimal recovery strategy to Chapter 10. The notations that have been introduced in Figures 11.4 and 11.5 are listed in Table 11.2.

Table 11.2. Notations introduced in Figures 11.4 and 11.5

assembly	production cost	disposal cost	recovery cost	recovery revenue
0	c_p^0		c_r^0 (disassembly)	
1	c_p^1	c_w^1		
2	c_p^2		c_r^2 (recycling)	r_r^2
3	c_p^3		c_r^3 (remanufacturing)	

If the recovery strategy is fixed, then the costs/revenues associated with the recovery/disposal are also fixed, and hence the values are fixed. Based on the results in previous sections, the following valuation method is proposed (see Teunter, 2001).

Valuation Method
The value of a produced or an as-good-as-new remanufactured assembly is the production cost (price). The value of a recoverable/disposable assembly is its net profit. The net profit is equal to revenue minus cost, where the revenue of remanufacturing is the value of a remanufactured assembly.

We will illustrate this method for the product structure depicted in Figure 11.4 and the recovery strategy depicted in Figure 11.5. For further discussion on how to determine the net profits of recoverable/disposable assemblies, we refer interested readers to Krikke et al., (1998).

Using Figure 11.4, the values for produced/recovered assemblies can easily be calculated starting with the components (bottom-up, in general). Similarly, given those values, Figure 11.5 can be used to calculate the net profits of recoverable/disposable assemblies, again starting with the components (bottom-up, in general). The calculations are done in Table 11.3.

For the simple inventory system depicted in Figure 11.1 without the disposal option, the proposed valuation method indeed results in the values and corresponding holding cost rates ($h_n = \alpha(c_p - c_r), h_r = h_p = \alpha c_p$) that were shown in Section 11.2 to roughly transform the average cost into the discounted cost. Unfortunately, it is not possible to analytically test the valuation method for complex inventory systems with (dis)assembly.

Teunter (2001) tests the valuation method for an inventory system with 3 levels (one product, two components, four parts) using simulation. For all examples that he considers, it turns out that the optimal (cost minimizing) strategies under the average cost approach (combined with the valuation method) and the discounted cost approach are (almost) identical. This provides some confidence that the valuation method indeed works. We do remark that only a single recovery strategy is considered and that all production and

Table 11.3. Values for recoverable and recovered assemblies associated with the product structure in Figure 11.4 and the recovery strategy presented in Figure 11.5, resulting from the proposed valuation method

| | value of assembly | |
assembly	produced/remanufactured	recoverable
1	c_p^1	$-c_w^1$
2	c_p^2	$r_r^2 - c_r^2$
3	c_p^3	$c_p^3 - c_r^3$
0	$c_p^0 + c_p^1 + c_p^2 + c_p^3$	$-c_r^0 + (-c_w^1) + (r_r^2 - c_r^2) + (c_p^3 - c_r^3)$

recovery lead times are assumed to be zero. The zero lead times allow a restricted focus on inventory strategies that are characterized by batch sizes alone, making the search for the optimal strategy manageable.

Of course, more (simulations) tests are needed before we can be confident that the proposed valuation method works in general. Performing such tests will be time-consuming, since it involves the simultaneous determination of inventory strategies for all assemblies of a product structure. But the effort is certainly worthwhile. Knowing the way to roughly include capital costs in an average cost model of a product recovery inventory system implies that the use of a more difficult discounted cost system can be avoided.

11.5 Implications for Multiple-source Systems

The previous sections have shown that a major complicating factor for valuing inventories in systems with product recovery is the two-source character of those systems. Production and recovery have different associated marginal costs. As a result of this cost difference, it is not straightforward to determine the 'right' value (and the corresponding holding cost rates) for a serviceable product. In fact, the reported findings indicate that the 'traditional' cost-price reasoning should not be used for systems with recovery. In this section, we discuss the implications for multiple-source systems in general.

There are several reasons for using multiple sources in systems without product recovery: e.g. capacity restrictions of the source with the lowest cost (or price), a large lead time for the source with the lowest cost (regular and emergency ordering), or the reduced lead time that results from order-splitting. See also Fearon (1993) for a discussion of advantages and disadvantages of multiple sourcing versus single sourcing.

Multiple source models have been studied by many authors, e.g. Hong and Hayya (1992), Lau and Zhao (1993), and Sculli and Wu (1981). However, to the best of our knowledge, none of them address the problem of determining the correct inventory value of an arbitrary product. They either assume the value to be fixed and given, or calculate it as a (weighted) average of the marginal costs for the different sources (i.e. the average cost price).

Based on the reported finding for product recovery systems, we expect that the simple approach of averaging marginal costs is not correct from a DCF point of view. However, the approach seems reasonable as long as the differences in marginal costs are small (say less than 10%). For many multiple-source systems, the cost differences will indeed be small, since sources will not be used if they are too expensive. This partly explains why the issue of inventory valuation has not been addressed in the literature. However, for models with a regular and an emergency replenishment option, the cost difference can be large (even more than 100%). So especially for these models, future research is needed on valuing inventories.

11.6 Conclusions and Directions for Further Research

Values of recoverable and recovered products (and components, parts, etc.) are relevant for accounting purposes and also for strategic and operations management decisions. In this chapter, we discussed their relevance for inventory control decisions. Inventory control is linked to product valuation through holding cost rates. For average cost (AC) models, which dominate inventory control, it is common use to include product values (multiplied by an interest rate) in the holding cost rates, thereby roughly correcting the mistake of having not discounted cash flows properly.

In Section 11.2, the average cost and the (annuity stream of the) discounted cash flow were compared analytically for some simple models. These models are all based on a simple inventory system, where returned products can either be remanufactured or disposed of. They differ in the type of inventory strategy considered. It appeared that transformation rates, i.e. a set of holding cost rates that approximately transforms the average cost into the discounted cash flow, indeed exist (for the considered models). Moreover, we observed that these rates are very different than those that would result from applying the traditional cost-price logic. For instance, that logic would result in equal holding cost rates for all returned non-serviceable products, but it turns out that different rates should be used for remanufacturables and disposables.

In Section 11.3, we offered intuitive cash flow explanations for the transformation rates. In short, remanufacturing and disposal have different associated revenues/costs, and hence lead to different values of returned products. Unfortunately, this implies that in order to know the right (AC) holding cost rate of a returned product, one has to know whether that product will be disposed of or remanufactured. This is not necessarily true for any inventory model, and we therefore discussed the limitations of using the transformation rates that were derived in Section 11.2.

Sections 11.2 and 11.3 only considered the simple inventory model where a returned product is either directly disposed of or directly remanufactured. In Section 11.4, more complex inventory systems with product (dis)assembly were considered. We proposed a method for determined transformation rates and discussed its limitations (the recovery strategy is fixed).

Testing the proposed valuation method is one area for further research. Other research directions are to analyze the effect of (returned) product values on strategic and operations management decisions other than inventory decisions, and to perform case studies. Finally, inventory valuation should be studied for more general multiple-source systems, especially systems with a regular and an emergency supply option (see Section 11.5).

Part IV

Supply Chain Management Issues
in Reverse Logistics

12

Coordination in Closed-Loop Supply Chains

Laurens G. Debo[1], R. Canan Savaskan[2], and Luk N. Van Wassenhove[3]

[1] Graduate School of Industrial Administration, Carnegie Mellon University, Schenley Park, Pittsburgh, PA 15213, USA, laurdebo@andrew.cmu.edu
[2] Kellogg School of Management, 2001 Sheridan Road, Evanston, IL 60208, USA, r-savaskan@kellogg.northwestern.edu
[3] Henry Ford Chair in Manufacturing, INSEAD, Boulevard de Constance, 77305 Fontainebleau, France, luk.van-wassenhove@insead.edu

12.1 Introduction

In any type of supply process, closed-loop or open-loop (Chapter 2), multiple decision-makers within a firm or in different firms are involved in decisions concerning the efficiency and the profitability of the supply process. For example, for consumer goods such as one-time use cameras (Kodak), the distributor network operates as the product return point due to the disposal convenience for the consumer. Kodak collects cameras from large retailers who also develop film for customers. The retailers are reimbursed both on a fixed-fee-per-unit basis and for transportation costs. Decision-makers in different companies control the return flow of cameras to the remanufacturing facility. Return flows of products can also be influenced by different decision-makers within a single firm. One example is tire retreading, a common recovery process for tires. When the tread of a tire is worn, the tire can no longer be used but the casing is often in good shape. Retreading is the process of replacing the worn tread by a new tread and reusing the casing. Typically, the demand of retreaded tires depends on the price discount that is given with respect to new tires. The supply of retreadable tires, however, depends on the technology chosen by the tire manufacturer. Hence, technology decisions made by the product design department, and marketing decisions made by the marketing department, determine the flows of retreaded tires.

Typically, the decision-makers' decisions interact and determine the total system performance. However, often decision-makers pursue their own local objectives only. Reverse logistic flows in a supply chain magnify the dependency among the decision makers' objectives at a second level.

In the examples cited above, it is obvious that in a decentralized decision-making environment, the equilibrium outcome will not be optimal from a systems perspective. Some degree of coordination is necessary in order to align the incentives of the individual decision-makers with the objective of achieving greater efficiency from the point of view of the overall supply chain.

Much of the recent literature in supply chain management is concerned with such coordination in an open-loop supply chain (Tayur et al., 2000). Researchers study incentive alignment using complex contracts, better decision-making through sharing of information, and integration of different functional domains.

On the other hand, the emerging literature on reverse logistics has focused primarily on single decision-maker problems within a single functional area (eg. production or inventory control).

In this chapter, our goal is to present research opportunities in the coordination of closed-loop supply chains and a summary of the research undertaken in that area during the REVLOG project. The research discussed will consider decision-making both at the firm level (i.e. decisions of a manufacturer and a retailer) and at the functional level (i.e. decisions of a marketing manager and a product design manager). To better position the REVLOG work, we first start by reviewing recent advances in reverse logistics as well as the work on coordination in open-loop supply chains.

12.2 Recent Topics on Coordination

A supply chain is a network of organizations that jointly add value to a product. Adding value to a product, from an operations management perspective, is delivering the right product at the right time, in the right quantity and with the right quality. The organizations in a supply chain are connected to each other by material, information, and financial flows. Within a supply chain, different decision-makers typically control these flows. The decision-makers in a supply chain belong either to the same organization, but have different functions (e.g. production, logistics, marketing, product design, etc.), or they belong to separate organizations (e.g. a component supplier, a retailer, a manufacturer, etc.). The total added value to the product depends on the collective actions of all decision-makers in the supply chain. Often, each decision-maker's incentive to add value depends on local objectives. As all decision makers exert externalities on each other via the network of flows that connects them (physical, informational, and financial), it is unlikely that pursuing local objectives will maximize the value added to the product.

Supply chain coordination aims at obtaining maximum added value in a system controlled by many decision-makers. Often, though, many obstacles stand in the way of reaching this goal. Researchers of supply chain coordination have identified and studied several of these obstacles and found ways to overcome them. Below, we summarize the most important ones.

Incentive Alignment

In supply chains involving different companies, typically a price is set between the buying company and selling company. Financial incentives of each company are clearly opposed, as the price determines the revenues for the selling company and the cost for the buying company. This results in suboptimal value delivery from the supply chain point of view. The costs for the retailer, when determining the optimal order quantity, is based upon the wholesale price quoted by the manufacturer. However, the manufacturer's production costs rather than the wholesale price determine the optimal order quantity from the point of view of the supply chain. Therefore, the retailer orders and sells less than the optimal quantity from the point of view of the supply chain. This is the well-known principle of double marginalization in the economics literature (Tirole, 1988).

Different contractual arrangements that coordinate decisions at the operational level in a supply chain have been proposed in the literature. A two-part tariff contract, in which the retailer pays a fixed fee upfront, but receives the products from the manufacturer at marginal cost, can coordinate this supply chain (Tirole, 1988).

Researchers typically study, by means of a game-theoretic model, the uncoordinated behavior in which each decision-maker maximizes his local objective (assuming that other decision-makers will do the same) and take the externalities caused by the network in which they are embedded into account. Next, they compare the uncoordinated with the coordinated supply chain performance, under which decision-makers jointly take decisions that maximize the total added value to the product. Finally, they construct a contract that aligns the financial incentives of the decision-makers in such a way that decision-makers pursuing their modified local objectives take actions that maximize the value added to the product.

For example, Pasternack (1985) studies how buy-back contracts can coordinate the material flow in a supply chain with uncertain market demand. A buy-back contract stipulates that any leftover inventory will be bought back by the manufacturer. This changes the financial incentives of the retailer. By choosing the buy-back price in the appropriate way, the order quantity of the retailer can coincide with the optimal order quantity for the supply chain.

Cachon (2003) gives an overview of optimal supply chain contracts including other features like retail price, effort to increase demand, competition between retailers, etc.

Information Sharing

Decision-makers need information from throughout the supply chain in order to maximize the total value added to the product. However, some major obstacles to full information dissemination in a supply chain remain, in spite of the growing popularity of Enterprise Resource Planning systems which make this technically possible. One of the most important consequences of the lack of timely information that lead to suboptimal value delivery in a supply chain is the bullwhip effect: a small perturbation on demand downstream can result in huge perturbations upstream. The bullwhip effect has been observed in many different industries and can be reduced by sharing downstream demand information with the whole supply chain (Lee, Padmanabhan and Whang, 1997).

Information sharing aims at improving supply chain performance by making information available throughout the supply chain. In this literature stream, researchers typically study 1) the value of information and 2) the implications of information asymmetry between two different decision-makers in the supply chain.

For example, researchers have studied the value of information in the well-known serial two-echelon inventory system, in which a supplier and a manufacturer each holds inventory of a product at different stages of the production process. Clark and Scarf (1960) studied the system-wide optimal inventory policy which depends on information about the inventory position at both stages of the supply chain. Chen et al. (2000) study the total supply chain cost decrease in a multi-echelon supply chain in which the upstream decision-maker gains access to information about the inventory position of the downstream decision-maker. A series of researchers studied asymmetry of inventory information in a two-echelon inventory system. Cachon and Zipkin (1999), Chen (1999), and Lee and Whang (1999) study the performance of the uncoordinated ('competitive') supply chain and discuss different transfer pricing schemes that have been proposed to coordinate such a two-echelon supply chain.

Cachon and Lariviere (1999) study the strategic behavior of a customer with superior knowledge about the market demand in a situation where the supplier has to decide upon capacity based on the customer's information. They determine which supply contract induces the customer to truthful information sharing. Corbett and Tang (1999) study the ability of contracts to coordinate the supply chain when one party does not know exactly the cost structure of the other party.

Functional Integration

Decision-makers in different functions within the same supply chain jointly determine the added value of the final products. However, objectives of different functional areas may differ from the overall objective of a supply chain. Functional integration aims at improving supply chain performance by coordinating decision-makers in different functions. For example, Porteus and Whang (1993) study the inefficiency resulting from the conflict of interest between a production manager (concerned about efficiency) and a marketing manager (concerned about customer satisfaction) and propose how the owner of the company should create incentives for both managers that will yield maximum returns. Lee et al. (1993) show how by choosing the appropriate product or process design significant reductions in buffer inventories can be achieved in the supply chain. They study how the added value to a product can be maximized by jointly determining product/process design and inventory management decisions.

12.3 New Perspectives on Reverse Logistics

As discussed in the other chapters of this book, the extant literature on reverse logistics and remanufacturing addresses research questions at several echelons of a supply chain. A large stream of papers considers firm-level issues related to the management of inbound remanufacturing and recovery processes (Chapters 2,5,6,7,8,9,10). At a higher level, the second stream of work deals with supply chain level problems relating to the design and the management of reverse logistics networks (Chapters 4,13,14,15). The present chapter builds on the mentioned papers but uses micro-economic and game theory to frame and investigate reverse logistics decisions, as these decisions are influenced by the conflicting incentives of independent supply chain partners.

Like forward distribution networks, in reverse logistics systems we find incentive, information, and functional coordination issues that result from pursuing locally optimal decisions by supply chain members. The challenge of inducing supply chain optimal solutions while preserving the decentralized decision making requires an indepth understanding of the following three types of flows in reverse logistics.

Material flows, which are affected by quantity, time, and place of returns, as well as their reuse potential, which often can only be assessed after product disassembly.

Information flows, which are crucial to reduce uncertainty in the timing and reusability of the returned product, as well as the time that the product spends in the market, etc.

Financial flows, which include buy-back clauses, disposal costs, and other end-of-use costs. While pricing for the original product is relatively straightforward, designing contracts specifying conditions under which the product will be taken back after use is more challenging, as the quality of the product is unknown and often difficult to observe. Yet, as Guide and Van Wassenhove (2001) argue, appropriate financial incentives are an important way of managing the physical return flow.

In Table 12.1, we summarize some recent work which examines reverse logistics problems as part of the total supply chain structure with an emphasis on pricing, incentive alignment, and information sharing.

Competition and the implications of market structure on incentives to remanufacture are investigated by Savaskan and Van Wassenhove (2000b) and Groenevelt and Majumder (2001). While Savaskan and Van Wassenhove

Table 12.1. A framework on recent closed-loop supply chain research

Competition	Value of Information Sharing	Functional Coordination/ Incentive Alignment
Focus: Prices		
Savaskan and Van Wassenhove (2000b)		
Focus: Quality / Quantity		
Groenvelt and Majumder (2001)	Toktay et al. (2000)	Debo et al. (2002), Ferrer (1996), Geyer and Van Wassenhove (2000), Corbett and DeCoix (2001)
Focus: SC Structure		
Savaskan and Van Wassenhove (2000b)		Savaskan et al. (2000a)

(2000b) model the retailer competition in prices with remanufacturing, Groen-evelt and Majumder (2001) focus on the conflict between third-party reman-ufacturers and original equipment manufacturers. Assuming the existence of an exogeneous mechanism that allocates returned units, Groenevelt and Ma-jumder (2001) develop an understanding of how competition in quantities re-manufactured will impact profitability of remanufacturing and manufacturing decisions for the OEM.

The impact of information has been examined by Toktay et al. (2000) by using a closed-loop queuing network which models forward and reverse logistics channel structures.

Coordination of product design decisions (such as durability) with mar-keting decisions (such as customer segmentation) is also studied in several papers. Ferrer (1996), Geyer and Van Wassenhove (2000), and Debo et al. (2001) are some recent examples. Corbett and DeCroix (2001) investigate a different type of incentive alignment and coordination problem. They examine contract structures that would induce efforts to minimize consumption of in-direct materials between a supplier and a manufacturer, focusing on reduction of consumption as well as its environmental consequences. They discuss selling versus leasing alternatives and their impact on incentives for different supply chain members. Finally, Savaskan et al. (2000a) model the choice of the supply chain structure and the coordination in prices with product remanufacturing.

We can also make a couple of general remarks by inspecting Table 12.1.

1. It is striking that in the reverse logistics literature the concept of price plays almost no role, while it is essential in most of the SC coordination

literature. It seems that pricing may also play a role in different reverse activities, e.g. as rebate prices in Savaskan and Van Wassenhove (2000b), or as discounts for remanufactured products that are differentiated from new products (Ferrer, 1996; Debo et al., 2001).

2. The well-established general recommendation in designing supply-chain incentive mechanisms is to attempt to make all parties internalize the full system-wide costs and benefits associated with their decisions. This requires the ability to offer more complex contracts than pure, spot-market, volume-based agreements. This could include two-part tariffs instead of simple linear prices, or even more complicated 'menus' of contracts from which the other party can choose the most appropriate. The presence of multiple flows of goods (for instance, a forward and a reverse flow) can in itself be used as a way to implement more sophisticated contracts.

3. Similarly, the notions of 'design for disassembly' or 'design for remanufacturing' or 'design for environment' are increasingly well established in the design world. For true closed-loop supply chains, one might expect all of these design concepts to be important, which clearly requires coordination between the design teams, logistics parties (both forward and reverse), remanufacturers, etc. (see Debo et al., 2002, and Geyer and Van Wassenhove, 2000, for examples).

4. While the literature on design for reuse provides insights into design issues for remanufacturing systems, there is a need to develop a more interdisciplinary approach to address issues for remanufacturing across several product generations.

We believe that there are two fundamental research questions that should constitute the basics of future research in this area:

- How can we use our learnings to date and the existing theory in forward supply chains to address issues faced in reverse logistics?
- How can the theory of reverse logistics help us improve our understanding of forward supply chain issues?

Because the forward supply chain literature is more developed, we expect that many lessons from this area can be of value for closed-loop supply chains, but the exact nature of the relationships between forward (or open-loop) and reverse (or closed-loop) supply chains still needs to be examined. The list of such research questions can easily be extended through critical thinking about reverse logistics as it relates to the overall supply-chain strategy of a company. In the next section, we describe Savaskan et al. (2000a), Savaskan and Van Wassenhove (2000b), and Debo et al. (2002), research undertaken during the REVLOG project and an attempt to study coordination issues in a closed-loop supply chain.

12.4 Reverse Logistics and Coordination Research During the REVLOG Project

In this section, we present recent work on coordination issues in reverse logistics. In particular, we present two streams of work, Savaskan et al. (2000a) and Savaskan and Van Wassenhove (2000b), and Debo et al. (2002), which investigate the interaction of forward and reverse channel decisions at the functional level, as well as at the firm level.

The first investigates the strategic implications of reverse channel choice on forward channel product pricing decisions in a monopolistic as well as in a competitive retail environment. The second investigates the economics of technology choice and market segmentation for remanufacturable products. For more details, the reader may also wish to consult the PhD dissertations of Savaskan (2002) and Debo (2002).

12.4.1 Reverse Logistics Channel Design and Coordination Issues

Motivation

The first stream of research looks at the reverse channel choice decision of a manufacturer and its implications on the forward channel product pricing decisions. More specifically, it addresses the following research questions.

- What is the optimal reverse channel structure in a closed-loop supply system from the points of view of the manufacturer, the retailer, and the consumer?
- How can the reverse channel be used for coordinating pricing decisions in the forward channel?

Three reverse channel formats are analyzed using game theoretic modeling. A centralized system in which the manufacturer assumes the product collection activity, a decentralized system in which the retailers act as the product return points, and an outsourced system in which a third party performs the product take-back activity.

The economic objectives of the forward and the reverse channel members are modeled under two market structures: a bilateral monopoly and a competitive retail market. The two settings yield different structural insights as they are based on differing market powers of the retailers. More precisely, in the bilateral monopoly case, the single retailer has full control over market demand, whereas in a competitive retailing environment each retailer impacts only his local market size by taking into account the competition from the other retailer.

The study of the two market structures reveals the implications of the strategic interaction in the distribution channel on the design of reverse logistics channels. Monopolistic settings focus on the vertical interaction between the retailer and market demand, and competitive retailing examines the horizontal interaction between competing retailers.

First, we present the basic modeling framework of the two models, and next discuss the insights on reverse channel choice and the use of the reverse channel as a coordination tool.

A Modeling Framework for Reverse Channel Choice in Monopolistic and Competitive Retail Markets

Consider a single manufacturer-single retailer, two-echelon supply chain. The manufacturer has incorporated product remanufacturing into her original manufacturing process and can reduce her unit production cost c_m at a rate of $\Delta\tau$, where Δ is the unit cost savings from product remanufacturing and τ is the return rate of used product. τ can be influenced by the investment decision of the reverse channel member. Hence, we assume a strictly convex cost function $C(\tau)$ to model the costs incurred by the agent assuming the product collection activity. The manufacturer transfers the new product at a wholesale price w to the retailer, and the retailer sells it at the price p. We make the standard assumption of a downward sloping demand function, i.e. the demand for the product is decreasing in its own price, p. We model the incentives to invest in product collection under three reverse channel structures: Model MC, in which the manufacturer assumes product collection; Model RC, where the retailer does the collection; and Model $3P$, in which a third party assumes the reverse channel responsibility. All supply chain members are profit maximizers and the manufacturer acts as a Stackelberg leader, i.e. anticipates the reaction function of the retailer and the third party. From the concavity of the objective functions of the channel members, the existence and the uniqueness of the equilibrium in prices and the collection effort are shown easily.[4]

In the MC model, the retailer solves $Max_p\ \Pi_R^M = (p - w)(\phi - \beta p)$ to determine how much to order as a function of the wholesale price charged by the manufacturer. Given the retailer's reaction function, the manufacturer determines the wholesale price and the collection effort, and therefore solves $Max_{w,\tau}\ \Pi_M^M = D(w)(w - c_m + \tau\Delta) - B\tau^2$ for w and τ.

In the RC model, the manufacturer chooses both w and the buy-back price b of a used product. Hence, given (w, b), the retailer solves $Max_{p,\tau}\ \Pi_R^R = (\phi - \beta p)[p - w] + b\tau(\phi - \beta p) - B\tau^2$. Given the retailer's reaction function in the forward and the reverse channel, the manufacturer determines (w, b), for which she solves $Max_{w,b}\ \Pi_M^R = D(w, b)(w - c_m + (\Delta - b)\tau(w, b))$.

In the $3P$ model, the third party decides on the investment in the reverse channel and therefore chooses τ to maximize $Max_\tau\ \Pi_{3P}^{3P} = b\tau D(p) - B\tau^2$.

---

[4] Π_j^i denotes the profits of the channel member i in model j. i takes the values of M, R, and $3P$, denoting the manufacturer, the retailer and the theird party. j is set equal to MC, RC and $3P$ denoting the manufacturer collecting, retailer collecting, and the third party collecting models.

The retailer chooses the price of the product, i.e. solves $Max_p \; \Pi_R^{3P} = (p - w)(\phi - \beta p)$. Given the reaction functions of the third party and the retailer, the manufacturer sets w and b to maximize $Max_{w,b} \; \Pi_M^{3P} = D(w) (w - c_m + \tau(b)\Delta)$.

When there is competition at the retail level, the demand faced by each retailer depends not only on his own price but also the price of the competitive store. More specifically, we consider the following demand structure for retailer i. $D_i(p_i \, , \, p_j) = \phi_i - p_i + \beta p_j \quad s.t. \quad 0 \leq \beta < 1 \quad i, j = 1, 2 \quad i \neq j.$

The retailers play a Bertrand pricing game. In this case, they assume product collection, so we also solve for the Nash equilibrium in each retailer's collection effort τ_i. We consider two settings, centralized collection by the manufacturer (Model MC) and the decentralized collection by the retailers (Model RC). Consistent with the monopolistic setting, the manufacturer is modeled as the Stackelberg leader. Given the outcome of the game played at the retail level, the manufacturer solves $Max_{w,\tau} \; \Pi_M^{MC} = (D_1(w) + D_2(w)) (w - c_m + \Delta\tau) - B\tau^2$ in the MC model and $Max_{w,b} \; \Pi_M^{RC} = D_1^{RC}(w,b) (w - c_m + (\Delta - b)\tau_1(w,b)) + D_2^{RC}(w,b)(w - c_m + (\Delta - b)\tau_2(w,b))$ in the RC Model, the retailer-collecting decentralized model.

In the next subsection, we discuss insights gained from the analysis of the game theoretic models presented above.

Insights on Market Power, Scale Economies and Reverse Channel Choice

When demand is price sensitive and there are fixed costs for investing in the product collection effort, the analysis of the bilateral monopoly case shows that unless there are collection efficiency differences between the reverse channel structures, the retailer is the most suitable decision-maker in the closed-loop supply chain to assume the product take-back activity. When the incentive to invest in product collection is driven by the size of the recycling/remanufacturing market[5], the ability to impact the size of the market most directly and efficiently (i.e. market power) becomes critical and therefore the retailer should assume the product take-back activity in the closed-loop supply system. When the retailer engages in collection, the manufacturer benefits from the expansion of the consumer demand and from the savings in investment costs, the retailer benefits from the buy-back payments he receives for returned units, and the customer benefits from a lower retail price for the new product and lower product-disposal and environmental costs.

When the manufacturer engages in product collection directly from the customers, it is shown that, due to the double marginalization effect in the forward channel, her incentives to invest in product take-back are smaller than the incentives of the retailer.

---

[5] Note that this follows from the fixed costs of investing in the collection effort.

Unless there are arguments based on cost efficiency in product collection, the third-party system proves to be the least preferred structure among the three. Note that since any buy-back payment allocated to a third party is a direct cost to the manufacturer in particular, and to the channel in general, the manufacturer only partially transfers the unit cost savings from remanufacturing to the third party, whereas she fully transfers them to the retailer in the retailer collecting system. With partial transfer, the incentives of the third party to invest in product collection are lower than the incentives of the retailer.

However, in practice, manufacturers do outsource their product take-back activities to third parties. In addition to the cost efficiency argument highlighted above, an important reason to use third-party logistics channels is that they can provide several value-adding product recovery services such as sorting, dismantling, and quality monitoring, which lead to a lower cost of remanufacturing for the producers. Hence, outsourcing of product collection can be economically justifiable considering the third party's capability to impact the amount of value recovered from a remanufactured product. Including these aspects in the modeling framework is certainly an interesting direction for future research and will provide a deeper understanding of the issues in reverse channel design.

In contrast to the bilateral monopoly case, Savaskan and Van Wassenhove (2000b) show that in the competitive retailing case the pareto-optimal reverse channel structure changes from the retailer-managed channel to the centralized-manufacturer-managed structure. When each retailer impacts only the size of his local market, more scale benefits are obtained by the centralization of product take-back by the manufacturer. Hence, the results of the monopolistic case are altered when the retailer no longer holds full market power. While the vertical interaction of the retailer with the market demand is no longer a significant factor for the reverse channel choice, it is shown that the decentralization of product collection has implications for the horizontal market interaction between the retailers. More precisely, the intensity of competition between the retail outlets is higher when product take-back is decentralized. Even though this interaction effect benefits the manufacturer in terms of a lower retail margin, it does not dominate the scale effect of a centralized system. Table 12.2 provides a comparison of the benefits to the manufacturer, the retailer, and the consumer in the pareto-optimal reverse channels under each market structure.

Insights on Coordination of Pricing Decisions Through Reverse Channels

The research also shows that the reverse channel in a closed-loop supply chain can be used as a new coordination tool by the manufacturer to attain the coordinated channel performance in a decentralized setting. In the bilateral monopoly case, the buy-back payments can be incorporated into the wholesale

Table 12.2. Comparison of optimal channel structures

BILATERAL MONOPOLY *Retailer should collect*	COMPETITIVE RETAILING *Manufacturer should collect*
Benefits for the Manufacturer:	
Higher sales volume	Scale economies in collection
Savings on investment cost	Higher return rate of products
Benefits for the Retailer:	
Indirect discount on the wholesale price	Externality due to reduction in unit production costs
	Savings on investment costs
Benefits for the Consumer:	
Lower retail price	Lower retail price
Lower product disposal and environmental costs	Lower product disposal and environmental costs

pricing scheme in the form of a two-part tariff. The two-part tariff consists of a wholesale price contingent on the return rate of used products from the market, and a fixed franchise fee, which represents both the rights to sell and to take back used products of the manufacturer. In the bilateral monopoly case, the optimal buy-back payment for the retailer is set equal to the unit cost savings from remanufacturing to maximize his incentives to engage in product collection. More specifically, one can easily show that the two-part tariff takes the form (F, t) where $t = c_m - (\Delta - b)\tau$ and $F = \Pi_{coordinated} - \Pi_{decentralized}^R$.

Buy-back payments play a very important role in channel coordination when the multi-retailer setting is considered. When retailers serve markets of different sizes, the manufacturer can attain the profits of a coordinated channel only if he can charge different wholesale prices to each outlet. However, in the U.S., such a practice is restricted by the Robinson Patman Act, which protects the retailers against price discrimination by the manufacturers. It is shown that the buy-back payments for used products provide a second degree of freedom for the manufacturer to differentiate the average wholesale price charged to each retail outlet, and thereby attain the coordinated channel profits in a decentralized setting. The coordinating choice of b and w found by the simultaneous solution of $p_i^{RC}(w^{*C}, b^{*C}) = p_i^{*C}$ for $i = 1, 2$, where p_i^{*C} denotes the retail price charged in a coordinated system.

The coordination aspect examined in this work considers the channel decisions of the manufacturer and the retailers. The consumer and his decision-making process is not part of the present framework. In that respect, coordination of product return decisions and sharing of benefits from remanufacturing among the channel members, including the customer, is certainly an interesting avenue for future research.

12.4.2 Marketing of Reverse Products and Technology Selection

Motivation

Here, we discuss the critical choice of product technology and its implications on market segmentation for remanufacturable products. To set the stage for discussion, we start by describing the industry context that inspired the study.

Tire retreading is a common recovery process for tires. There is anecdotal evidence that retreaded tires have suffered for a long time from a reputation problem (Préjean, 1989). Furthermore, they are valued less than new tires and are sold at a discount price to the lower end of the market. Not all worn tires are retreadable, and thus the manufacturer is constrained in the sales of retreaded tires by the supply of retreadable tires. This is a key characteristic of remanufacturable products. Furthermore, manufacturers can typically influence the fraction of used products that are remanufacturable. Indeed, a tire manufacturer can increase the fraction of worn tires that are retreadable by increasing the amount of steel cord.

Hence, there is an interesting interaction between the choice of product technology (i.e. durability) and the size of the manufactured and the remanufactured segments in the market. Note that pricing decisions determine the demand of new and remanufactured products and production technology decisions determine the supply of remanufactured products. Thus those two decisions jointly determine the profits of the manufacturer.

We observed the same interdependency in other industries. For example, Bosch patented an Electronic Data Logger (EDL), a chip recording the peak load and temperature of a power tool during usage. The EDL has to be mounted on every new power tool. Reading this data from a used power tool allows for the increase of the fraction of used products that can be remanufactured. Furthermore, remanufactured power tools are sold at lower prices than new power tools, as they are valued less by consumers. Thus, technology decisions (i.e. adding the EDL) determine the supply of remanufacturable products while pricing decisions determine the demand for remanufactured products.

Motivated by these observations and similar observations in other industries, Debo et al. (2002) study two factors that make remanufacturing difficult: remanufactured products are valued less than new products and not all used products can be remanufactured. The manufacturer's levers to encourage demand for remanufactured products and to generate supply of remanufacturable products are typically the pricing strategy for new and remanufactured products and the technology selection for new products, respectively.

More specifically, Debo et al. address the following questions that a firm considering remanufacturing faces.

- Is it worth it to produce a more expensive remanufacturable product, knowing that the remanufactured product can only be sold to the lower end of the market?

- How does the optimal technology depend on the market characteristics?
- What drives the pricing strategy?
- How do subsidies for remanufactured products change the composition of the product portfolio?
- How do product and market characteristics determine the incentives to collect used products?

According to the market segmentation literature (Mussa and Rosen, 1978), the market is segmented with a high- and a low-quality product. However, in this literature these products are independent from one another. This is not the case when we consider new products and remanufactured products, because the remanufactured product only exists due to historical sales of new products.

The operations literature has mainly focused on production control, inventory management, and the design of networks with forward and reverse flows. In these papers, pricing is typically not considered. Ferrer (1996) studies market segmentation with new and remanufactured products, but assumes that technology is exogeneously determined. The contribution of the research of Debo et al. is to address jointly technology selection and market segmentation in a remanufacturing context and to develop insightful structural results.

The Model

The technology choice in Debo et al. (2001) is modeled by means of $c_n(q)$, which are the unit production costs of the new product as a function of $q \in [0, 1]$, which is the fraction of used products that is remanufacturable. q is referred to as the level of remanufacturability and is a strategic decision vaiarable of the manufacturer. $c_n(q)$ is assumed to be convex and twice continuously differentiable. The per-unit remanufacturing cost of a used product is c_r.

Each consumer is characterized with a parameter $\theta \in [0, 1]$, which is its willingness-to-pay for a new product. It is assumed that the willingness-to-pay for a new product is distributed on $[0, 1]$ according to a function F, where $F(\theta)$ denotes the volume of consumer types in $[0, \theta]$ and is a strictly increasing and continuous function with $F(0) = 0$ and $F(1) = 1$. The willingness-to-pay of consumer type θ for a remanufactured product is $(1 - \delta)\theta$, with δ the 'perceived depreciation' of the remanufactured product. p_n and p_r denote the prices of new and remanufactured products, respectively. The net utility that a consumer of type θ derives from buying a new product, a remanufactured product, and no product are denoted by $\theta - p_n$, $(1 - \delta)\theta - p_r$, and 0, respectively. The manufacturer's goal is to maximize average profit over an infinite time horizon by setting the remanufacturability level q and a sequence of prices for new and remanufactured products, $\{p_t = (p_{n,t}, p_{r,t}) \ \forall t \geq 1\}$. A constant level of remanufacturability in all periods is assumed since it is the initial technology choice that determines this value for all subsequent periods.

It is easy to show that in any period t, consumers with a willingness-to-pay for a new product between $\frac{p_{n,t}-p_{r,t}}{\delta}$ and 1 will prefer to buy a new product, while consumers with a willingness-to-pay for a new product between $\frac{p_{r,t}}{1-\delta}$ and $\frac{p_{n,t}-p_{r,t}}{\delta}$ will prefer to buy a remanufactured product.

Therefore, the resulting period-t demand volumes of remanufactured and new products are $r_t = F\left(\frac{p_{n,t}-p_{r,t}}{\delta}\right) - F\left(\frac{p_{r,t}}{1-\delta}\right)$ and $n_t = 1 - F\left(\frac{p_{n,t}-p_{r,t}}{\delta}\right)$, respectively. As no more remanufactured products can be sold in each period than the available supply of remanufacturable products, we have a constraint: $\sum_{k=1}^{t}(qn_{k-1} - r_{k-1}) \geq r_t$. Initially, no used products are available; $r_0 = 0$ or $p_{r,0} = (1-\delta)p_{n,0}$. In every period, the prices are set such that demand is non-negative; $0 \leq \frac{p_{r,t}}{1-\delta} \leq \frac{p_{n,t}-p_{r,t}}{\delta} \leq 1$. The manufacturer's infinite-horizon average-profit maximization problem is

$$\max \lim_{N \to \infty} \left(\sum_{t=0}^{N} \pi(p_{n,t}, p_{r,t}, q)\right) / (N+1)$$

subject to

$$0 \leq q \leq 1$$
$$\sum_{k=1}^{t}(qn_{k-1} - r_{k-1}) \geq r_t \quad \forall t \geq 1$$
$$(1-\delta)p_{n,0} = p_{r,0}, p_{r,0} \leq 1$$
$$0 \leq p_{r,t} \leq (1-\delta)p_{n,t}, p_{n,t} - p_{r,t} \leq \delta \ \forall t \geq 1,$$

where

$$\pi(p_{n,t}, p_{r,t}, q) = (p_{r,t} - c_r)r_t + (p_{n,t} - c_n(q))n_t.$$

Under the appropriate assumptions of $c_n(q)$ and $F(\theta)$ (see Debo et al., 2001), the solution to infinite-horizon average-profit criterion allows one to restrict the analysis to steady-state policies, policies that reach a stationary point (p_n^*, p_r^*) after a finite number of stages. Debo et al. determine the optimal stationary solution (q^*, p_n^*, p_r^*). Below, we summarize the main insights that can be obtained from the model.

Insights Obtained From the Model

Debo et al. find that, all else equal, it is optimal to produce a remanufacturable product if the production costs of the single-use product are high, the perceived depreciation of a remanufactured product is low, the remanufacturing costs are low, or the required effort to make a single-use product remanufacturable is low. In other reports on remanufacturing, similar observations are made, though not derived from a formal theoretical framework.

The optimal technology depends on the market characteristics. Debo et al. find that a manufacturer chooses a higher level of remanufacturability in a market with a high concentration of consumers with a high reservation price, and a low concentration of consumers with a low reservation price. The reason is that relatively more consumers are interested in remanufactured products and fewer are interested in a new product. Therefore, all else equal, the manufacturer has to supply more remanufacturable products with fewer new products and therefore chooses a higher level of remanufacturability. The retread business press provides anecdotal evidence supporting this finding: retread markets in India, for example, are more important than retread markets in North America or Europe.

Debo et al. also consider a legislator contemplating subsidizing remanufactured products in order to encourage demand for remanufactured products and to lower demand for new products that would cause more environmental hazard. A subsidy corresponds to an exogenous decrease in remanufacturing costs. Analysis shows that it may be possible that demand for new products will increase, exactly the opposite of what the legislator is expecting.

12.5 Conclusion

Recently, lots of creative thinking has been put into making forward supply chains from suppliers to distributors more cost efficient and responsive to market needs. Now, most companies realize that getting the product into the hands of the consumer is not the end of the story but the beginning of a new era of challenges, challenges which require more forward thinking into issues concerning the design of reverse logistics channels to handle product returns from customers. The research on coordination of forward and reverse logistics decisions is still in its infancy stage. The current sporadic output of papers on the coordination of closed-loop supply chains shows that there is still a need for a general framework which provides a better linkage to the coordination research on forward supply chains. To provide that vital linkage, we believe any framework on reverse logistics research should have the following fundamental questions as its basis.

1. *How can the insights from forward supply chain research be used to address issues regarding reverse logistics channels?*

For instance, what does the bullwhip effect mean in a reverse logistics context, what are the costs associated with it, and how can information distortions in the reverse channel be reduced? Can a manufacturer make use of postponement and part modularity to improve the responsiveness and the flexibility of the reverse channel operations? What process and product changes are required for that?

2. *How does the theory of reverse logistics change and help our understanding of the management of forward supply chains?*

For instance, given the fact that the manufacturer can manipulate the salvage value of a product via remanufacturing, how does this impact the capacity choices and the pricing decisions in the forward supply chain? Can the manufacturer use pricing decisions in the reverse channel to further coordinate the forward supply chain prices?

While the REVLOG project has produced answers to some questions, there is still a growing need to enrich our understanding and build a more general and integrated theory of closed-loop supply chains.

13

Long–term Analysis of Closed-loop Supply Chains

Patroklos Georgiadis, George Tagaras, and Dimitrios Vlachos

Department of Mechanical Engineering, Aristotle University of Thessaloniki,
P.O. Box 461, 54124 Thessaloniki, Greece,
geopat@eng.auth.gr, tagaras@auth.gr, vlachos1@auth.gr

13.1 Introduction

Most of the previous chapters, especially in Parts II and III of the book, concentrate on analyses of closed-loop supply chains (CLSC) at the operational level, and they are confined to rather specific issues, which are extremely important from an operational and tactical point of view. This chapter introduces a longer term, strategic perspective into the analysis of closed-loop supply chains based on a quantitative approach.

Long-term strategic management issues of reverse logistics systems have not been adequately analyzed in the past, possibly because of the difficulty in handling the variety of involved factors and the complexity of their interdependencies. A notable exception is the work of Thierry et al. (1995), which systematically describes the implementation steps of a copier recovery strategy. Although the contribution of Thierry et al. is valuable, it does not delineate a specific formal quantitative analysis. The purpose of this chapter is to show how the methodological tool of System Dynamics (SD) can be employed to assist the analysis of long-term strategies. by means of quantifying the anticipated effects of alternative strategic choices. Some other approaches for long-term analysis are discussed in Chapter 15.

Forrester (1961) introduced the SD approach in the early 60s as a modelling and simulation methodology for analysis and long-term decision making in dynamic industrial management problems. Since then, SD has been applied to various business policy and strategy problems. There are already some publications using SD in supply chain modelling, but all of them refer to forward logistics. Forrester included a model of a supply chain as one of his early examples of applying SD methodology. Towill (1995) uses SD in supply chain redesign to gain additional insight into system dynamics behavior and particularly into the underlying causal relationships. The output of the proposed approach is a collection of effective industrial dynamics models of supply chains. Minegishi and Thiel (2000) use SD to improve the knowledge of the complex logistic behavior of an integrated food industry. They present a generic model and some practical simulation results applied to the field of poultry production and processing. Hafeez et al. (1996) describes the analysis and modelling of a two-echelon industry supply chain that services the construction industry, using an integrated System Dynamics framework. Simulation results are used to compare various re-engineering strategies. Sterman (2000) presents two case studies where SD methodology is used to model reverse logistics problems. In the first one, Zamudio-Ramirez (1996) analyzes part recovery and material recycling in the US auto industry to assist the industry in thinking about the future of enhanced auto recycling. In the second one, Taylor (1999) concentrates on the market mechanisms of paper recycling, which usually lead to instability and inefficiency in flows, prices, etc.

The application of SD in all these papers shows that System Dynamics can indeed be a useful tool for long-term analysis of traditional (forward) supply chains. It remains to be seen in the subsequent paragraphs exactly how this tool can be applied to supply chains involving reverse logistics as well.

13.2 Systems Dynamics Methodology

The structure of a system in SD modelling is described using causal-loop or influence diagrams. A causal-loop diagram consists of variables connected by arrows denoting the causal influences among the variables. The major feedback loops are also identified in the diagram. These loops are either positive feedback (reinforcing) or negative feedback (balancing) loops. In a positive feedback loop, an initial disturbance leads to further change, suggesting the presence of an unstable equilibrium. Figure 13.1 represents the causal loops for a simplified inventory planning and control system. The variables of an influence diagram are connected through arrows that are called influence lines and represent influence between variables. The direction of the influence lines is the direction of the causation. The sign (+) or (-) at the upper end of the influence lines shows the sign of the effect. When the sign is (+), the variables change in the same direction; otherwise they change in the opposite direction. A feedback loop is a succession of causes and effects such that a change in a given variable travels around the loop and comes back to affect the same variable. The polarity of a feedback loop is obtained by the algebraic product of individual signs around the loop and is represented by (+) and (-) signs. If an initial increase (or decrease) in a variable in a feedback loop eventually results in an increasing (or decreasing) effect on the same variable, then the feedback loop is identified as a positive feedback loop. If an initial increase in a variable eventually results in a decreasing effect on the same variable or vice versa, then the feedback loop is identified as a negative feedback loop. The signed clockwise cycles in Figure 13.1 represent feedback loops with positive (+) or negative (-) feedback. Actual serviceable inventory and production rate are the variables that determine the internal environment of the system, while sales determine the external environment. Loop 1 that consists of production

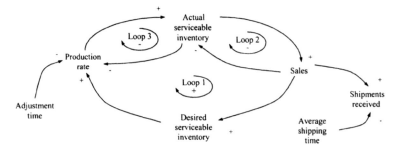

Fig. 13.1. An example of a causal loop (influence) diagram

rate, the actual serviceable inventory, sales, and the desired serviceable inventory is a positive feedback loop. An increase in the production rate will increase the actual serviceable inventory, which may in turn increase sales. Increased sales will cause an increase in the desired serviceable inventory, which leads to an increase in the production rate. If the system consisted of only this loop, the production rate would grow indefinitely. Of course, this cannot be true in the real world. Negative feedback loops limit such growth. A negative feedback loop exhibits goal-seeking behavior: after a disturbance, the system seeks to return to an equilibrium situation. In the previous example, an increase in sales will decrease the actual serviceable inventory, which may in turn decrease sales (Loop 2). In addition, an increase in the production rate will increase the level of actual serviceable inventory, which will lead to a decrease in the production rate (Loop 3).

Causal loop diagrams play two important roles in system dynamics studies. First, during model development, they serve as preliminary sketches of causal hypotheses. Second, causal loop diagrams can simplify the representation of a model. The structure of a dynamic system model contains level and rate variables. The level (state) variables are the accumulations within the system, and their values over time describe the state of the system. The rate variables represent the flows, which alter the state of the system. For example, the actual serviceable inventory in Figure 13.1 is a level variable, while the production rate and sales are rate variables. Every rate variable constitutes a decision point and the modeler must specify precisely the decision rule determining the rate. Modelers must make a sharp distinction between decision rules and the decisions they generate. Decision rules are the policies and protocols specifying the decision process, while decisions are the outcome of this process. The production rate at time t in the previous example could be determined using the following decision rule, which adjusts the actual level of serviceable inventory until it is equal to the desired level:

$$\text{Prod. rate } (t) = \frac{\text{desired level of inventory}(t) \text{ - actual level of inventory}(t)}{\text{adjustment time}}.$$

In this decision rule, the adjustment time is a decision variable which refers to the time required to close the gap between the desired and the actual inventory levels. Aggressive correction actions require small values of adjustment time, while more conservative actions require greater values.

Figure 13.2 presents an adaptation of Towill's (1995) general SD methodology. The whole procedure is divided into two phases. The first phase is the qualitative analysis of the system. During this phase, the influence diagram of the model is built and is then transformed to a flow diagram. The second phase is the quantitative analysis of the system. During this phase, the flow diagram is translated to a simulation program and is then verified and validated. The program is executed for alternative 'what-if' scenarios and the results are analyzed.

Specifically, the first phase starts with the identification of the system objectives. The objective leads to the identification of the key variables and the system boundaries. The connection patterns between these variables are drawn using the list extension method (Coyle, 1978). The list extension method is a technique which helps to start an influence diagram, facilitates stopping with the aid of a closure test, and focuses attention on the purpose of the model.

The qualitative conceptual phase ends with the construction of the flow diagram, which represents the model structure and the interrelationships among the variables. The flow diagram is constructed using building blocks (variables) categorized as levels, flows, delays, converters and constants. Level variables (symbolized by rectangles) are the state variables of the system. Flow variables (symbolized by valves) are the rates of change in level variables and they represent those activities which fill in or drain the level variables. Delays (represented by circles with a square) introduce the time delay in material or information channels. Although in principle delays exist in all flow channels, in the SD modelling approach only the delays that have a significant contribution to system behavior are introduced to the flow diagram. Converters (represented by circles) are intermediate variables used for auxiliary calculations. Constants (represented by rhombuses) are the model parameters. These parameters are either control variables (e.g. the adjustment time in Figure 13.1) or constants that represent system parameters (e.g. average shipping time). Finally, the connectors, represented by simple arrows, are the information links representing the causes and effects within the model structure, while the double-line arrows signify product flows. Double lines across the arrows indicate a delayed information or material flow. Figure 13.3 shows the flow diagram of the influence diagram presented in Figure 13.1. The delay in this diagram determines the shipping time of goods to the market.

The flow diagram is a graphical representation of the model mathematical formulation. The embedded mathematical equations are divided into two main categories: the level equations, defining the accumulations within the system through the time integrals of the net flow rates, and the rate equations, defining the rate of change of the levels. For example, the mathematical

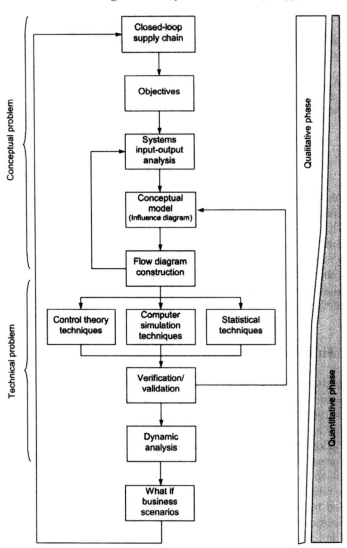

Fig. 13.2. System dynamics methodology for closed-loop supply chain modelling

form of the actual serviceable inventory at time t is the following.

$$\text{Actual serviceable inventory}(t) = \int_0^t [\text{Production rate}(t) - \text{Sales}(t)]dt$$

$$+\text{Actual serviceable inventory}(t_0)$$

The quantitative phase of the methodology begins with the development of the dynamic simulation model using specialized software. Nowadays, high-

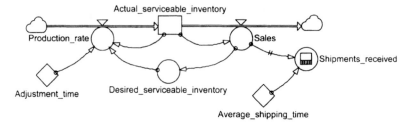

Fig. 13.3. An example flow diagram

level graphical simulation programs (such as i-think® and Powersim®) support this phase. Then, the simulation model is verified and validated. During that step it is likely to return to and correct the conceptual modelling in order for the model to accurately represent the system. Then we run the model and log the dynamic behavior of the variables. The final step is to use the model to design and evaluate new decision rules and strategies that might be applied in the real system. This can be done analyzing the sensitivity of the model by examining the results of 'what-if' scenarios.

13.3 A Holistic Approach

The integration of the forward and reverse flow channels transforms the 'one-way' structure of the traditional supply chain networks to closed-loop networks. For the different forms of reuse (direct reuse, re-manufacturing, repair, recycling) (Georgiadis et al., 2002), the main flows and the major loops of such closed-loop logistic networks are depicted in Figure 13.4. The solid lines represent the forward channel while the dashed lines represent the reverse channel. Four loops characterize the structure of the system. The first loop refers to the direct reuse. *Reusable packages* such as bottles, pallets, or containers are transported back to the *original producer* and possibly after a cleaning and minor maintenance are reused for packaging purposes. The second loop refers to the added value recovery process that includes the re-manufacturing and the repair forms of reuse. *Used products* are transported to producers and after an *added value recovery process*, *reusable products* that include good-as-new products or B-class products are produced. The last two loops refer to recycling. Recyclable materials are transported to the recyclers and after a *material recovery process* they are used by the *original producers* (Loop 3 - the outer loop) or the producers in the *added value recovery process* (Loop 4).

Several actors are involved in the above closed-loop system: suppliers, original producers, value added recovery producers, distributors, users, collectors, and recyclers (Fleischmann et al., 1997). Actors may be members of the forward channel (e.g. manufacturers, retailers, and logistics service providers),

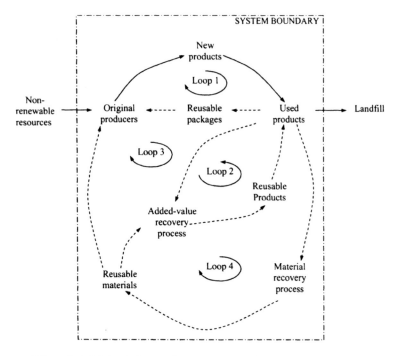

Fig. 13.4. Major influence loops in a closed-loop logistics network

private third parties (e.g. secondary material dealers, material recovery facilities, and added value recovery facilities), or members of the public sector (e.g. government, municipality). The motivation for the participation of original producers and/or specialized third parties in an integrated closed-loop logistics network may be economical, ecological, or both. Reverse inbound flows are economically attractive when the value still incorporated in a used product is greater than the cost of the required reverse activities. This value is called value gain. Positive value gain signifies an economically attractive operation.

Ecological motivation is expressed via state environmental legislation, holding, for example, the original producers responsible for the entire product life cycle or imposing a percentage of recycling. The goal is to reduce both the disposal rate of the used products and the usage rate of nonrenewable resources. Moreover, customer expectations urge companies to reduce the environmental burden of their products and a 'green' image has become an important marketing element that forces the original producers to take into account environmental aspects. Figure 13.5 contains the influence diagram of the described ecological motivation. The diagram consists of two major negative loops. Loop 1 and Loop 2 illustrate the need for stringent environmental legislation as either the pollution level of the environment, or the nonrenewable

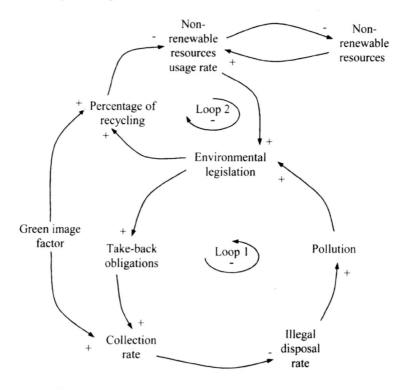

Fig. 13.5. Influence diagram of the ecological motivation

resource usage rate, or both are increased. Specifically the level of *pollution* in the environment forces governments to establish *environmental legislation* that imposes *take-back obligations* of reused products. *Take-back obligations* increase the collection rate and that causes a decrease in *disposal rate* that leads to a decrease in the environmental *pollution*. In addition, an increase in the *non-renewable resources usage rate* makes governments impose a *percentage of recycling*, which leads to a decrease in the *non-renewable resources usage rate*.

Bringing together the forward and reverse flow channels, the type of recovered items, the forms of reuse, the involved actors, and the reuse motivations, we can have a comprehensive view of the entire forward/reverse logistics network. Thus, the System Dynamics methodology allows a holistic approach that can be used to develop dynamic models including both quantitative and qualitative variables (e.g. users' environmental consciousness), time delays for each activity (e.g. collection time, delivery time), and uncertainty in variables (e.g. the timing of the return of used products). The objective of this modelling approach is twofold. The first objective is to understand the dynamic behavior of an integrated forward/reverse logistics network by evaluating the effects of shocks imposed by the external environment to the system (e.g. a new state regulation), or the magnitude of influences between internal elements of the system (e.g. the effect of collection rate to remanufacturing capacity). The second objective is to develop a powerful simulation tool for long-term policy design and evaluation in a real closed-loop supply chain. The investigation of new decision rules, strategies, and structures that might be applied in the real world can be performed from the point of view of a single company, a joint venture, or an industry sector. It is also possible to design and evaluate public policies that aim at securing the viability of reverse channels. Examples of long-term decisions that can be investigated include the following.

- The degree of centralization in the reverse channel. For example, how centralized should the collection activity be?
- The number of levels in the reverse channel. For example, should the re-manufacturing process and the original production be integrated in one type of facility or should they be carried out in different locations?
- Capacity planning in the forward or reverse channel. For example, what are the capacity adjustment policies for an original producer or a reman-ufacturer that improve the system performance?
- Inventory management in forward/reverse channels. For example, what are the appropriate levels of inventories of new products, as-good-as-new products, B-class products, etc.?

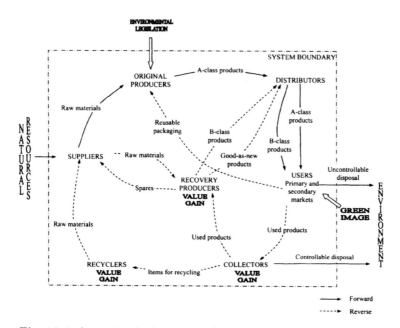

Fig. 13.6. Actors involved and main flows in a closed-loop supply chain

- The degree of partner cooperation. For example, what initiatives for reverse distribution must be taken by a single company or a group of companies to achieve a more efficient design and operation of the closed-loop supply chain?
- Public policy design. For example, what type of policy can induce efficient cooperation among government, businesses, and consumers?

Figure 13.6 summarizes the holistic approach towards the quantitative long-term analysis of CLSC by means of System Dynamics. The main flows among the involved actors and the influences of the economical and ecological motivations are depicted. From this general approach, system dynamics models that are targeted at a specific system objective can be developed using the methodology presented in the previous section. An application of this procedure is described in the next section.

13.4 A Specific Model

Georgiadis and Vlachos (2003) have developed a model to study the long-term behavior of a reverse logistics network for product recovery of a single product under various 'ecological awareness' influences. Their model refers to products from the tire manufacturing sector, but it is more generally applicable to reverse logistics networks for products that may be reused without disassembly. The design parameter is the capacity of remanufacturing facilities. The environmental influences incorporated are

- state environmental protection policies, i.e. environmental legislation which affects the collection procedure directly or indirectly, and state campaigns to reinforce proper collection of reusable products, and
- the 'green image' effect on demand realization.

The influence diagram of the model is depicted in Figure 13.7. The system under study includes the following operations: Supply, Production, Distribution, Use, Collection, Inspection, Remanufacturing, and Waste disposal. The level variables (shown in capital letters in Figure 13.7) and their notation are the following:

L_{rm}: Inventory of raw materials
L_i: Serviceable inventory
L_{di}: Distributor's inventory
L_u: Used products (at the end of their use)
L_{ud}: Uncontrollably (directly) disposed products
L_c: Collected products
L_r: Reusable products
L_d: Controllably disposed products

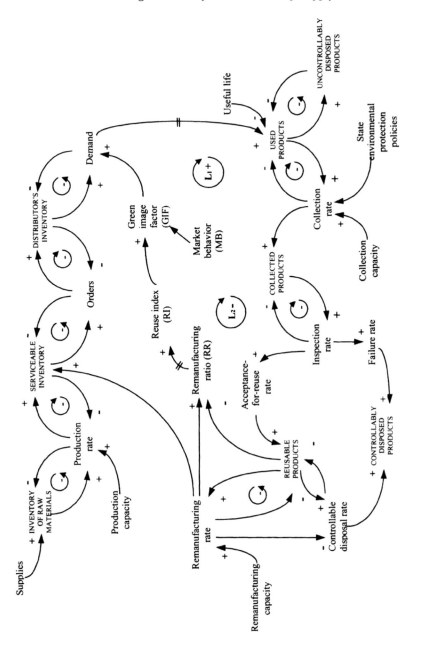

Fig. 13.7. Influence diagram of a forward-reverse logistics system with remanufacturing

The rate variables and their notation are the following:

R_p: Production rate
R_o: Orders
D: Demand
R_c: Collection rate
R_i: Inspection rate
R_a: Acceptance-for-reuse rate
R_f: Failure rate
R_d: Controllable disposal rate
R_r: Remanufacturing rate

Moreover, the following notation is used for the capacities:

C_p: Production capacity
C_c: Collection capacity
C_r: Remanufacturing capacity

Starting from the upper left corner of Figure 13.7, raw materials L_{rm} are furnished by external suppliers. R_p depletes L_{rm}, subject to the production capacity C_p. The rest of the influence diagram is constructed in the same way. We can see that L_i contains new (R_p) and remanufactured products (R_r), which then satisfy the distributors' orders R_o. R_o increase L_{di}, which is depleted by the customer demand D. Used products (L_u) at the end of their current use are either directly (uncontrollably) disposed (L_{ud}) or collected (R_c) for reuse. The reverse channel starts with the collection and inspection procedures. L_c is increased by R_c and decreased by R_i. The inspected products are either accepted-for-reuse (R_a) or rejected (R_f). The latter are disposed controllably, while L_r are either remanufactured (R_r) or controllably disposed (R_d), subject to remanufacturing capacity C_r.

The *environmental protection policies* are modelled with an influence line from 'State environmental protection policies' to collection rate R_c. According to the model, environmental protection legislation or campaigns affect the fraction of used products that are properly collected and handled. This influence may be direct in the form of a take-back obligation for the original producers or the merchandisers. An indirect influence may be the institution of penalty for each uncontrollably disposed product, which may take the form of a take-back fee included in the product price, to be paid back when the product is returned to a proper collection point. In this case, the influence of the penalty on R_c depends on the costs of the alternative options for handling used products. An environmental protection campaign also has an indirect effect on the volume of products that are collected for reuse. The dependency can be modelled using an S-shaped curve, typical in SD modelling, which represents the inertia of the customer to no or extensive protection measures.

The *green image effect* on product demand depends on the market awareness that the specific producer adopts product recovery and reuse (Reuse

Index-RI). One quantitative measure for a producer's dedication to reuse is the remanufacturing ratio (RR) defined as the fraction of remanufactured products (R_r) to L_r. RI expresses the opinion of the market (customers) for the specific firm and it generally changes slowly. The values of RI over time are determined by smoothing and delaying past values of RR. The delay time expresses the time needed for the market to be aware of the remanufacturing activities of the producer.

Moreover, the way RI affects demand D depends on the specific market and the product characteristics. For example, customers are more sensitive to the green image of some products, the production or disposal of which has pronounced negative environmental effects. The effect of RI on D will be larger for these products. The effect will also be larger in markets where customers have developed environmental consciousness. To model this influence, two variables are used: Green Image Factor (GIF) and Market Behavior (MB). GIF is determined from RI according to the impact of the qualitative variable MB, which expresses the market and product characteristics. GIF is the main driver of the demand as a function of time $D(t)$, which is calculated using 13.1.

$$D(t) = D(t - dt) * [1 + GIF(t)] \tag{13.1}$$

Equation 13.1 implies that GIF may be interpreted as the percentage change in product demand per time unit. Therefore, GIF may be either positive or negative. Positive values are obtained for a firm that supports remanufacturing and for this reason experiences an increase in demand, while negative values are obtained for a firm with a low degree of reuse, leading to a decrease in its demand. The magnitude of the influence depends upon the variables that regulate the 'green image' influence. RR is a ratio taking values between 0 and 1. The extreme values of RI are also 0 and 1. The form of dependency between RI and GIF depends on the reaction of a specific market to the reuse activities.

The loop between R_p and L_i is a negative feedback loop, which regulates the actual inventory to a desired level, readjusting the R_p over time. In total, there exist nine similar negative feedback loops. There are two more feedback loops (L1 and L2), the mechanisms of which reflect the 'green image' effect on the product demand. Specifically, an increase in demand sequentially increases L_u, R_c, L_c, R_i, R_a, and L_r through the positive influences of these variables. These dependencies are common for Loops L1 and L2. An increase in L_r will increase RR through the increase in R_r if we follow Loop L1, but it will decrease RR through the direct negative influence line if we follow Loop L2. Then both loops connect the RR to D, following the same positive influence path through RI and GIF. L1 is a positive feedback loop (all influences around the loop are positive), which would cause an uncontrollable increase in D if demand weren't controlled by the negative feedback Loop L2 (one negative influence around the loop).

The mathematical formulation of the product recovery chain is the following. The product recovery chain is described by three level variables, which are functions of time: $L_c(t)$, $L_r(t)$, and $L_d(t)$. $L_c(t)$ stands for the accumulation of used products that arrive at the collection facilities. The mathematical expression of $L_c(t)$ is

$$L_c(t) = \int_0^t [R_c(t) - R_a(t) - R_f(t)]dt + L_c(0) , \qquad (13.2)$$

where $R_c(t)$ is the collection rate at time t, $R_a(t)$ is the acceptance-for-reuse rate at time t, and $R_f(t)$ is the failure rate at time t. $R_c(t)$ represents the rate of increase in collected products through the collection activities and is given by

$$R_c(t) = \min\{C_c, p_1 * L_u(t)\} , \qquad (13.3)$$

where C_c is the collection capacity, p_1 is the percentage of collected products, and $L_u(t)$ denotes the used products at time t. $R_a(t)$ and $R_f(t)$ are the rates of decrease in collected products through the acceptance-for-reuse rate and the failure rate respectively, and they are given by

$$R_a(t) = (1 - p_2) * L_c(t) \qquad (13.4)$$

$$R_f(t) = p_2 * L_c(t) , \qquad (13.5)$$

where p_2 is percentage of inspected products that fail and therefore they are controllably disposed. The inventory of reusable products, $L_r(t)$, is computed from

$$L_r(t) = \int_0^t [R_a(t) - R_r(t) - R_d(t)]dt + L_r(0) , \qquad (13.6)$$

where the remanufacturing rate $R_r(t)$ and the controllable disposal rate $R_d(t)$ are

$$R_r(t) = \min\{C_r, L_r(t)\} \qquad (13.7)$$

$$R_d(t) = \max\{L_r(t) - R_r(t), 0\} , \qquad (13.8)$$

where C_r is the remanufacturing capacity. Finally, $L_d(t)$ is calculated from $R_d(t)$ and $R_f(t)$ as follows:

$$L_d(t) = \int_0^t [R_f(t) + R_d(t)]dt + L_d(0) . \qquad (13.9)$$

To solve the above system of equations, we have to determine the initial conditions at time $t = 0$ for all level variables.

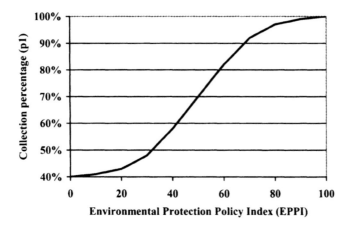

Fig. 13.8. Relationship between environmental protection policy index and collection percentage

The above model can be used to answer the following questions:

a. What are the product stocks and flows and their dynamic response?
b. What is the environmental protection policies' impact on reverse flows?
c. How do different market behaviors in recognizing the green image of a firm affect product demand?
d. What are the reverse channel capacities (remanufacturing, collection) that maximize or maintain a specific demand level?

Numerical Example

Assume that there are no stockouts in the forward channel and that the collection and inspection capacities are infinite. The initial demand is 100 items per week. Figure 13.8 depicts the dependency of the collection percentage p_1 on the environmental protection policy. The latter is expressed by the environmental protection policy index (EPPI), which takes values from 0 to 100, where 0 refers to the absence of such policy and 100 refers to the introduction of stringent policies that almost eliminate the uncontrollable disposal. The collection percentage for the first case (no policy) is assumed equal to 0.40. The dependency takes the form of an S-shaped curve, defined by

$$p_1 = \frac{a}{1 + e^{b*EPPI}} , \qquad (13.10)$$

where a, b are scale parameters.

As already mentioned, the magnitude of RI influence on GIF depends upon the market sensitivity to the reuse activities. This sensitivity depends on the type of market, the product and the promotion of the 'green image'. To model different market behaviors, we may use four alternative curves, as

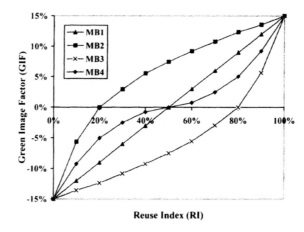

Fig. 13.9. Relationship between GIF and RI for various market behaviors

shown in Figure 13.9. The relationship under the first market behavior (MB1) is proportional. Under the second market behavior (MB2), the market and therefore GIF responds quickly for low levels of RI (sensitive market), while in MB3 the response is slow and becomes more acute for high levels of RI. Finally MB4 combines MB2 and MB3. The relationship takes the form of an S-curve (sensitive market for very poor or very good performance). We assume that the maximum and minimum values of the GIF have the same magnitude, which corresponds to a 15 percent maximum annual demand change (increase or decrease) due to the 'green image' effect.

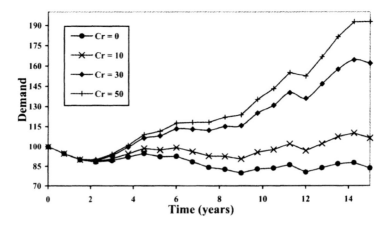

Fig. 13.10. Demand distribution over time for four levels of C_r

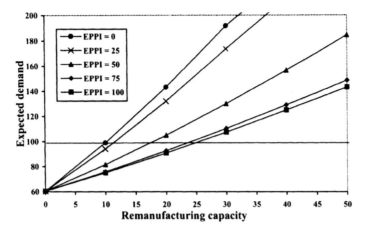

Fig. 13.11. Expected demand as a function of C_r under five $EPPI$ levels

The mean useful life of the product, i.e. the time period between the purchase and the end-of-use time, is 40 weeks, and the average time for the market to realize the change in remanufacturing ratio is almost 1 year (52 weeks).

The simulation program was developed using Powersim v2.5c. The model developed has 7 levels, 10 rates, 7 auxiliary variables, and 7 constants.

To obtain the dynamic (transient) behavior of the system variables, the model was simulated for various sets of input parameters. Specifically, all possible combinations of five $EPPI$ levels (0, 25, 50, 75, and 100), four market behaviors (MB1, MB2, MB3, and MB4), and six remanufacturing capacity levels C_r (0, 10, 20, 30, 40, and 50 units per week) were simulated, for a total of 120 scenarios. The simulation horizon was set to 15 years and the time step to 1 week, assuming that this time is longer than the lead time of the flows we have not modelled as delays. Figure 13.10 depicts the demand distribution over time for various levels of remanufacturing capacity under the first market behavior (MB1) and an $EPPI$ level of 25. As expected, higher remanufacturing capacity improves the 'green image', which eventually leads to demand increase. In cases of low remanufacturing capacity, the demand trend is negative. Moreover, there is an upper limit for C_r (50, for this example), above which an increase in capacity does not change the demand response. This threshold value is the capacity that can serve all reusable products.

Figure 13.11 shows the long-term effect (after 15 years) of the $EPPI$ level and the C_r on the expected demand (moving average for a one-year period). For constant C_r, a higher $EPPI$ decreases the expected demand. This can be explained if we consider that higher $EPPI$ levels lead to increased collection and increased L_r, which, since C_r is constant, will decrease RR, GIF, and

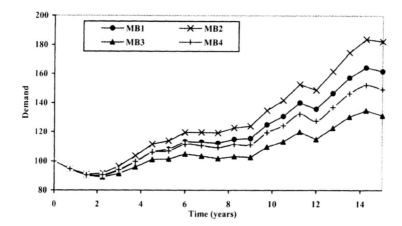

Fig. 13.12. Demand distribution over time for four different market behaviors

finally D. In the same figure, we also notice that the impact of a change in $EPPI$ on expected demand is more pronounced at higher levels of C_r.

Finally, the effect of market behavior is illustrated in Figure 13.12. The scenario shown in Figure 13.12 refers to C_r equal to 30 items per week and an $EPPI$ level of 25. For this specific example, but also under all combinations of C_r and $EPPI$ considered, D is the highest for MB2, while for MB3 it exhibits the worst performance.

13.5 Discussion and Outlook

The System Dynamics methodology presented in the previous sections helps modelers to develop dynamic models which can be used as a long-term decision tool, through the evaluation of specific strategic policies and structures in closed-loop supply chains. The main management areas in which SD models facilitate decision making are capacity planning, inventory management, production planning, environmental management, and cooperation between actors. More specifically, the SD models predict the dynamic behavior of flows, inventory levels, and economic results in closed-loop supply chains under alternative 'what-if' scenarios.

The ability to evaluate the performance of various subsystems and of the entire logistics network greatly facilitates decision making, since it is possible to identify policies, strategies and structures that optimize the behavior of the total system. For example, the remanufacturer can determine the capacity expansion strategy which maximizes profit, taking into consideration the integrated forward-reverse logistics network. Moreover, the system dynamics approach allows loop analysis and the identification of causes and effects between system variables. Furthermore, the dynamic model of a CLSC allows comprehensive description and analysis of the long-term system operation under alternative environmental protection policies concerning take-back obligation, proper collection campaigns, and green image effect.

In conclusion, the proposed SD methodology is a valuable and powerful experimental tool, which may help analyze many business and operational scenarios and questions about the long-term operation of product recovery networks. Thus, the approach may prove useful to policy makers and decision makers dealing with the complex, long-term strategic management problems of reverse logistics, as well as to researchers in environmental management.

LCA as a Tool for the Evaluation of End-of-life Options of Spent Products

Costas P. Pappis, Stavros E. Daniel, and Giannis T. Tsoulfas

Deptartment of Industrial Management and Technology, University of Piraeus, 80 Karaoli & Dimitriou Str., 18534 Piraeus, Greece, pappis@unipi.gr, sdaniel@unipi.gr, tsoulfas@unipi.gr

14.1 Introduction

The extension of the traditional supply chain not only raises significant logistics and operational issues, but also environmental ones. Reverse logistics is doubly connected with environmental management issues.

- It aims for the recovery of used products and materials with obvious environmental gains. The life cycle of used products or used parts is extended instead of being ended in the waste stream. In addition, certain materials are recycled and this at least prevents natural resources from being exhausted.
- Like any logistics operation, it should be subjected to environmental performance appraisal for the identification of the associated impacts on the environment and the opportunities for improvement. Indeed, the appropriate design and control of the logistics network may minimize the environmental impact.

Therefore, the planning of the extended supply chain has to deal also with the environmental aspects involved, as the total performance of the network is a function of all its characteristics, which are in a continuous interaction. Certainly, in addition to the environmental ones, there are also other criteria that have to be taken into account for effective decision-making, such as economic, social, political, and technical issues, which are sometimes at odds with one another. Relevant priorities change from time to time and that affects the relative significance of the criteria. It is quite clear, though, that the environmental requirements for manufacturers and, more generally, for the parties involved in the extended supply chain are getting even stricter. Meeting these requirements is not only a condition for doing business, but also a competitive advantage against companies in the same sector. Certainly, there are also many opportunities for direct business profit by using the environmental parameter in the design equation, an additional motivation for the companies.

Relevant legislation in developed countries derives from the social demand for environmental protection, and the perspective that the manufacturers should be made responsible for their products 'from cradle to grave' tends to dominate. However, while nobody would argue against the need for a global approach of the situation, it can be claimed that the efforts made so far are rather limited. Developed countries or leader firms, mainly, have established environmental policies for product recovery. The issue of how effective these policies are is debatable.

In this paper, the environmental aspects of reverse logistics are discussed and Life-Cycle Analysis (LCA) is introduced as a quantitative approach for measuring the environmental impacts of logistics activities. In order to illustrate LCA, the method is applied in the case of Starting, Lighting and Ignition (SLI) batteries recovery. In addition, Analytic Hierarchy Process (AHP) and LCA Polygon, which assist LCA-based decision-making, are presented and applied.

14.2 Closed-loop Supply Chain Management: The Environmental Perspective

The extension of the traditional supply chain may help companies meet market and customer expectations, legal and regulatory requirements, and commercial and economic demands. Undoubtedly, suppliers, manufacturers, and importers, together with customers, all have significant roles to play in activities related to the recovery and/or proper disposal of products that have completed their life cycle, and of the packaging materials associated with them. In addition, the appropriate selection of the activities regarding the production and the usage phases of products is necessary in order to minimize the net environmental impact of the extended supply chain. Therefore, the role of cleaner production is to provide an integrated preventative environmental strategy applied to processes, products, practices, and services to increase eco-efficiency and to reduce risks for humans and the environment.

Every product generated, transported, used, and discarded within the supply and reverse chains causes certain impacts on the environment. It is necessary, however, that all life-cycle stages of a product are identified in order to establish or optimize environmental policies. It is important to emphasize that each link in the extended supply chain not only has some environmental impact, but it also affects the activities that follow. For this reason, the analysis should be made for each activity separately, as well as for the whole system.

For the introduction of the environmental parameter in the design of extended supply chains several practices are followed. The extent to which these practices attribute significance to environmental issues varies. Environmental management standards such as ISO 14000 and EMAS provide guidance for developing environmentally friendly organizational systems, although individual operations must adapt corporate policies to site-specific risks (Rondinelli and Vastag, 1996). They provide a structured approach to planning and implementing environmental protection measures. Environmental management systems (EMS) can be the first step towards environmental improvement, as they enable organizations to benchmark their environmental performance and then regularly evaluate their performance and improvement. To develop an EMS, an organization has to assess its environmental impacts, set targets to reduce these impacts, and plan how to achieve the targets.

Several OR methods and tools are employed either to identify the type and intensity of environmental problems (observation and analysis) or to assist environmental planners to effectively cope with these problems (solution identification). The main topics investigated in the reverse chain concern, from an OR perspective, are measurement of environmental parameters, estimation of the systems' response to pollution, and identification of effective management practices. In this case, OR studies focus mainly on the exploration of the damages induced and the environmental management practices to be applied

- to natural recipients (air, water, soil, ecosystems) and
- to the global environment.

In terms of an integrated overview of a product's impact on the environment, aspects of the environmental planning in the traditional supply chain have to be taken into consideration. In particular, focus should be drawn to issues such as the activities' products and by-products, the technology used, the location concerned, the risks associated with the system's operation, etc. In this case, OR research primarily attempts to investigate the possibility of increasing the environment-related benefit from changes

- in the product itself (kind, material, synthesis, operational characteristics, etc.),
- in the technology applied (production process, transportation, equipment, kind and quantity of energy and raw materials, etc.),
- in the location of facilities and the layout of the productive chain, and
- in other decision parameters.

Many quantitative methods which aim either to measure the environmental performance of policies or to provide a decision-support tool have been proposed. Furthermore, many methods have been adapted by companies and have been adjusted to their own specific needs. The most widely used methods follow.

- LCA
- Environmental Impact Assessment
- Ecological Risk Assessment
- Material Flow Analysis
- Life-Cycle Cost Analysis

- Ecological Footprint
- System of Economic and Environmental Accounting
- Environmental Auditing
- Input-Output Analysis
- Strategic Environmental Assessment
- Energy Analysis
- Location Analysis
- Cost-Benefit Analysis

Some of these methods deal also with other managerial aspects and try to provide integrated and efficient support to decision-making. However, the different dimensions which have to be taken into consideration pose certain constraints in this case, as does the complexity of the examined systems.

The most widely used method for assessing the environmental impacts of processes is LCA. Its nature and capabilities also suggest that it is the appropriate tool to apply in the environmental analysis of supply chains, as will be discussed in the following chapter.

14.3 Life-cycle Analysis: Quantitative Tools of Analysis

LCA is utilized as a quantitative tool to promote sustainable development by assessing the environmental impacts caused by a product during its entire life cycle (Lee, 1998). Although it has been used in some industrial sectors for about 20 years, LCA has received wider attention and methodological development only since the beginning of the 1990s, when its relevance as an environmental management aid in both corporate and public decision making became more evident. Examples of this include incorporation of LCA in ISO 14000, EMAS, and the EC Directive on Integrated Pollution Prevention and Control (IPPC), which require companies to have a full knowledge of the environmental consequences of their actions, both on- and off-site (Azapagic, 1999). Furthermore, it is used by governments, e.g. when establishing eco-labeling criteria for certain product groups or when defining mandatory reuse or recycling quotas (Scholl and Nisius, 1998).

LCA visualises the environmental and resource consequences of the choices in materials and processes in the manufacture of a product (Ong et al., 1999) and effectively maps out the entire life cycle of a product, beginning from raw materials extraction, processing, manufacture, transport, use, maintenance, recovery, and, finally, disposal. This shift in focus is related to a preventative strategy for improving environmental performance along the whole product chain. The new strategy is based on a set of new approaches to environmental problems:

- A life-cycle approach, where environmental problems are evaluated from a 'cradle to grave' perspective, from raw material acquisition to final waste treatment.

Table 14.1. LCA phases according to the framework developed by SETAC and ISO

SETAC	ISO
Goal Definition and Scoping	Goal and Scope Definition
Inventory Analysis	Inventory Analysis
Impact Assessment	Impact Assessment
Improvement Assessment	Interpretation

- A systems approach, where all processes and activities along the product chain which are necessary for the existence and functionality of a product are assessed within a traditional material and energy balance.
- A holistic approach to environmental problems, where, at least in theory, all types of problems are evaluated, and the main problems related to a given system are defined (Hanssen, 1998).

The Society for Environmental Toxicology and Chemistry (SETAC) was, in 1990, the first to initiate activities to define LCA and develop a general methodology for conducting LCA studies. Soon afterwards, the International Organization for Standardisation (ISO) started similar work on developing principles and guidelines on the LCA methodology. Although SETAC and ISO worked independently of each other, a general consensus on the methodological framework between the two bodies has started to emerge, with the difference being in the level of detail only. In this paper, the terminology suggested by ISO will be used, as it seems to be gaining acceptance among practitioners. According to both approaches, LCA studies comprise four phases, which are presented in Table 14.1.

Because of its holistic approach to system analysis, LCA is becoming an increasingly important decision-making tool in environmental management. Its main advantage over other, site-specific, methods for environmental analysis lies in broadening the system boundaries to include all burdens and impacts in the life cycle of a product or a process, and not focusing on the emissions and wastes generated by the plant or manufacturing site only (Azapagic and Clift, 1999).

14.3.1 Goal and Scope Definition

In the first step of LCA, it is vital that the study team

- identifies the decisions or applications for which the study results will be used,
- determines what information is needed for those decisions or applications and what part of this information can be provided by LCA, and
- defines the goal and scope of the study.

The scope of the study describes the system to be studied and directs how much information is to be collected, in what categories, and to what level of detail and quality. In addition, the study boundaries, assumptions, simplifications, and limitations are defined. An important task in this LCA phase concerns the selection of the reference product and reference data. Finally, the choice of geographical borders of a system (Lindfors et. al., 1995a, 1995b) and the definition of time horizon are crucial issues regarding the system's boundaries.

14.3.2 Inventory Analysis

Inventory analysis is perhaps the most important stage of LCA (Wenzel and Hauschild, 1998; Hauschild and Wenzel, 1998). The performance of this phase is essential for the realization of the next phases of LCA (i.e. Impact Assessment and Interpretation).

Inventory begins with a flow diagram and process tree, where all the relevant processes and stages of the predefined system are outlined. Aggregation of data is of primary importance in this stage of LCA. Some basic issues in inventory analysis concern the data sources (private versus public), the choice between the use of average versus case-specific data, the kind of measurements (internal versus external), the specification of data (publicity, age and frequency of data, changes of technology), and, finally, data availability (Miettinen and Hamalainen, 1997; SETAC, 1991). Furthermore, the classification of data in categories is of particular importance because it is the first step regarding the quantitative part of LCA. This classification leads to the determination of impact categories that can provide a more structured environmental analysis, since it is easier to identify their relation to certain environmental problems.

The next step concerns the modeling of the system, where calculation of mass and energy entering or exiting in every stage of the system takes place. In particular, mass and energy balances for each stage of the system are examined (SETAC, 1991) from where total balances would occur in order to evaluate the total environmental impact of a product.

Finally, all the inputs and outputs of the system concerning mass or energy are listed in the Inventory Table. This table is the eco-profile of the process examined.

14.3.3 Impact Assessment and Aggregation of Results

In the Impact Assessment (IA), the inventory is translated into potential contributions to various impacts within the main groups of predefined impact categories. During this phase, one also attempts to identify related hazards, thus assisting manufacturers to prioritize areas for action in order to get the best results for their investments (Curan, 1993; Berkhout, 1995; Lee et al., 1995). The IA qualifies the inventory as a basis for decisions in the case of comparison between products. It can also reveal the need to collect additional data, that is it may lead back to the Inventory Analysis. This assessment may be quantitative and/or qualitative and should address effects such as environmental and health considerations, habitat modification, and other impacts (SETAC, 1991; SETAC, 1993; NORD, 1995).

For the evaluation of environmental effects, several different methods have been used (Hanssen et. al., 1994; Krozer, 1998; Hertwich et al., 1997):

- the health hazard scoring (HHS) system,
- the material input per service-unit (MIPS),
- the Swiss ecopoint (SEP) method,
- the sustainable process index (SPI),
- the environmental priority strategies (EPS),
- the Society of Environmentally Toxicology and Chemistry's life-cycle impact assessment (SETAC LCA) method,
- the method of critical volume,
- the Eco-indicator method, and
- the Tellus method.

The method used to get a more accurate interpretation of parameters which influence the eco-balance, and which probably has gained the widest acceptance, is the Environmental Design of Industrial Products (EDIP) method adapted to SETAC LCA IA method (Hertwich et al., 1997). EDIP includes a tool for the design and construction of products in light of environmental, working environment, and resource considerations.

The next step is to aggregate all the environmental impact categories included in an IA method. Two different techniques may be used in order to provide a single index for each policy (Daniel et al., 2003b). The two methods that have been chosen belong to the category of multi-criteria decision-making tools that can be applied when it is necessary to take into consideration criteria of different natures for the purpose of comparing policies. Indeed, decision-making in LCA is usually faced with such situations where decisions have to be taken under an uneven set of criteria.

Analytic Hierarchy Process (*AHP*)

For each policy, based on the various criteria, an environmental score is computed for the impact assessment data and the associated environmental impacts. The AHP method (Saaty, 1986, 1988a, 1988b) is applied to decision making in order to assist decision makers to describe the general decision operation by decomposing a complex problem into a multi-level hierarchic structure of objectives, criteria, sub-criteria, and alternatives. It is employed to weigh the criteria according to the relative importance attached to each criterion by the decision maker. These computed environmental scores and assigned weights are next combined to produce an environmental score representing the environmental merit of the policy. In order to obtain the true values of the priorities, the supermatrix approach is suggested. A matrix is created which is composed of the weights of actions according to the criteria and of the weights of criteria according to the actions (Harker, 1987). However, the number of paired comparisons required by this approach often places a limitation on the actual size of the matrix. A simple solution is to rescale the weights of criteria in such a way as to undo the effects of the normalization which takes place when the local weights of the actions are determined. This rescaling takes place with the introduction of two kinds of criteria weights (Giangrande, 1994): intrinsic weights and specific weights.

An intrinsic weight *(iw)* expresses a scaling constant that reflects the importance that the decision maker ascribes to a criterion regarding the goal, based on his system of values.

The specific weights *(sw)* measure the discriminatory power of the criteria and depend on the kind of normalization applied to the local weights of the actions.

For the elimination of the negative effects of the normalization of the local weights of the actions, a calculation of the rescaled weights of the criteria *(w)* by multiplying the intrinsic weights by the specific weights and normalizing the products is done as it appears in Formulas 14.1-14.3. The principle of hierarchic composition is then applied with these assessed weights.

$$w_{c1} = \frac{iw_{c1} \cdot sw_{c1}}{\sum\limits_{i=1}^{n} iw_{ci} \cdot sw_{ci}} \tag{14.1}$$

$$w_{c2} = \frac{iw_{c2} \cdot sw_{c2}}{\sum\limits_{i=1}^{n} iw_{ci} \cdot sw_{ci}} \tag{14.2}$$

$$\cdots$$
$$\cdots$$

$$w_{cn} = \frac{iw_{cn} \cdot sw_{cn}}{\sum\limits_{i=1}^{n} iw_{ci} \cdot sw_{ci}} \tag{14.3}$$

where w_{ci} : rescaled weights of the criteria
iw_{ci} : intrinsic weights
sw_{ci} : specific weights

LCA Polygon

Georgakellos (1997) proposed the LCA polygon as a technique which aims to contradistinguish the results that are reached from the Inventory Analysis. Daniel et al. (2003b) applied this method to the IA phase. Impact categories are described in a radial system of axes. In a hypothetical system of n impact categories, a regular n-sided polygon is formed, the edges of which are inscribed in a circle. Each radius ending on an edge of the circle is a measuring axis for each impact category. The point where the axes meet corresponds to a value of 0. The values corresponding to the edges of the circle are, by definition, the normalized maxima (with a value equal to 1) for each category and correspond to the environmental policies for the reduction of environmental pollution. Thus, to every impact category corresponds a value in $[0,1]$. Each of the axes expresses different natural values and thus has different individual characteristics (scale and units). The actual values for different impact categories are given for the corresponding axes, forming a new n-sided polygon. The examination of management policies, or the comparison of such policies, leads to the formation of alternative polygons in the same radial system. The environmental efficiency of each policy is then described by comparing the areas of the two polygons. The larger the area, the worse the environmental profile of the policy. The arrangement of the n axes in the polygon influences the total value of the area surface. This problem may prove to be crucial, especially when approximate values are compared. For this reason, the areas for all the possible triangles and different impact categories arrangements are calculated and then the average area is calculated.

The average area of the LCA polygon is calculated by:

$$E' = \frac{1}{2} \cdot \sin\left(\frac{360}{n}\right) \cdot \left\{ n \cdot \left[\frac{2 \sum\limits_{\substack{i,j=1 \\ i<j}}^{n} R_i R_j}{[n(n-1)]} \right] \right\} \qquad (14.4)$$

where E' : the area of the LCA polygon
$\quad\quad n$: the number of the impact categories
$\quad\quad R_{i,j}$: the values for the impact categories.

Obviously, the average area of the LCA polygon is independent of the arrangement of the impact categories, and is thus more objective. The area of a regular n-sided polygon inscribed in a circle of a radius R is calculated by Formula 14.5.

$$E = \frac{1}{2} \cdot n \cdot R^2 \left(\frac{360}{n}\right) \qquad (14.5)$$

The index E_{LCA} may be used to describe the ratio of the average area of the LCA polygon E to the area of the regular polygon.

$$E_{LCA} = \left(\frac{E'}{E}\right) \cdot 100\% \qquad (14.6)$$

14.3.4 Interpretation

In the Interpretation phase of LCA, alternative policies may be ranked and opportunities for the reduction of the environmental burdens may be identified and valuated. The procedure of interpretation is further elaborated and there is no complete actual framework for this LCA phase. As mentioned earlier, the term 'Interpretation' instead of 'Impovement Assessment' is used by ISO because, after a lot of discussion, it became evident that the improvement of a product or a service along its life cycle is not simply a matter of LCA application. In real life, when decision makers have to decide on an improvement for a product or a service, they will take into account other criteria, such as economic ones, in addition to the environmental recommendations produced in previous LCA phases.

Almemark et al. (2000) proposed a framework as a first step for the implementation of this phase of LCA. This framework focuses on data quality and practically provides a means for a final evaluation of the first step of LCA, i.e. Goal and Scope Definition, after having performed the Inventory Analysis and the Impact Assessment phases.

Tsoulfas et al. (2002a) proposed two approaches that may be used in this LCA phase. The first approach is Collaborative Decision Making (CDM), which may provide a means for a well-structured decision-making process. Usually, CDM is performed through debates and negotiations among a group of people, where conflicts of interest are inevitable and support for achieving consensus and compromise is required. Opinions about the relevance or value of a position when deciding an issue may differ. Decision makers may have arguments for or against alternative solutions. In addition, they have to confront the existence of insufficient or too much information. Such situations are always met in LCA. Karacapilidis and Papadias (2001) proposed HERMES, a system that augments classical decision making approaches by supporting argumentative discourse among decision makers. HERMES focuses on aiding decision makers in reaching a decision not only by efficiently structuring the discussion, but also by providing reasoning mechanisms for it.

The second approach is to define indicators which can assist decision-making as they reveal improvement possibilities as well as the degree of improvement. The first indicator is the 'distance from target' indicator, which is given by the formula

$$DT_i = 1 - \frac{ICT_i}{ICV_i}, \quad i = 1, 2, \dots, N, \tag{14.7}$$

where ICT_i : The i-th Impact Category Target
$\quad\quad ICV_i$: The i-th Impact Category Value
$\quad\quad N$: the number of the impact categories.

If, $0 \leq DT_i \leq 1$, DT_i shows the percentage distance between the observed value and the goal value. If $DT_i < |DT_i|$, DT_i shows the excess percentage above the target.

The second indicator is the 'improvement margin' indicator and is given by the formula

$$IM_{ij} = 1 - \frac{BATEP_{ij}}{ATEP_{ij}}, \quad i = 1, 2, ..., N \text{ and } j = 1, 2, ..., k_i, \quad (14.8)$$

where $BATEP_{ij}$: the j-th Best Available Technique's Environmental Performance for the i-th impact category

$ATEP_{ij}$: the j-th Applied Technique's Environmental Performance for the i-th impact category

N : the number of impact categories

k_i : the number of activities that contribute to the i-th impact category.

Although the above mentioned tools can assist in defining a robust framework that may be applied in the Interpretation phase, they do not provide a complete answer to the questions that have to be dealt with in this part of LCA. Indeed, a lot of research efforts must be done in this direction, so that thorough and complete LCA studies can be conducted.

14.4 LCA Applied to Starting, Lighting and Ignition Batteries Recovery: A Case Study

14.4.1 The SLI Batteries' Reverse Chain

In the sequel, the LCA methodology presented in Section 3 is applied to the case of SLI batteries in order to illustrate the method. Due to the fact that there is no complete framework for the Interpretation phase, the analysis is limited to the remaining LCA phases.

The work presented is based on data (Daniel et al., 2003a) from a medium-size company, which is a private enterprise and whose main activity is lead recycling, which is extracted mainly from SLI batteries (from here on batteries) or, in small percentage, from leaden conduits. This enterprise is located in Aspropirgos, Greece, about 20 Km from Athens.

Two alternative waste management policies are studied and compared. The first policy deals with the recovery chain, that is, the flow of used products from consumers to recovery facilities. The second policy deals with the disposal chain, in which used products are carried to landfills.

Reverse Logistics Network

At the beginning, the batteries are usually deposited at a car electrician's shop, where cars are brought for battery replacement. In rare cases, an individual himself replaces the used battery. This means that most of the used batteries enter the reverse flow chain instead of ending up in a landfill. The batteries containing their liquids are stored in columns in the electrician's shop. Imported batteries have no liquids, because of the international legislation restrictions on transportation.

Collectors buy used batteries from the car electricians' shops and transfer them using pick-up trucks. Then they forward them to wholesalers trading used materials. Sometimes the collector forwards the batteries directly to the recycling unit. Collectors do not store batteries.

Wholesalers store used batteries in yards, together with other used materials (brass, iron, aluminum. etc.), for some time, and then transport them with trucks to the recycling unit. The batteries are unloaded next to the breaker, and then they are loaded into the breaker using a small lifting machine. The casing of some truck batteries is made of bakelite. Such batteries are broken manually and their lead is pushed to the next stages of the production process, while the bakelite is transferred to a landfill. During disaggregation, lead oxides, plastic, paper, and battery liquids are extracted. The lead oxides and plastic are further processed in the unit but the paper is disposed of as waste.

Lead oxides are stored next to a vessel of water and are wet. For this reason, they remain in outdoor storage sites until they dry and then are forwarded to the furnace for the remaining processes. In the furnace, apart from the lead oxide, some ancillary materials are also inserted. Pure lead is produced in fluid form. Pure lead is placed in molds and, after that, is driven outdoors and transferred to grids of a smaller size. Then it is stored in columns until its sale (Tsoulfas et al., 2002b).

Disposal

Disposal of used batteries concerns an alternative channel that a used battery may go through if it is not collected at the different storage points (car electricians' shops or wholesale collection points) and then transported to the recycling units. Disposal comprises two basic procedures, namely collection/transportation and landfilling or dumping. It should be noted that uncontrolled dumping of used batteries is not a usual practice.

14.4.2 Goal and Scope Definition

The intention of the study is to analyze and compare the reverse supply chain and the disposal chain of batteries and to evaluate the potential environmental implications associated with the logistics activities taking place in them. The final goal of this study could be to support decision makers in taking into consideration the environmental dimension when dealing with the management of batteries that have ended their life cycle.

The functional unit which has been used for this study is 1 ton of lead. The data references and sources are presented in Table 14.2.

The definition of the system boundaries regarding the selected processes and life-cycle stages or the selected geographical boundaries of the product

Table 14.2. Data sources-data references

Product system & Process type	Data specificity			Data source type					Comments
	Product specific	Site specific	General	1	2	3	4	5	
Collection									
Road transportation of used batteries			✓			✓		✓	Literature (CEC, 1991)
Sea transportation of used batteries			✓			✓		✓	Literature (Georgakellos, 1997)
Transportation of ancillary materials			✓					✓	Literature (CEC, 1991)
Disaggregation									
Separation		✓		✓					Direct measurements
Remanufacturing/Processing/Formation									
Furnace		✓			✓				Empirical estimation
Cleaning basin		✓			✓				Empirical estimation
Air control system			✓		✓				Empirical estimation
Distribution									
Road transportation of lead grids			✓			✓		✓	- Literature (CEC, 1991) - Estimation
Sea transportation of lead grids			✓			✓		✓	- Literature (Georgakellos, 1997)
Internal distributions			✓					✓	- Literature (CEC, 1991) - Estimation
Notes									

1. Measurements
2. Computations (e.g. mass balance, input data, etc.)
3. Extrapolation of data from similar process type or technology
4. Extrapolation of data from different process type or technology
5. Estimate

Product-specific data: concern specific processes of batteries' reverse chain
Site-specific data: concern data from actual sites in the product system of batteries' reverse chain
General data: all the others

system is determined by the description of the logistics operations given in Section 14.4.1.

Some important remarks for the simplification of the study are taken into consideration. These concern the transportation process, which takes place in two stages of the reverse supply chain, namely, at the collection stage (transportation of used batteries) and at the distribution stage (distribution of the final products), that is, lead grids to the end users. It also appears in the collection stage of the disposal chain.

By using various approaches (cut-off methods or 5% rule), some less important processes at the processing stage have been ignored. This assumption does not influence the results significantly. More specifically, it has been assumed that storage of the ancillary materials and all the material movements in the processing unit cause negligible environmental effects (negligible emissions and minimal energy consumption).

Exclusion of the reuse stage of the final products of batteries' reverse supply chain was imperative due to the lack of reliable data in terms of undetermined secondary lead uses.

Regarding the geographical borders, a significant percentage of used batteries is imported from countries where the environmental legislation may not harmonize with that of the EU. This entails possible damage to the environment due to the toxicity of batteries. The situation may deteriorate depending on the means of transportation.

Finally, in this study, the allocation of the environmental exchanges is excluded as long as the examined waste-management policies are focusing on a single service, as does the production of secondary lead.

14.4.3 Inventory Analysis

An assumption made regarding the inventory analysis concerns the implementation of the conservation law of mass. It is supposed that the total mass of lead entering the system, mainly contained in used batteries, is equal to the mass leaving the system in the form of grids (product), slag, or lead dust. Similarly, it is supposed that during transportation, either after the collection of used batteries or at the stage of distribution of recycled lead to the end users, no increase or reduction of the mass of lead is incurred. The same assumption holds for the ancillary materials, so there is no change in the mass of the ancillary materials at their procurement and transportation phases.

Based on the above assumptions, and using the symbols appearing in Figure 14.1, we have the following, where

$m_{i,ii,iii,iv,v,vi}$	: the mass of lead in the corresponding stages (i, ii, iii, iv, v, vi)
m	: the mass of lead to be reused by the end user
$m_{a,b,c}$	: the mass of the ancillary materials in the corresponding stages (a, b, c)
m_L	: the mass of lead which enters the ecosystem in the vicinity of the plant
m_s	: the mass of slag leaving the system
α	: the percentage of slag, which corresponds to the amount of net metal lead produced
β	: the percentage of lead emitted to the ecosystem in the vicinity of the plant
f	: the percentage of the ancillary materials required for the production of lead mass m

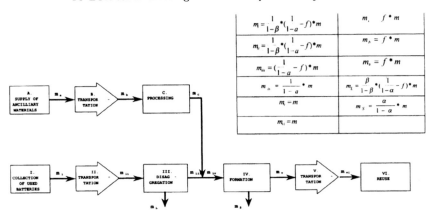

Fig. 14.1. Mass balance in the batteries' reverse chain

The total energy consumed in the system is the sum of energy consumed at every system's stage. Thus

$$E_{TOT} = E_i + E_{ii} + E_{iii} + E_{iv} + E_v + E_{vi} + E_a + E_b + E_c . \quad (14.9)$$

The energy consumed at each stage is the product of the specific energy and the mass leaving it. The specific energy u_j is defined as the energy consumed per unit of mass. Therefore:

$$E_j = u_j \cdot m_j \quad , \quad (14.10)$$

with $j = i, ii, iii, iv, v, vi, a, b, c$. E_j is expressed in MJ, u_j in MJ/ton and m_j in tons. From (14.9) and (14.10) we conclude that

$$E_{TOT} = (u_i + u_{ii}) \cdot \{\frac{1}{1-\beta} \cdot (\frac{1}{1-\alpha} - f)\} \cdot m$$
$$+ u_{iii} \cdot (\frac{1}{1-\alpha} - f) \cdot m + u_{iv} \cdot (\frac{1}{1-\alpha}) \cdot m + u_v \cdot m$$
$$+ u_{vi} \cdot m + (u_a + u_b + u_c) \cdot f \cdot m \quad . \quad (14.11)$$

Equation (14.11) holds also for the consumption Q_{TOTi} of each ancillary material required, the only difference being that, instead of u_j, we have g_j, i.e. the specific mass. The latter is defined as the mass of the required ancillary materials (or airborne, waterborne, and solid waste produced) per unit mass, and instead of E_j, we have Q_{ji}.

All the inputs and outputs of the system concerning mass or energy are listed in the Inventory Table (Table 14.3).

Table 14.3. Inventory analysis for the recovery processes of *SLI* batteries

		Collection/ Transportation	Disaggregation	Remanufacturing	Distribution/ Transportation	Total
Power demands						
Electrical energy	KWH		57.14	64.36		121.5
Liquid fuels (oil, etc.)	MJ	78.36		5539.3	230.2	5847.86
Other fuels						0
Resource consumption						
Carbon (C)	kg			142.86		142.86
Oil	kg	1.68		119	4.95	125.63
Water	m³		0.5			0.5
Iron (Fe)	kg			171.43		171.43
Antimony (Sb)	kg			-7.14		-7.14
Copper (Cu)	kg			-1.2		-1.2
Sodium chloride (NaCl)	kg			71.43		71.43
Sodium carbonate (Na₂CO₃)	kg			142.86		142.86
Sodium hydroxide (NaOH)	kg			1.6		1.6
Calcium oxide (CaO)	kg		9.52			9.52
Sulfate (S)	kg			0.8		0.8
Sodium nitrate (NaNO₃)	kg			0.8		0.8
Airborne emissions						
Carbon dioxide (CO₂)	kg	5.5	24.1	27.43	1.7	58.73
Carbon monoxide (CO)	g	88.23		87.14	382.3	557.67
HC	g	0			0	0
HCHO	g	0			0	0
Sulfate dioxide (SO₂)	g	0.01	84.85	969.1	0	1053.96
Nitrogen oxides (Nox)	g	74.35	28.4	162.7	194.8	460.25
Particulates	g	7.33	10.81	606.12	20.87	645.13
Volatile Organic Compounds (VOC)	g	26.72			63.2	89.92
Lead (Pb)	g			349.4		349.4
Liquid effluents						
Acid (H₂SO₄)	g					0
Lead (Pb)	g					0
Solid wastes						
Bulk waste (plastics & paper)	kg		452.6			452.6
Slag & ashes	kg			1484.3		1484.3
Hazardous waste (Lead (Pb))	kg		33.33			33.33

14.4.4 Impact Assessment and Aggregation of the Results

In the present study, the 'distance to target' principle is mainly applied for valuation, based on the political targets as defined by the relevant regulations for each environmental impact category, which hold globally, in the EU, and in Greece. It must be noted that the exchanges with the environment from the various processes in the systems examined were assessed and the most significant resources consumption, ecological impact, and impact on the working environment were designated on the basis of the weighting criteria applied in the EDIP method (Wenzel et al., 1997).

Table 14.4. The results of the Impact Assessment phase

Impact category	Normalization reference unit for 1990 $E_{W(j)}$ kg/pers/year (W/EU/GR)	Actual waste production $E_{W(j)}$ Kg/pers/year		Normalized waste NW(j) pers⁻¹ $NW(j) = \dfrac{E_{W(j)}}{E_{WGR(j)}}$		Weighted factor WEP(j)	Weighted waste j WW(i) $WW(j) = WEP(j) \cdot NW(j)$	
		RC	D	RC	D		RC	D
Global warming	$8.7 \cdot 10^3$ (W)	$7.85 \cdot 10^6$	$3.98 \cdot 10^5$	$1.7 \cdot 10^5$	$8.65 \cdot 10^9$	1.43	$2.43 \cdot 10^5$	$1.24 \cdot 10^3$
Stratospheric ozone depletion	0.202 (W)	0.0	0.0	0.0	0.0	23	0.0	0.0
Photochemical ozone formation	18 (EU)	9508.8	10941	$1.6 \cdot 10^6$	$3.4 \cdot 10^5$	1.28	$2.05 \cdot 10^6$	$4.35 \cdot 10^5$
Acidification	83 (EU)	$7.0 \cdot 10^4$	$2.7 \cdot 10^6$	$2.6 \cdot 10^6$	$1.0 \cdot 10^4$	1.186	$3.1 \cdot 10^6$	$1.186 \cdot 10^4$
Nutrient enrichment	298 (GR)	$5.24 \cdot 10^4$	$6.05 \cdot 10^4$	$1.76 \cdot 10^5$	$2.03 \cdot 10^5$	1.2	$2.1 \cdot 10^5$	$2.44 \cdot 10^5$
Human toxicity (HT)	*combination (GR)	*combination	*combination	$1.33 \cdot 10^2$	$8.73 \cdot 10^6$	2.8	$3.72 \cdot 10^2$	$2.44 \cdot 10^5$
Ecotoxicity (ET)	*combination (GR)	*combination	*combination	$5.21 \cdot 10^3$	$1.72 \cdot 10^3$	2.3	$1.2 \cdot 10^2$	$3.95 \cdot 10^3$
Persistent toxicity (PT)	*combination (GR)	*combination	*combination	$2.9 \cdot 10^2$	0.228	2.5	$7.25 \cdot 10^2$	0.57
Bulk waste	2500 (GR)	0.326	0.326	$1.3 \cdot 10^4$	$1.3 \cdot 10^4$	1.1	$1.43 \cdot 10^4$	$1.43 \cdot 10^4$
Slag & ashes	235 (GR)	1.07	0.0	$4.55 \cdot 10^3$	0.0	1.1	$5 \cdot 10^3$	0.0
Hazardous waste	20.7 (GR)	0.024	1.32	$11.6 \cdot 10^4$	0.064	1.1	$1.27 \cdot 10^3$	0.07
Radioactive Waste	0.035 (GR)	0.0	0.0	0.0	0.0	1.1	0.0	0.0
Carbon	570 (W)	$1.94 \cdot 10^4$	0.0	$3.4 \cdot 10^7$	0.0	0.0058	$1.66 \cdot 10^9$	0.0
Iron	100 (W)	$2.33 \cdot 10^4$	0.0	$23.3 \cdot 10^7$	0.0	0.0085	$19.8 \cdot 10^9$	0.0
Oil	590 (W)	$3.12 \cdot 10^4$	$2.21 \cdot 10^4$	$5.29 \cdot 10^7$	$3.74 \cdot 10^7$	0.023	$1.22 \cdot 10^8$	$8.61 \cdot 10^9$
Lead	0.64 (W)	0.0	$1.36 \cdot 10^3$	0.0	$2.13 \cdot 10^3$	0.048	0.0	$1.02 \cdot 10^4$
Copper	1.7 (W)	0.0	$1.62 \cdot 10^6$	0.0	$9.53 \cdot 10^7$	0.028	0.0	$26.6 \cdot 10^9$
Antimony	0.0 (W)	0.0	$9.73 \cdot 10^6$	0.0	$9.73 \cdot 10^6$	1.0	0.0	$9.73 \cdot 10^6$

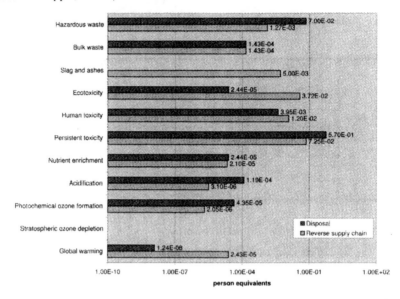

Fig. 14.2. Weighted ecological impact potentials of batteries reverse supply and disposal chain

1990 has been chosen to be the year of reference and the total global population of 5.29 billion people (World Resources Institute, 1990) has been used in order to calculate the person-equivalents for each environmental impact category. The application of the EDIP method leads to the results presented in Table 14.4.

The final results of this impact assessment are presented in Figures 14.2 and 14.3. The weighted impact potentials expressed in person equivalents for each environmental category are shown quantitatively and graphically.

In order to provide a single index for each waste-management policy so that comparisons between the policies are eased, the aggregation of the environmental impact categories is done using the AHP and LCA Polygon methods presented in Section 3.

Application of the *AHP*

In order to apply AHP in the final results of LCA, it is necessary to make the following assumptions.

- The evaluation of the alternative waste-management policies, which simultaneously take the value of zero following some criteria, either due to lack of inventoried data of substances that contribute to the corresponding environmental impacts or of reliable data, are not included in the analysis.

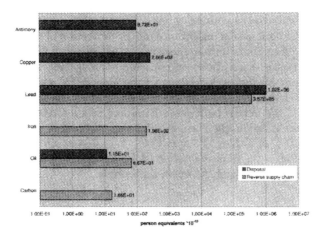

Fig. 14.3. Weighted resources consumption impact potentials of batteries' reverse supply and disposal chain

- The weights of the criteria that are not included in the analysis are not taken into account because they affect the weights of other criteria and, consequently, the final decision.
- The intrinsic weights of the impact categories are computed based on the weighting, taking into consideration the EDIP method (Daniel and Pappis, 2003).
- The principle of hierarchic composition takes place for the overall estimation of each waste-management policy in the three main impact categories (ecological impact, resources consumption, and impact on the working environment) as described in the EDIP method. In a similar way with criteria, the main impact category of 'impact on the working environment' is not included in the computation due to lack of data.
- The impact categories of the EDIP method are considered to be of equal importance.

The application of AHP starts with the itemization of the normalized values of each waste-management policy (reverse supply chain (RC) and disposal chain (D)) at the stage of IA. The criteria refer to the sub-categories of the impact categories and include 'environmental impact' and 'resources consumption' and are presented in columns 'STEP 0 RC' and 'STEP 0 D' of Tables 14.5 a and b, respectively.

Next, the values of the criteria are further normalized so the corresponding sum for the sub-category is equal to 1. The results are presented in columns 'STEP 1 RC' and 'STEP 1 D' of Tables 14.5 a and b.

The next step is the calculation of the intrinsic weights and the specific weights of the criteria. The weight of each criterion is a result of the distance-to-target method and the political targets. The calculation of intrinsic weights (iw) is based on their actual importance for the protection of the environment. The corresponding contribution (positive or negative) of the whole waste-management policy is not taken into account. The specific weights (sw) are calculated based on the performances of the alternative waste-management policies with respect to the particular criteria, that is, the average (or total) performance of the actual policies with respect to the criteria. The overall weights are calculated using Formula 14.3 and the results are presented in columns 'STEP 2' of Tables 14.5 a and b.

The next step is the final evaluation of the alternative waste-management policies, which is done using the following formula.

$$A_{ij} = \sum a_i \cdot w_{cj} \tag{14.12}$$

where a_i : the alternative waste-management policies
$\quad\quad$ w_{cj} : the overall weights of the alternative waste-management policies i according to criterion j
$\quad\quad$ A_{ij} : the final evaluation of the waste-management policy i according to criterion j.

Application of Formula 14.12 leads to the results presented in columns 'STEP 3 RC' and 'STEP 3 D' of Tables 14.5 a and b.

In the final step, the main impact categories are evaluated following the same process, taking into account the corresponding weights according to Formula 14.13. The main impact categories (environmental impacts and resources consumption) are considered to be of equal importance, thus $w_k = 0.5$.

$$E_i = \sum A_{ij} \cdot w_k \tag{14.13}$$

where A_{ij} : the alternative waste-management policies
$\quad\quad$ w_k : the weight of the main impact category k
$\quad\quad$ E_i : the overall evaluation of the waste-management policy i.

For each waste-management policy [(RC) and (D)] the results are the following:

$$E_{RC} = 0.1659 \cdot 0.5 + 0.2421 \cdot 0.5 = 0.2040 \Rightarrow E_{RC} = 20.40\%$$
$$E_D \;= 0.8341 \cdot 0.5 + 0.7579 \cdot 0.5 = 0.7960 \Rightarrow E_D \;= 79.60\%$$

Application of the LCA Polygon

In order to apply the method of the LCA polygon in the final results of LCA, it is necessary to make the following assumptions.

Table 14.5. Application of AHP - (a) Ecological Impacts and (b) Resources Consumption

(a)

	STEP 0 RC	STEP 0 D	STEP 1 RC	STEP 1 D	Criterion weight	STEP 2		w_{cj}	STEP 3 RC	STEP 3 D
						Intrinsic Weights (iw_{cj})	Specific Weights (sw_{cj})			
Global warming	1.70E-05	0.99949	8.65E-09	0.00051	1.43	0.089	4.90E-05	0.00003	3.15E-05	1.60E-08
Photochemical ozone formation	1.60E-06	0.04494	3.40E-05	0.95506	1.28	0.080	1.02E-04	0.00006	2.65E-06	5.63E-05
Acidification	2.60E-06	0.02534	1.00E-04	0.97466	1.186	0.074	2.95E-04	0.00016	3.99E-06	1.53E-04
Nutrient enrichment	1.76E-05	0.46438	2.03E-05	0.53562	1.2	0.075	1.09E-04	0.00006	2.73E-05	3.15E-05
Persistent toxicity	2.90E-02	0.11284	2.28E-01	0.88716	2.5	0.156	7.40E-01	0.83128	9.38E-02	7.37E-01
Human toxicity	1.33E-02	0.99934	8.73E-06	0.00066	2.8	0.175	3.83E-02	0.04821	4.82E-02	3.16E-05
Ecotoxicity	5.21E-03	0.75180	1.72E-03	0.24820	2.3	0.144	1.99E-02	0.02062	1.55E-02	5.12E-03
Slag and ashes	4.55E-03	1.00000	0.00E+00	0.00000	1.1	0.069	1.31E-02	0.00648	6.48E-03	0.00E+00
Bulk waste	1.30E-04	0.50000	1.30E-04	0.50000	1.1	0.069	7.48E-04	0.00037	1.85E-04	1.85E-04
Hazardous waste	1.16E-03	0.01780	6.40E-02	0.98220	1.1	0.069	1.88E-01	0.09274	1.65E-03	9.11E-02
								A_{IJ}	16.59%	83.41%

(b)

	STEP 0 RC	STEP 0 D	STEP 1 RC	STEP 1 D	Criterion weight	STEP 2		w_{cj}	STEP 3 RC	STEP 3 D
						Intrinsic Weights (iw_{cj})	Specific Weights (sw_{cj})			
Carbon (C)	3.40E-07	1.00000	0.00E+00	0.00000	0.0058	5.21E-03	1.18E-04	0.00001	1.34E-05	0.00E+00
Oil	2.90E-07	0.98312	4.98E-09	0.01688	0.023	2.07E-02	1.02E-04	0.00005	4.52E-05	7.77E-07
Iron (Fe)	2.33E-06	1.00000	0.00E+00	0.00000	0.0085	7.63E-03	8.09E-04	0.00013	1.34E-04	0.00E+00
Lead (Pb)	7.43E-04	0.25861	2.13E-03	0.74139	0.048	4.31E-02	9.95E-01	0.93370	2.42E-01	6.92E-01
Copper (Cu)	0.00E+00	0.00000	9.53E-07	1.00000	0.028	2.52E-02	3.31E-04	0.00018	0.00E+00	1.81E-04
Antimony (Sb)	0.00E+00	0.00000	9.72E-06	1.00000	1	8.98E-01	3.37E-03	0.06593	0.00E+00	6.59E-02
								A_{IJ}	24.21%	75.79%

- The principle of hierarchic composition applies also in the *LCA* polygon
  method for the overall estimation of each waste-management policy with
  respect to the three main impact categories (ecological impact, resources
  consumption, and impact on the working environment) as described in the
  EDIP method. For the same reasons as in the *AHP* method, the main
  impact category of 'impact on the working environment' is not included.
- The impact categories of the *EDIP* method are considered to be of equal
  importance.

The results of the method are presented in the form appearing in Figure 14.4,
which shows the *LCA* polygon of the reverse supply chain in the case of Ecological Impacts. Similar polygons are obtained for the other impact categories.
The areas of the polygons define the environmental score of the alternative
waste-management policies. The results are summarized in Table 14.6.

The final step is to apply the principle of hierarchical composition (Formula
14.14). The main impact categories (environmental impacts and resources
consumption) are considered to be of equal importance, thus $w_k = 0.5$.

$$E_i = \sum E_{LCAi} \cdot w_k \qquad (14.14)$$

Fig. 14.4. LCA polygon of the reverse supply chain - Ecological Impacts

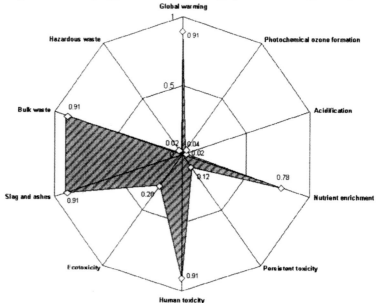

where E_{LCAi} : the ratio of the average area of the LCA polygon E to the
area of the regular polygon

w_k : the weight of the main impact category k

E_i : the overall evaluation of the waste-management policy i.

For each waste-management policy [(RC) and (D)] the results are the following.

$$E_{RC} = 0.15 \cdot 0.5 + 0.27 \cdot 0.5 = 0.21$$
$$E_D = 1.87 \cdot 0.5 + 5.51 \cdot 0.5 = 3.69$$

These results clearly show that, from an environmental point of view, it is preferable to set up a reverse logistics network instead of having the batteries landfilled. It should be noted that, in the case examined, the company has an economic gain from its recovery activities (otherwise it would not do business in this industry). However, there are cases where the environmental and economic profiles of the relevant activities are contradictory and it is difficult to aggregate these profiles in order to support effective decision making.

14.5 Conclusion and Discussion

Businesses have many bottom-line incentives to include the environmental parameter in their supply chain management considerations. Many firms take the view that, rather than maximizing profits within a fixed set of environmental constraints, it is better to modify these constraints in order to gain a competitive advantage (Tsoulfas and Pappis, 2001). The environment is introduced as a new criterion for differentiating among products, extending to the development of environmental standards or labels (Faucheux and Nicolai, 1998). For starters, an environmental approach can lead to internal cost savings from using energy more efficiently, producing less waste and recycling materials. Second, it can open new markets, as consumers are attracted to companies that have invested in green technologies. Third, opportunities abound to assist other companies and communities, both at home and in emerging markets, to use energy, water, and by-products more efficiently and to find beneficial uses for wastes (International Institute for Sustainable Development, 2000).

The development of environmental management standards and methods for measuring the environmental impacts of supply chains has provided the

Table 14.6. Percentage area of the examined waste-management policies according to the LCA polygon method

	Index E_{LCA} (%)	
Waste-Management Policy	Environmental impacts	Resources consumption
Reverse Supply Chain	0.15	0.27
Disposal Chain	1.87	5.51

necessary means for businesses to adapt to the new scene. In addition, the introduction of Reverse Logistics practices not only is connected with economic profit due to the exploitation of materials and parts that have ended their life cycle, but also with major contributions regarding environmental protection. LCA is the most promising quantitative method for the assessment of environmental impacts. LCA may reveal the overall loss/gain incurred by recovery practices and may be used as a basis for making comparisons of different such practices. Especially in the case of logistics activities, the concept of identifying the life-cycle stages is similar to any other kind of analysis from an economic/business perspective. Thus, there are definitely grounds for an integrated analysis that could encompass and aggregate all the aspects needed in order to achieve business targets.

In most cases, an LCA comparison between materials, products, or processes has a main goal, identification of the 'best one' (Hanssen, 1998). Each phase has its own sources of uncertainty. Goal and Scope Definition determine the system boundaries, an essential step towards a comprehensive and practicable LCA. The inventory data set is characterized by various data elements, typically containing inherent uncertainties, which ultimately undermine the certainty of the LCA results.

The examination of different IA methods raises certain questions regarding the interpretation of the results, since there is ground for contradictory conclusions, which is equivalent to the inability to draw *any* conclusions. This is mainly due to the different impact categories that are examined in each method and also to the different targets that these methods use to mark their limits. It follows that the results of any LCA study should not be generalized for any similar case study because of the specific conditions and assumptions that apply to each one of them (Tsoulfas et al., 2002b). It is suggested (ISO 14042: 2000(E)) that LCA studies should be implemented using as many IA methods as possible.

Moreover, an LCA inevitably portrays a simplification of the real life cycle of a product: it collapses many life-cycle steps into a few broad stages to make analysis tractable (Andrews and Swain, 2000).

These methodological gaps complicate LCA and limit its acceptance. A universal methodology has to be adopted and a number of aspects need to be further worked out to overcome some of the barriers mentioned above. The ISO is now standardizing the approach to LCA. The proposed ISO standards are based on SETAC's approach but with several modifications (Ong et al., 1999).

OR Models for Eco-eco Closed-loop Supply Chain Optimization

Jacqueline M. Bloemhof-Ruwaard[1], Harold Krikke[2], and Luk N. Van Wassenhove[3]

[1] Rotterdam School of Management / Faculteit Bedrijfskunde, Erasmus University Rotterdam, P.O. Box 1738, 3000 DR Rotterdam, The Netherlands, JBloemhof@fbk.eur.nl
[2] Tilburg University, CentER Applied Research, Department of Operations Research, P.O.Box 90153, 5000 LE Tilburg, The Netherlands, Krikke@uvt.nl
[3] Henry Ford Chair in Manufacturing, INSEAD, Boulevard de Constance, 77305 Fontainebleau, France, luk.van-wassenhove@insead.edu

15.1 Introduction

Growing concern for the environment has led to a range of new environmental policies for various industries in which the recovery of waste is an essential element. The formulation of waste policies is a complicated matter for a number of reasons. First, the determination of environmental impact of supply chain processes (including recovery) is complex and partly subjective in its interpretation. Second, and related to this, the relative environmental impact of waste recovery, as opposed to other supply chain processes, may vary strongly by type of industry. Third, environmentally friendly processing of waste is often costly, especially when taking an end-of-pipe perspective. Fourth, there may be unexpected side effects of waste policies, such as (illegal) trading.

Lifset and Lombardi (1997) point out that the eco(nomy)-eco(logy) contradiction can only be solved by adopting a product life-cycle approach and that the Original Equipment Manufacturer (OEM) is the most appropriate candidate to manage that product life cycle. Theoretical advantages of (Extended) Producer Responsibility or EPR (adapted from Lifset and Lombardi, 1997) are

- to generate funds for End-Of-Life (EOL) processing,
- to give incentives to optimize recovery characteristics of the product,
- to tab expertise and product information for recovery operations, and
- to set and assign collection and recovery targets with clear settlements.

Product life-cycle management and hence policy measures have a product focus. However, the actual environmental load results from the supply chain processes forwarding the product from raw material to the final consumer.

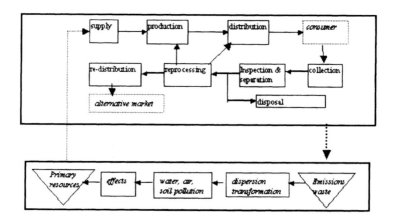

Fig. 15.1. Relating closed-loop supply chain and environmental chain

Figure 15.1 shows the relationship between the supply chain (including re-covery operations) and the environmental chain. Optimization of closed-loop supply chains can be based on costs and environmental impacts. Supply chain costs are directly linked to supply chain processes. In order to measure the supply chain environmental effects, we use the observation that the supply chain is linked to the environmental chain. All supply chain activities gen-erate emissions and waste which, via diverse processes in the environmental chain, eventually have an impact on our natural resources. In turn, these natural resources are necessary inputs for the supply chain.

Reductions in the environmental load can result either directly from im-proving the processes or by changing the product design through which a lower usage of processes or the usage of other less polluting processes is en-abled. Hence, adopting a life-cycle perspective implicitly means concurrent optimization of product design and (forward and recovery) supply chain pro-cesses.

A lot of literature is available on critical success factors, including prod-uct material composition, internal reuse at manufacturing, energy friendly manufacturing, low energy products, reducing weight of products, disassem-bly, return quality, collection rates, and recovery technology. There is also a debate regarding whether or not sector-wide solutions are better than sys-tems run by individual companies. Sector-wide policies may be carried out by individual companies or by collective nationwide systems. According to De Koster et. al (2002), collective systems have their merits in terms of visibility, economies of scale, and ease of implementation but also tend to be expensive and not necessarily environmentally friendly. An eco-eco product and supply chain design is the result of many factors and there is little consensus about their relative importance. This is not surprising. In our opinion it is impossible to denote a strict hierarchy on success factors, since they depend on goals set, product parameters, market characteristics, geography, and so on. Therefore, success factors need to be analyzed on a case-by-case basis. However, a set of basic questions can be posed:

- How can we measure the environmental impact of the circular supply chain as objectively as possible?
- How do direct economic cost functions behave compared to environmental impact functions?
- Who are the shareholders and what are their objectives?
- What are the most important factors and hence decision variables?
- Which constraints need to be taken into account?
- To what extent does recovery contribute to lowering the overall environmental impact of the supply chain?
- Can recovery be done in concurrency with economic objectives?
- Are there any indirect economic consequences, both positive and negative?
- Is there evidence that either a sector-wide or individual solution yields better results?

Given the intrinsic complexity of these matters, we advocate the use of OR models to support this kind of decision-making. Looking into decision support models available in the literature, we see on the one hand cost (economical) models for supply chain optimization and on the other Life-Cycle Assessment or analogous ecological models aiming at product design improvements. However, a common framework from which integrated eco-eco models can be developed seems to be lacking. The aim of this chapter is threefold:

1. To develop an eco-eco modelling framework for the evaluation of EU policy.
2. To develop OR models for illustrative case studies on both individual and sector-wide systems.
3. To reflect on the lessons learned from EU policy.

In section 15.2, we discuss recent EU policy and different EPR systems (individual and sector-wide). Section 15.3 deals with OR models and literature regarding eco-eco optimization for closed-loop supply chains. We present two case studies: one on the pulp and paper sector and one on consumer electronics. Section 15.4 presents a case study on refrigerator life-cycle management, based on individual recovery decisions, whereas Section 15.5 deals with the pulp and paper case, based on a sector-wide approach for recovery. Section 15.6 ends with discussions and conclusions.

15.2 EU Policy

Policy and trade are increasingly influenced by the European Union. As part of this, EU member states have to follow guidelines for implementation of European environmental policies. A number of so-called priority waste streams have been defined, like packaging (1994/1999) and Vehicles (2000). Special directives are in force, e.g. the directive on Waste of Electrical and Electronics Equipment (WEEE) (submitted 2001), the directive on restriction on Hazardous Substances (RoHS) and the one on Eco-design of End-use Equipment (EEE). In preparation are directives for demolition waste, tires, health-care waste, and hazardous waste. There are also special directives to reduce or prevent landfill and incineration (1994/1999) and reduce or ban waste transportation (1994). Once the directives are in force, EU member states must implement legislation on 'producer responsibility' within 18 months.

The paper and pulp sector in Europe is one of the most polluting, but it has made significant efforts to reduce its environmental impact. EU policy for this sector consists of the following elements (see Gabel et. al, 1996) .

- Eco-labelling plans have been proposed and partly introduced to identify ecologically responsible pulp and paper products.
- Eco-taxes on products that fail to meet environmental requirements.
- Facility licences. In order to operate, pulp and paper mills will need to meet emissions criteria.
- Mandated recovery and secondary fiber contents. A large part of the pulp and paper volumes in Europe is used for packaging. For packaging, the EU directive prescribes recovery percentages of 65%. Also, new paper and cardboard packages need to be more and more of recycled fiber.

In June 2001, the EU Environmental Ministers agreed in principle to enter the special directives into force. It boils down to the following targets for member states.

- Minimize the use of dangerous substances and the number of plastics, and some materials (e.g. mercury) must be phased out (RoHS).
- Promote design for recycling, (no targets) (EEE).
- Producers should take responsibility for EOL phases of their products, encode new products for identification, and provide information to the processors for accurate recycling (WEEE).
- Appropriate systems for separate collection of WEEE should be put in place and private households are entitled to return for free.
- Producers have to set up and finance appropriate systems in order to ensure accurate processing and reuse/recycling of WEEE and are responsible, at the latest, from the point of collection at municipalities.
- Collection targets of 4 kg per head of population yearly. Collection services must be offered by retailers (trade-ins) and municipalities. Goods must be collected separately and pre-processed such that optimal recovery is possible (WEEE).
- Recovery quotas are set between 70-90% of the goods collected (depending on the product category). Hazardous contents must be removed at all times and processed according to strict prescriptions. Reuse in the original

supply chain is encouraged, however only for plastics are targets set (5% of weight of newly produced consumer electronics) (WEEE).

15.3 OR Models for Eco-eco Optimization

In this overview, we discuss reverse supply chain cost models, closed-loop (forward and reverse) supply chain cost models, and LCA-oriented models. We do not discuss general location-allocation models, since they are not directly applicable to the problem description (compare also Table 4.1 of Chapter 4). Table 15.1 gives an overview of the most important models and shows that none of the models has all the modelling features we are looking for. Below, we summarize the major findings of our literature study.

In reverse supply chain cost models only the reverse chain is considered. The problem is generally represented as a directed graph, where the sources identified are collection sites, the sinks are either disposal sites or reuse locations, and the arcs are transportation links (see also Figure 4.1 of Chapter 4). Sets of feasible locations for different types of facilities are given and an optimal choice must be made. The collected items can be processed via various routings in the graph. Optimization occurs on supply chain costs, generally fixed and variable costs for facilities and variable costs for intermediate transportation. Mixed Integer Linear Programming (MILP) is the most commonly used technique. Sometimes, locations are imposed, resulting in Linear Programming (LP) problems. From a mathematical point of view, the models are pretty standard. Some papers describe nice branch and bound and heuristic algorithms and some model extensions like economies of scale with concave cost functions and multi-period variants. This type of model is discussed more extensively in Chapter 4.

Closed-loop supply chain cost models are similar to reverse supply chain cost models, but simultaneously optimize the forward and reverse network. Locations may serve both as sink and source, and nodes and transportation links may serve both as part of the forward and reverse network, increasing efficiency but also system complexity. Because goods flows are literally closed, one needs additional decision variables: the amount of reuse in the original supply chain or the number of containers per location. As a result, the structure and formulation of these models is generally somewhat more intricate, but mathematically the same remarks apply as for the reverse supply chain cost models. MILP models will be suitable in situations with stable network designs, using scenario analysis to examine the results of possible changes in the external parameters. If a closed-loop supply chain has to built from scratch, with a high degree of dynamics in the structure, then it is worthwhile to examine Systems Dynamics, as is mentioned in Chapter 13.

Life-cycle analysis (LCA) can be defined as an input-output analysis of resources or materials and energy requirements in each phase of the life cycle of a product. Although commonly acknowledged that a full life cycle assessment

Table 15.1. Literature study on model characteristics in supply chain optimization

Authors	Type of Supply Chain	Decision Variables
	Reverse SC Cost Models	
Spengler et al., 1997	rev., open-loop	loc-allocation
Ossenbruggen, 1992	rev.. open-loop	allocation
Barros et al.. 1998	rev., open-loop	loc-allocation
Krikke et al., 1999	rev., closed-loop	loc-allocation
Louwers et al., 1999	rev., open-loop	loc-allocation
Ammons et al., 1997	rev., open-loop	loc-allocation
Jaramayan et al., 1997	rev., open- and closed-loop	loc-allocation
Shih, L., 2001	rev., open-loop	network design
Brouwers and Stevels, 1995	rev., open- and closed-loop	product design
	Closed-loop SC Cost Models	
Berger and Debaille, 1997	forw. + rev., closed-loop	loc-allocation
Kroon and Vrijens, 1995	forw. + rev.. closed-loop	loc-allocation
Thierry, 1997	forw. + rev., closed-loop	allocation
Chen et al., 1993	forw. + rev., closed-loop	product design
Fleischmann et al.. 2001	forw. + rev.. open- and closed-loop	loc-allocation
	LCA-oriented Models	
Sasse et al., 1999	rev., open-loop	loc-allocation
	multi-objective with simple LCA (energy and waste)	
Berger et al., 1998	rev., open-loop	loc-allocation
	flexible objectives: costs and environment, LCA	
Caruso et al., 1993	rev., open-loop	loc-allocation
	multi-objective with environmental indicators	
Bloemhof-Ruwaard, 1996	forw.+ rev., closed-loop	allocation and product mix
Daniel et al., 1999	rev., open-loop	product design
Guelorget et al., 1993	forw. + rev., open-loop	product design

is the appropriate way to assess the environmental profile of a product (see also Chapter 14), many practical burdens with regard to data collection and result interpretation remain. By definition, LCA only considers environmental issues. In reality, economic and technical issues cannot be ignored in any decision. In practice, simplifying methods are used for measuring the supply chain environmental impact. The methods used differ slightly per case study, but boil down to defining a (small number of) environmental indicators, based on LCA methodologies. Given LCA's complexity, it becomes regular practice to make use of quantitative physical indicators, e.g., emission, waste, and energy use, instead of a full LCA (see Azzone, 1996).

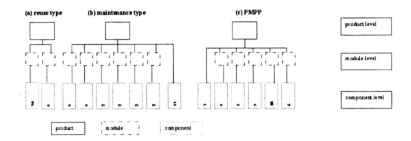

Fig. 15.2. Disassembly structures of the three alternative product designs

15.4 Case: Refrigerators

Umeda et al. (2000) present a case study on refrigerator life-cycle management. The case study was carried out by a team from Tokyo University that obtained R&D data from a large Japanese OEM of consumer electronics. Product modularity is optimized for three alternative product designs: reuse, maintenance, and PMPP, consisting of modules, components, and materials (in decreasing order of hierarchy). The designs are equal in terms of quality and functionality and utilize the same 25 components (see Figure 15.2) of the same materials. However, because the products have different modular structures, they have different costs and environmental impact functions (measured by energy and waste), and have differently feasibility for recovery and disposal options at various levels of disassembly.

In the original work by Umeda et al. (2000), the supply chain network is assumed to be fixed. Krikke et. al (2002) concurrently optimize the supply chain network and product design choices, assuming the product design parameters to be given. Moreover, logistical data and supply chain cost functions are added. Krikke et. al (2001) describes all data gathering. Our logistics data set comprises locations in Europe, and we assume that the Japanese manufacturer wishes to sell and recover the refrigerators in Europe. Demand is known at the level of 10 domestic agents, which are spread over Europe. All domestic agents will take any of the three types of refrigerators (these are considered identical in terms of final quality). Mixed strategies are allowed, i.e., we may serve different agents with different product designs. Decisions will be taken both on the network structure and product design (three types). The network decisions are numbers, capacities, and locations for supply chain process facilities combined with allocation of goods flows in the system. Goods flows represent intermediate flows between facilities, inbound supply flows, outbound market deliveries, inbound return flows, and outbound disposal/thermal disposal flows. Inbound and outbound flows relate to supply points (i.e., suppliers of raw materials), market locations, and disposal/thermal disposal locations. Allocation implicitly means choosing recovery options (product repair, module

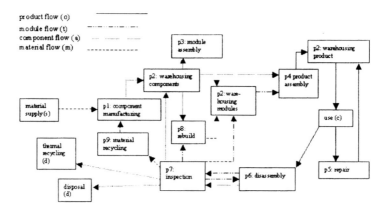

Fig. 15.3. Closed-loop supply chain for the refrigerator case

rebuild, etc.). The most important parameters are rate of return, recovery feasibility (read return quality), and recovery targets (read legislation). We measure optimization criteria as follows. Costs include all fixed and variable costs related to the installment and operations of the processes as well as the disposal sinks and supplies from the sources. Supply chain environmental impact is measured by energy use and waste volume.

We present a Mixed Integer Linear Programming model to support solving this optimization model. Figure 15.3 represents the closed-loop supply chain system. The sources of the system are the supplier locations, from where materials are supplied to the first process (component manufacturing). Here, components are made out of raw materials. Ten suppliers for all materials are available in Europe. After component manufacturing, modules and products are assembled. Modules are assembled from components only and products from components and/or modules, depending on the product design option. Components, modules, and final products are stored at the warehouse. The final products are sold to agents. Agents are the only nodes in the system serving as sink and source. All products are returned after use within the planning period. They can be repaired and delivered to the warehouse for resale. They can also be disassembled, after which modules are inspected and rebuilt. In the rebuild process, non-reusable components in the modules are replaced. Rebuilt modules are sent to the warehouse. Modules can also be further disassembled, after which components are inspected and sent to the warehouse (for direct reuse), disposed of thermally or 'normally' or recycled at a material level. Note that a single common type of facilities is used for inspection and disassembly of products and modules. Materials recycled re-enter the supply chain via component manufacturing.

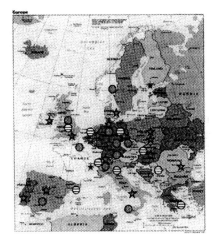

Fig. 15.4. Facility locations for all supply chain processes

In total, 20 potential facility locations are available. Transportation is done by truck. There are 10 disposal sites, of which 5 are energy recovery installations. Note that the system is consumer driven and all forward processes are 'pulled' by consumer demand. The reverse chain is 'pushed' by the discarding of products by that same consumer. We assume that a closed-loop supply chain for Europe must be set up from scratch. Figure 15.4 represents facility locations for all supply chain processes.

15.4.1 The Model

Let us first introduce some notation. Amounts are all in kilograms. Costs are denoted in euro/kg., and Energy is denoted in GwH/kg.

Index sets:

$\mathcal{O}$ $= \{o_1, \ldots, o_3\}$ $=$ set of possible product designs
$\mathcal{A}$ $= \{a_1, \ldots, a_{25}\}$ $=$ set of components
$\mathcal{T}$ $= \{t_1, \ldots, t_{13}\}$ $=$ set of modules
$\mathcal{M}$ $= \{m_1, \ldots, m_4\}$ $=$ set of materials
$\mathcal{S}$ $= \{s_1, \ldots, s_{10}\}$ $=$ set of supply points of raw materials
$\mathcal{K}$ $= \{k_1, \ldots, k_{10}\}$ $=$ set of customer locations, primary markets
$\mathcal{D}$ $= \{d_1, \ldots, d_{10}\}$ $=$ set of disposal and thermal recycling locations
$\mathcal{P}$ $= \{p_1, \ldots, p_9\}$ $=$ set of processes (forward or return)
$\mathcal{I}$ $= \{i_1, \ldots, i_{20}\}$ $=$ set of facility locations
$\mathcal{F}$ $= \{f_1, \ldots, f_3\}$ $=$ set of performance indicators (costs, energy, waste)
$(\mathcal{P}, \mathcal{I})$ $=$ set of processes $p \in \mathcal{P}$ assigned to locations $i \in \mathcal{I}$

Parameters:

$\mathcal{I}(p)$ = set of feasible locations i for p

wi_{ma} = amount of material input m for manufacturing 1 kg of component a

wa_{at} = amount of component a for assembling 1 kg of module t

wr_{at} = amount of component a for rebuilding 1 kg of module t

wc_{ao} = amount of component a for assembling 1 kg of product design o

wm_{to} = amount of module t for assembling 1 kg of product design o

uk_{kp} = fraction of return flow of customer k feasible for process p

um_{tp} = fraction of released modules t feasible for process p

ud_{ad} = fraction of released components a feasible for disposal at location d

up_{ap} = fraction of component a feasible for process p

ro_{ot} = amount of module t released by disassembling product design o

rm_{ta} = amount of component a released by disassembling 1 kg of module t

rp_{oa} = amount of component a by disassembling 1 kg of product design o

rc_{am} = amount of material m from recycling 1 kg of component a

vk_k = demand finished product at customer location k

vd_d = maximum demand from external secondary market d

ω_f = weight assigned to objective f

ϵ_f = scaling factor for deviation variable Z_f

T_f = target value for performance indicators f

β = a very large number

σ = rate of return

cm_{msi} = costs of supplying material m from supply point s for facility location i

$ci_{ii'}$ = (internal) transportation costs from location i to location i'

cc_{ik} = transportation costs from location i to customer k

cd_{id} = transportation costs from location i to disposal location d (including linear processing costs)

cpc_{api} = costs of processing component a by process p at location i

cpm_{tpi} = costs of processing module t by process p at location i

cpp_{opi} = costs of processing product design o by process p at location i

FIX_{pi} = annualized setup costs plus fixed costs for installing process p at location i

em_{msi} = energy use of supplying material m from supply point s to facility location i

$ei_{ii'}$ = energy use of transporting goods from location i to location i'

ec_{ik} = energy use of transporting goods from location i to customer c

ed_{id} = energy use of transporting goods from location i to disposal d (including processing)

cpc_{api} = energy use of processing component a by process p at location i

epm_{tpi} = energy use of processing module t by process p at location i

epp_{opi} = energy use of processing product design o by process p at location i

Decision variables:

X^{re}_{mspi} = (external) raw material flow m from s to (p, i)

$X^{ri}_{mpip'i'}$ = (internal) raw material flow m from (p, i) to (p', i')

$X^{ci}_{apip'i'}$ = (internal) component flow a from (p, i) to (p', i')

$X^{m}_{tpip'i'}$ = module flow t from (p, i) to (p', i')

$X^{gi}_{opip'i'}$ = (internal) goods flow of type o from (p, i) to (p', i')

X^{gc}_{opik} = product flow of type o from (p, i) to customer k

X^{gr}_{okpi} = reverse product flow of type o from customer k to (p, i)

X^{cd}_{apid} = component flow a from (p, i) to disposal/thermal recycling locations d

X^{g}_{opi} = product design o processed at (p, i)

X^{c}_{api} = component flow a processed at (p, i)

X^{m}_{tpi} = module flow t processed at (p, i)

Y_{pi} = boolean indicating assignment of process p to location i

Z_f = deviation variable for performance indicator f

Now the problem studied can be formulated in (simplified) mathematical terms in the following way. Within the (single) planning period considered, vk_k refrigerators can be sold at each market k, from where each agent serves customers in the area. To this end, refrigerators have to be manufactured from materials m, components a, and modules t, where the product structure depends on the product design o. Raw materials are supplied from supplier locations s (sources). Subsequently the components, modules, and products can be manufactured in the supply chain by processes p. Products are used and discarded at markets k (which serve both as sinks and as a source) and returned to the logistic network where several recovery processes p can be applied. Unrecovered return flows end at sink d, for either disposal or thermal recycling. Supply chain processes p can be installed at locations i. Together they serve as nodes (p, i) in the network. For each process p, locations i can be chosen from a set $\mathcal{I}(p)$. Network design choices involve location choices Y_{pi} and goods flows X, where location choices incur a fixed cost FIX_{pi} and variable processing costs. Transportation costs for goods flows are linear. Energy use is linear to volumes. Residual waste is the total volume of all components disposed of at sinks d. The objective is to optimize total economic costs ($f1$), energy use ($f2$), and residual waste ($f3$). In a multiple-criteria model, relative deviations Z_g to a priori set target values T_f are minimized, where different weights can be assigned to the criteria.

$$\min \quad w_f \times Z_f \tag{15.1}$$

s.t.

Definitions and logical constraints

$$\sum_{mspi} cm_{si} X^{re}_{mspi} + \sum_{apipi'} ci_{ii'} X^c_{apip'i'} + \sum_{tpip'i'} ci_{ii'} X^m_{tpip'i'}$$

$$+ \sum_{opip'i'} ci_{ii'} X^{gi}_{opip'i'} + \sum_{opik} cc_{ik} X^{gc}_{opik} + + \sum_{okpi} cc_{ik} X^{gr}_{okpi}$$

$$+ \sum_{apid} cd_{id} X^{cd}_{apid} + \sum_{mpip'i'} ci_{ii'} X^{ri}_{mpip'i'} + \sum_{api} cpm_{api} X^c_{api}$$

$$+ \sum_{tmi} cpm_{tpi} X^m_{tpi} + \sum_{opi} cpp_{opi} X^g_{opi} + \sum_{pi} FIX_{pi} Y_{pi} - \epsilon_1 Z_1 = T_1 \tag{15.2}$$

$$\sum_{mspi} em_{si} X^{re}_{mspi} + \sum_{apipi'} ei_{ii'} X^c_{apip'i'} + \sum_{tpip'i'} ei_{ii'} X^m_{tpip'i'}$$

$$+ \sum_{opip'i'} ei_{ii'} X^{gi}_{opip'i'} + \sum_{opik} ec_{ik} X^{gc}_{opik} + \sum_{okpi} ec_{ik} X^{gr}_{okpi}$$

$$+ \sum_{apid} ed_{id} X^{cd}_{apid} + \sum_{mpip'i'} ei_{ii'} X^{ri}_{mpip'i'} + \sum_{api} epm_{api} X^c_{api}$$

$$+ \sum_{tmi} epm_{tpi} X^m_{tpi} + \sum_{opi} epp_{opi} X^g_{opi} - \epsilon_2 Z_2 = T_2 \tag{15.3}$$

$$\sum_{apid} X^{cd}_{apid} - \epsilon_3 Z_3 = T_3 \tag{15.4}$$

$$Y_{pi} = 0, 1 \qquad\qquad\qquad \forall p, \forall i \in \mathcal{I}(p) \tag{15.5}$$

$$\sum_o X^g_{opi} + \sum_a X^c_{api} + \sum_t X^m_{tpi} \le \beta Y_{pi} \qquad \forall p, \forall i \in \mathcal{I}(p) \tag{15.6}$$

$$X^{re}_{mspi}, X^{ri}_{mpip'i'}, X^{ci}_{apip'i'}, X^{mi}_{tpip'i'}, X^{gi}_{opip'i'}, X^{gc}_{opik},$$

$$X^{gr}_{okpi}, X^{cd}_{apid}, X^g_{opi}, X^c_{api}, X^m_{tpi} \ge 0 \qquad \forall p, s, f, d \tag{15.7}$$

Balance equations forward

$$\sum_{oi} X^{gc}_{opik} = vk_k \qquad\qquad \forall k, p = p2 \tag{15.8}$$

$$\sum_k X^{gc}_{opik} = X^g_{opi} \qquad\qquad \forall o, i \in \mathcal{I}(p), p = p2 \tag{15.9}$$

$$X^g_{opi} = \sum_{p'=p4,p5} \sum_{i'} X^{gi}_{op'i'pi} \qquad \forall o, i \in \mathcal{I}(p), p = p2 \tag{15.10}$$

$$X^g_{opi} = \sum_{i'} X^{gi}_{opip'i'} \qquad\qquad \forall o, i \in \mathcal{I}(p), p = p4, p' = p2 \tag{15.11}$$

$$\sum_o wm_{to} X^g_{opi} = \sum_{i'} X^{mi}_{tp'i'pi} \qquad \forall t, i \in \mathcal{I}(p), p = p4, p' = p2 \tag{15.12}$$

$$\sum_o wc_{ao} X^g_{opi} = \sum_{i'} X^{ci}_{ap'i'pi} \qquad \forall a, i \in \mathcal{I}(p), p = p4, p' = p2 \tag{15.13}$$

$$X^m_{tpi} = \sum_{i'} X^{mi}_{tpip'i'} \qquad\qquad \forall t, i \in \mathcal{I}(p), p = p2, p' = p4 \tag{15.14}$$

$$X^c_{api} = \sum_{p'=p3,p4,p8} \sum_{i'} X^{ci}_{apip'i'} \quad \forall a, i \in \mathcal{I}(p), p = p2 \tag{15.15}$$

$$X^m_{tpi} = \sum_{p'=p3,p7,p8} \sum_{i'} X^{mi}_{tp'i'pi} \quad \forall t, i \in \mathcal{I}(p), p = p2 \tag{15.16}$$

$$X^m_{tpi} = \sum_{i'} X^{mi}_{tpip'i'} \quad \forall t, i \in \mathcal{I}(p), p = p3, p' = p2 \tag{15.17}$$

$$\sum_t wa_{at} X^m_{tpi} = \sum_{i'} X^{ci}_{ap'i'pi} \quad \forall a, i \in \mathcal{I}(p), p = p3, p' = p2 \tag{15.18}$$

$$X^c_{api} = \sum_{p'=p1,p7} \sum_{i'} X^{ci}_{ap'i'pi} \quad \forall a, i \in \mathcal{I}(p), p = p2 \tag{15.19}$$

$$X^c_{api} = \sum_{i'} X^{ci}_{apip'i'} \quad \forall a, i \in \mathcal{I}(p), p = p1, p' = p2 \tag{15.20}$$

$$\sum_s X^{re}_{mspi} + \sum_{i'} X^{ri}_{mp'i'pi}$$
$$= \sum_a wi_{ma} X^c_{api} \quad \forall m, i \in \mathcal{I}(p), p = p1, p' = p9 \tag{15.21}$$

Balance equations reverse

$$\sum_{i'} uk_{kp} X^{gc}_{op'i'k} \geq \sum_i X^{gr}_{okpi} \quad \forall o, c, p = p5, p6, p' = p2 \tag{15.22}$$

$$\sum_{i'} \sigma X^{gc}_{op'i'k} = \sum_{p5,p6} \sum_i X^{gr}_{okpi} \quad \forall o, c, p' = p2 \tag{15.23}$$

$$X^g_{opi} = \sum_k X^{gr}_{okpi} \quad \forall o, i \in \mathcal{I}(p), p = p5, p6 \tag{15.24}$$

$$X^g_{opi} = \sum_{i'} X^{gi}_{opip'i'} \quad \forall o, i \in \mathcal{I}(p), p = p5, p' = p2 \tag{15.25}$$

$$\sum_o ro_{ot} X^g_{opi} = \sum_{i'} X^{mi}_{tpip'i'} \quad \forall t, i \in \mathcal{I}(p), p = p6, p' = p7 \tag{15.26}$$

$$X^m_{tpi} = \sum_{i'} X^{mi}_{tp'i'pi} \quad \forall t, i \in \mathcal{I}(p), p = p7, p' = p6 \tag{15.27}$$

$$um_{tp} X^m_{tpi} \geq \sum_{i'} X^{mi}_{tpip'i'} \quad \forall t, i \in \mathcal{I}(p), p = p7, p' = p2, p6, p8 \tag{15.28}$$

$$X^m_{tpi} = \sum_{p'=p2,p6,p8} \sum_{i'} X^{mi}_{tpip'i'} \quad \forall t, i \in \mathcal{I}(p), p = p7 \tag{15.29}$$

$$X^m_{tpi} = \sum_{i'} X^{mi}_{tp'i'pi} \quad \forall t, i \in \mathcal{I}(p), p = p8, p' = p7 \tag{15.30}$$

$$\sum_t wr_{at} X^m_{tpi} = \sum_{i'} X^{ci}_{ap'i'pi} \quad \forall a, i \in \mathcal{I}(p), p = p8, p' = p2 \tag{15.31}$$

$$X^m_{tpi} = \sum_{i'} X^{mi}_{tpip'i'} \quad \forall t, i \in \mathcal{I}(p), p = p8, p' = p2 \tag{15.32}$$

$$X^m_{tpi} = \sum_{i'} X^{mi}_{tp'i'pi} \quad \forall t, i \in \mathcal{I}(p), p = p6, p' = p7 \tag{15.33}$$

$$\sum_o rp_{oa} X^g_{opi} + \sum_t rm_{ta} X^m_{tpi}$$
$$= \sum_{i'} X^{ci}_{apip'i'} \quad \forall a, i \in \mathcal{I}(p), p = p6, p' = p7 \tag{15.34}$$

$$X^c_{api} = \sum_{i'} X^{ci}_{ap'i'pi} \quad \forall a, i \in \mathcal{I}(p), p = p7, p' = p6 \tag{15.35}$$

$$up_{ap'} X^c_{api} \geq \sum_{i'} X^{ci}_{apip'i'} \quad \forall a, i \in \mathcal{I}(p), p = p7, p' = p2, p9 \tag{15.36}$$

$$ud_{ad} X^c_{api} \geq X^{cd}_{apid} \quad \forall a, i \in \mathcal{I}(p), \forall d, p = p7 \tag{15.37}$$

$$X^c_{api} = \sum_{p'=p2,p9} \sum_{i'} X^{ci}_{apip'i'}$$
$$+ \sum_d X^{cd}_{apid} \quad \forall a, i \in \mathcal{I}(p), p = p7 \tag{15.38}$$

$$\sum_a \sum_i X^{cd}_{apid} \leq vd_d \quad \forall d, p = p7 \tag{15.39}$$

$$X^c_{api} = \sum_i X^{ci}_{ap'i'pi} \quad \forall a, i \in \mathcal{I}(p), p = p9, p' = p7 \tag{15.40}$$

$$\sum_a rc_{am} X^c_{api} = \sum_i X^{ri}_{mpip'i'} \quad \forall m, i \in \mathcal{I}(p), p = p9, p' = p1 \tag{15.41}$$

Explanation: Equation (15.1) states the objective function. Equations (15.2)–(15.4) define the deviation variables optimized in (15.1). Equations (15.5)–(15.6) specify requirements for the Boolean variables being 1 if a facility is open and 0 if not. A facility is open if the goods flows totalized over a, t, and o are positive. Equation (15.7) specifies non-negativity requirements for all goods flows. Equations (15.8)–(15.21) make sure that for each facility in the forward chain input and output are balanced. Note that in (15.8) customer demand triggers warehousing and pulls all other processes via (15.9)–(15.21) up to material supply. Equations (15.22)–(15.41) balance input and output for all reverse facilities. The return flow is triggered by (15.22), (15.23), and (15.24), from where all reverse processes are 'pushed' via (15.24)–(15.41) until the point of reuse. In time, the balance equations are 'activated' sequentially; first (15.8) and (15.9) (sales), then (15.22)–(15.41) (returns), and finally (15.8)–(15.21) (manufacturing). Of course, this is not relevant for the algorithm, but it is relevant in view of the assumptions made in the problem definition.

Now given the model formulation, we have several ways not only to find the 'optimal' solution, but also to manipulate and analyze system behavior. First, the objective function is formulated as a form of parametric analysis. The weights assigned can be varied one by one or in couples, which enables us to test solution sensitivity to the importance of one or more optimization criterion in the MILP version of the model. Second, when fixing the Booleans and weights, we obtain an LP version of the model. The righthand side coefficients representing the business targets set might give some interesting insights. The shadow prices (optimal value of the dual variables) indicate the decrease of the total objective value when loosening a business target. So if we relax, for example, the cost target, the cumulative reduction of energy and waste equals the sum of cost increase and total objective value decrease (both positive by definition). Of course this only goes within the range that the current basis remains optimal. One can use the 100% rule to calculate whether this is the case. Third, reduced cost of the non-basic variables indicate the extent to which the coefficient can or needs to be reduced before it enters the basis. If, for example, one wishes to stimulate product reuse and its corresponding variable is non-basic, its reduced cost indicates the unit reduction of cost (or environmental impact) necessary to make this option interesting. However, this needs to be offset against necessary investments. Of course, with coefficients, parametric analysis is also possible.

15.4.2 Results and Lessons Learned

Results are extensively reported by Krikke et. al (2002). In this study, runs were split into two major parts: optimization and sensitivity analysis. Below we discuss the major findings of the study.

In the first part, solutions are optimized, and optimal values for product design and location- allocation are determined. These scenarios are optimized with the full MILP model. The relative importance of product design and network design was analyzed for different weights assigned to the optimization criteria. The supply chain network structure has the most impact on costs, whereas the product design has the most impact on energy and waste. A centralized supply chain network by far outperforms a decentralized supply chain network in terms of costs. Product design, using modularity as the main instrument, can reduce opposing behavior of costs, energy use, and waste functions. Overall, it appears that design PMPP is dominant over the other designs, regardless of location-allocation choices. Product design PMPP enables solutions with the most synergy between supply chain costs, energy use, and waste, whereas the other two have more conflicting effects on the three optimization criteria. Reuse at a component/module level is the most beneficial recovery option. When recovery feasibility goes down, material recycling and thermal disposal are the second-best choices. Product repair does not come into the picture in this case. Energy and waste are the most conflicting objectives over a wide range, due to the fact that thermal disposal increases waste, but reduces energy use. It also appears that low-cost solutions have nearly optimal environmental impact, but small improvements in energy and waste are often costly.

In the second part, we do a sensitivity analysis. Based on results of the first part, two solutions are selected in which the zero-one variables in the MILP model are fixed to obtain an LP model. The robustness of these solutions is tested to parameter settings reflecting recovery targets, rate of return, and recovery feasibility. Recovery feasibility and rate of return appear to be far more important than both product design and location-allocation. Optimal recovery feasibility and maximal rate of return lead to objective function values one can only dream of, while in the case of, for example, 10% rate of return and 10% return feasibility (hence 1% reusability of original sales) results are pretty bad. Under realistic assumptions, the robust solution with nine facilities outperforms a so-called efficient solution (with seven facilities) due to its flexibility in applying various recovery options. A suggested management strategy would be a centralized, robust supply chain network in Dresden combined with the PMPP product design, with maximal incentives for consumers to return 100% of the products in optimal condition. Recovery targets based on proposed EU legislation have an ambivalent impact, as they reduce waste but increase energy use and costs. EU legislation and management solutions should enhance a high rate of return and high recovery feasibility in order to maximize component and module reuse next to setting recovery targets. Krikke et. al (2002) conclude that generalization of results from the case is limited. Results will depend strongly on the product and model/data at hand. Dynamic product designs in turbulent markets obviously make reuse problematic. More case evidence from consumer electronics and from other sectors is necessary. However, some of the conclusions do have fairly robust general value, in particular the case suggests that a life-cycle-based approach yields better results than merely optimizing an end-of-pipe solution only. Also, the concurrent eco-eco modelling of supply chain and product designs is applicable to other cases.

15.5 Case: Pulp and Paper

The paper and pulp sector is a very large and capital-intensive part of industry. Apart from oil and gas, it is the biggest industry measured in terms of world trade. For some countries, pulp and paper trade is essential for the economy. For example, the export of Swedish pulp and paper covers 20% of total export value of that country.

The principal environmental impacts of the European pulp and paper sector are associated with (1) its consumption of raw materials, including virgin resources, (2) its emissions to air and water during the production of paper and pulp (especially pulp bleaching, a very polluting process), and (3) the existence of waste paper.

European policy makers seem persuaded of the benefits from recycling, but many analysts consider these benefits ambiguous, because recycling may reduce some environmental impacts while increasing others. The emphasis on recycling also raises suspicion that environmental policy may be manipulated to serve (national) industry and trade policy objectives or to shift environmental problems from one place or type to another. Since different activities and places involved in the life cycle of pulp and paper products are linked, actions to reduce environmental impacts in one place may have serious implications for activities that occur elsewhere. A very high imposed percentage of recycling, which helps reduce Western Europe's solid waste problem, would adversely affect the Nordic industry. Scandinavia, Europe's main producer of virgin pulp, would be unable to generate sufficient recycled paper by recovery from its relatively low population. The only solution would be to import waste paper, which would generate an environmental impact of its own, particularly from the energy used for transportation. France, Spain, the UK, and other countries have had to face considerable flows of untreated paper at very low prices from Germany, where high recycling standards were decreed without the necessary treatment capacity. Such mishaps could lead EU Member States to adopt protectionist measures.

Policy choices should be rational and consistent. A scientific basis for preferring one technology over another can be found in an explicit evaluation system. Such a system is also essential for understanding the rationale for policy choices and verifying the consistency of environmental policies. Apart from evaluation, environmental optimization is necessary to search for efficient and robust policies. The possibility of different environmental optima based on some set of assumptions implies that policy makers should search for robust development pathways for the sector through time. The evaluation approach we suggest is described in more detail in Bloemhof et al. (1996).

The overall model combines a linear programming network flow model of the European pulp and paper sector with life-cycle analysis. LCA provides a tool for tracking flows of materials through the life cycle from one place or process to another. It combines objective information about the environmentally relevant characteristics of processes with a subjective evaluation of the relative importance of different types of environmental damage. The environmental index that results from the LCA is used as a coefficient to build the objective function for the linear programming model. The output of this model is a life-cycle configuration with its corresponding environmental impacts. The configuration entails a pattern of production, consumption, waste management, and transportation. The flows imply the geography and the technologies of the sector, including levels of raw materials demand, production, and international trade.

Economic criteria are not explicitly included in the model so, formally, this case is an example of eco(logical) optimization. However, most ecological models apply only an LCA. Here, the model is extended by a network flow model, very suitable for economic optimization. Implicitly, economic drivers play a role in the model as follows. The network flow model minimizes, among others, (environmental) transportation costs, and these costs are calculated as a positive constant times the number of kilometers. Economic transportation costs are calculated in the same way, so minimizing environmental costs will point in the same direction as minimizing economic transportation costs. Moreover, the model can easily be extended to a multiple objective one, as presented in section 15.4, optimizing both economical and environmental costs.

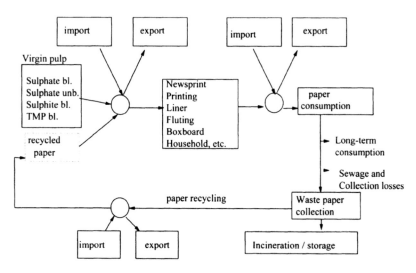

Fig. 15.5. Network flow model for Scandinavia

Figure 15.5 gives a graphical representation of the life-cycle optimization for Scandinavia. Virgin pulp can be produced or bought to be used for consumption or export. Each paper type needs a specific furnish of pulp types. Recycled pulp can only be used in the paper-making process if it is available in the region itself. Waste paper will be disposed of in the region itself, or used for the making of recycled pulp within Europe. Each region will decide on a level of virgin pulp production, paper production, and paper recycling that is consistent with minimizing environmental impact of the life cycle as a whole. The optimization can be represented by a network flow model.

15.5.1 The Model

This section contains the mathematical formulation of the network flow model. First we introduce the notation, next the linear programming model.

Indices:

$i, j, k \in I$ = indices for the six regions
$v \in V$ = index for the four virgin pulp types
$p \in P$ = index for the seven paper types

The set $I^+ := I \cup \mathrm{NA}$ relates to the OECD-Europe regions plus North America.

Decision variables:

VP_{iv} = virgin pulp production of type v in region i
VT_{ijv} = virgin pulp transportation of type v, from region i to region j
VD_{iv} = virgin pulp demand in region i of type v
RP_i = recycled pulp production in region i
RD_i = recycled pulp demand in region i
PT_{ijp} = paper transportation of paper type p, from region i to region j
WT_{ij} = waste paper transportation, from region i to region j
WI_i = waste paper incinerated in region i
WP_i = waste paper 'production' for recycling in region i
WD_i = waste paper demand for recycling in region i
Λ_{ip} = share of recycled pulp in furnish of paper type p in region i

Exogenous variables:

vs_i = virgin pulp wood supply in region i
pp_{ip} = paper production of paper type p in region i
pd_{ip} = paper demand for paper type p in region i
ws_i = waste paper supply for recycling or incineration in region i
pe_{ip} = paper export from region i to non-EU countries
pi_{ip} = paper import from non-EU countries to region i

Other parameters:

ev_{iv} = environmental impact of virgin pulp production of type v in region i
er_i = environmental impact of recycled pulp production in region i
ei_i = environmental impact of incineration of waste paper in region i
et_{ij} = environmental impact of transportation from region i to region j
ρ_p = total share of pulp in the inputs of paper type p (yield of paper from pulp)
μ_{vp} = furnish rate of pulp type v with regard to total virgin pulp share in paper product p
λ_p^{max} = maximum share of recycled pulp in the furnish of paper type p
δ_p = long-term consumption rate for paper type p
σ_p = sewage rate of waste paper originating from paper type p
γ_{ip} = collection rate of waste paper originating from paper type p in region i
ω = estimated fiber yield of waste paper

Waste paper supply for either recycling or incineration in region i is defined as follows:

$$ws_i := \sum_{p \in P} (1 - \delta_p)(1 - \sigma_p) \times \gamma_{ip} \times pd_{ip}$$

The paper recycling problem (PR) can be formulated as a linear program:

$$z_{PR} = \min \sum_{i \in I^+} \sum_{v \in V} ev_{iv} VP_{iv} + \sum_{i \in I} er_i RP_i + \sum_{i \in I} ei_i WI_i +$$

$$\sum_{i \in I^+} \sum_{j \in I} et_{ij} \sum_{v \in V} VT_{ijv} + \sum_{i \in I} \sum_{j \in I} et_{ij} \left(\sum_{p \in P} PT_{ijp} + WT_{ij} \right) \quad (15.42)$$

subject to

$$\sum_{v \in V} VP_{iv} \leq vs_i \qquad\qquad i \in I^+ \qquad (15.43)$$

$$\sum_{i \neq j \in I^+} VT_{ijv} + VP_{jv} = VD_{jv} + \sum_{k \neq j \in I} VT_{jkv} \qquad j \in I, v \in V \,(15.44)$$

$$RP_j = RD_j \qquad\qquad j \in I \qquad (15.45)$$

$$VD_{iv} = \sum_{p \in P} \mu_{vp}(1 - \Lambda_{ip}) \times \rho_p \times pp_{ip} \qquad i \in I, v \in V \,(15.46)$$

$$RD_i = \sum_{p \in P} \Lambda_{ip} \times \rho_p \times pp_{ip} \qquad\qquad i \in I \qquad (15.47)$$

$$\Lambda_{ip} \leq \lambda_p^{max} \qquad\qquad i \in I, p \in P \,(15.48)$$

$$pi_{jp} + \sum_{i \in I, i \neq j} PT_{ijp} + pp_{jp}$$

$$= pd_{jp} + \sum_{k \in I, k \neq j} PT_{jkp} + pe_{jp} \qquad j \in I, p \in P \,(15.49)$$

$$ws_i = WI_i + WP_i \qquad\qquad\qquad\qquad i \in I \qquad (15.50)$$

$$\sum_{i \in I, i \neq j} WT_{ij} + WP_j = WD_j + \sum_{k \in I, k \neq j} WT_{jk} \qquad j \in I \qquad (15.51)$$

$$RP_i = \omega \times WD_i \qquad\qquad\qquad\qquad i \in I \qquad (15.52)$$

The objective (15.42) is to minimize environmental impacts of virgin pulp production (both in Europe and North America), recycled pulp production, waste paper incineration, and transportation. Constraints (15.43) are the capacity constraints for wood pulp. Constraints (15.44) and (15.45) define flow conditions for virgin pulp and recycled pulp. Constraints (15.46) define the allowable share of virgin pulp in order to satisfy quality conditions for paper types. Constraints (15.47) define the allowable share of recycled pulp in order to satisfy quality conditions for paper types. Constraints (15.48) represent the natural bound on the share of recycled pulp in the overall furnish of paper products. Constraints (15.49) define flow conditions for paper. Constraints (15.50) define the destination of collected waste paper to be either incineration or recycling. Constraints (15.51) define flow conditions for waste paper. Constraints (15.52) consider the yield from waste paper for recycled pulp.

15.5.2 Results and Lessons Learned

The solutions suggest a small contribution of virgin pulp (15%) and a large contribution of recycled pulp (85%). These percentages are quite different from current contributions (on average 60% virgin pulp and 40% recycled pulp). The large amount of recycled pulp requires considerable transportation of waste paper. Comparing current technologies on the basis of our evaluation methodology, the average environmental index of recycled pulp is indeed much lower than the one for virgin pulp. This result justifies the current policy emphasis on recycling. The potential environmental gain from energy recovery in waste incineration is insufficient to offset the difference between the environmental impacts of virgin and recycled pulp production. A shift to this optimal solution would have major implications for trade and industry. This is inevitable because the optimal solution involves maximum recycling while the current recycled pulp share in aggregate paper production is only 40%.

Interestingly, were one to shift to the optimal life-cycle configuration (thus shifting the mix of virgin and recycled pulp use in the sector), Scandinavia would supply all of Europe's residual need for virgin pulp. Scandinavia would produce and use virgin pulp to export paper with high virgin pulp furnishes (graphic products). The major consuming countries would, in turn, focus on recovering these pulps from post-consumption wastes, making paper and board grades that would allow the complete substitutability of virgin pulp by recycled pulp (paper and board products). Essentially, the optimal solution entails specialization of production and of products because it minimizes the environmental impacts of transportation.

A policy to increase progressively the use of recycled pulp may be simplistic. Given the relative environmental indices of the existing process technologies, maximal recycling is profitable. But, would this profit still hold if the environmental indices for the unit processes were different from those that now apply? How much would they have to differ for the optimum life-cycle configuration to be inconsistent with maximum recycling?

Results show that there is a limit to the environmental improvement that can be achieved through recycling, and that more improvement can be made via cleaner virgin pulping technologies. A comparable analysis is done for incineration. These results indicate the degree to which environmental optimality depends on the environmental performance of the different technologies and thus on prospective technological progress. The relative balance between the environmental impacts of recycling, incineration, and virgin pulp production is critical to the determination of the optimum recycling level and, by implication, to whether recycling is the best route to improved environmental performance. The results also point to the technological improvement needed if clean pulp production plus incineration were to prove a preferred route compared to higher recycling percentages.

The model can easily be used to do scenario analysis based on European policy making. Interesting conclusions based on the results from the scenario analyses follow.

- *Single-minded focus on recycling is myopic.*
  A high level of recycling will deliver short-term environmental impact reductions as long as full energy recovery is not available. Investment in technology and capacity for energy recovery has a large potential for environmental impact reductions, without disturbing international trade relations. Apart from energy recovery, a higher level of recycling in those regions with a large supply of waste paper gives a further improvement of the environmental impact of the sector.
- *Command and control instruments limit environmental impact reductions.*
  We demonstrate that policies such as regulated levels of recycled pulp are less effective than flexible policies. The potential for environmental impact reduction is reduced by regulation. Regions with a large production of high-quality graphic papers should be able to focus on cleaner virgin pulp production.
- *Policies on paper recycling have major international trade implications.*
  A high recycling level can easily lead to the transportation of waste paper from regions with a high population to Scandinavia. The dominant place of Scandinavian virgin pulp will be replaced by recycled pulp produced in Germany or France. This will have high impacts on forestry (renewable resource).

- *National legislation can easily result in protectionist actions.*
  National mandated levels of recycling have a large impact on market shares of other regions. National legislations can also lead to an increase in total transportation flows.
- *Relocation of the paper industry has major economic implications.*
  A reduction of environmental impact of 4–8% is possible with relocation. However, the paper production configuration changes considerably and converges to the configuration of paper consumption as the recycling level increases.

15.6 Discussion and Conclusions

This chapter focuses on an eco-eco modelling approach for the evaluation of EU policy. The importance of modeling both economic and ecological effects is demonstrated in a literature study. Operations Research models have been developed for both an individual system of Waste of Electrical and Electronics Equipment (WEEE) and a sector-wide system for paper recycling.

The WEEE is a very heterogeneous sector and although some general lessons apply to all products, the practical implications differ strongly. Therefore it is very difficult to formulate general rules of thumb for the optimal setup of EPR systems. However, quantitative studies can evaluate the impact of different strategies for particular waste streams on a case-by-case basis. From this starting point, one can formulate targeted public policies, focusing on success factors for that particular waste stream (e.g. effective collection). At the same time, we note that the problems at hand are multidisciplinary. For example, observing the importance of the rate of return parameter in the model is worthwhile, and hence collection should be the prime focus. Support from other disciplines, e.g. behaviorial sciences or communications experts, is needed to give any sign of how to actually stimulate consumers to return. Nevertheless, advanced planning based on eco-eco OR models can contribute significantly to the implementation of life-cycle-based producer responsibility in consumer electronics and other sectors.

The paper and pulp results (originating from 1996) were not really heard at the time by EU people. However, five years later in the *Financial Times*, an article looking at current trends in the paper industry in Europe confirmed exactly what has been predicted in the study. This clearly shows the usefulness of these eco-eco OR models, especially the sensitivity analysis and scenario analysis.

Both cases show that it remains difficult to obtain a true integration between ecological and economical concerns. The refrigerator example optimizes one objective function subject to target constraints on the other objectives. The paper and pulp example does not have explicit cost parameters in the objective function. Therefore, we can conclude that combining ecological and economical concerns in MILP models is very well possible, but a true integration of these concerns is still elusive in practice.

16

Information and Communication Technology Enabling Reverse Logistics

Angelika Kokkinaki[1], Rob Zuidwijk[2], Jo van Nunen[2], and Rommert Dekker[1]

[1] Rotterdam School of Economics, Erasmus University Rotterdam, P.O. Box 1738, 3000 DR Rotterdam, The Netherlands,
kokkinaki@few.eur.nl, rdekker@few.eur.nl

[2] Rotterdam School of Management / Faculteit Bedrijfskunde,
Erasmus University Rotterdam, P.O. Box 1738, 3000 DR Rotterdam,
The Netherlands, rzuidwijk@fbk.eur.nl, elaan@fbk.eur.nl

16.1 Introduction

In this chapter, we examine how Information and Communication Technologies (ICT) are being used to support reverse logistics; ICT is the term widely used in a European context instead of IT, which is commonly used in American publications. Focusing on the ICT infrastructure, this chapter does not follow a quantitative approach as does the rest of the book. Nonetheless, the topics covered in this section outline how ICT systems enable and support the quantitative approaches presented in other chapters of this book. Furthermore, this chapter provides a roadmap to the reader about what aspects of reverse logistics are implemented and what remains to be addressed in the future.

Most ICT systems for reverse logistics have been developed to address needs in a specific sector (i.e. decision-making on different recovery options of returns, designing a product for optimal end of use recovery, etc.) or to cover the reverse logistics requirements of a particular company. Thus, in our attempt to present this area systematically we need to develop a framework of reference first.

For that reason, we go back to the essentials of reverse logistics where the recurring theme regarding reverse logistics is that they include *processes* related to the recovery of *products* with the objective to facilitate reintroduction of returns into a *market*.

Based on these three keywords, we have identified that ICT systems for reverse logistics have indeed attempted to address one or more of the issues related to

1) product data, that is, data regarding the condition and configuration of the returns;

2) process facilitation, and more specifically supporting operations of reverse logistics; and

3) redistribution to the market, in particular attempts to consolidate the fragmented marketplaces.

Product data are essential for efficient handling of returns. However, returns are plagued by a high degree of uncertainty regarding some of their important attributes (i.e. place of origin, timing and quality standards). Since product data of returns are rarely available, ICT systems have been developed to trace the required information through the systems that were used in the original production phase of each return or to retrieve these critical data through monitoring and, in some cases, reverse engineering methods.

ICT systems developed for the control and coordination of reverse logistics processes assist in the decision-making for the recovery options of returns (reuse, remanufacturing, recycling) and support administrative tasks related to returns handling that contribute to more efficient returns management.

Upon completion of the required recovery operations, used parts and products need to be forwarded to the market. Interestingly, markets for such parts or products are highly fragmented. In recent years, with the expansion of e-commerce applications, several attempts have been made to consolidate markets for returns through the creation of specialized e-marketplaces.

In view of these three main underlying themes, we develop a three-dimensional space (Figure 16.1) with axes of reference *Products, Processes,* and *Marketplaces,* with the aim to systematically analyze how ICT systems process information flows related to these aspects. First, we identify the constraints imposed upon these information flows. Namely, we discuss the impact of uncertainty about products, processes, and marketplaces and demonstrate with a concrete example the value of information with respect to control and planning of reverse logistics. Finally, using this three-dimensional space as a roadmap, we examine each ICT system shown in Figure 16.1 in detail.

This chapter is structured as follows. Section 16.2 examines uncertainty issues and Section 16.3 outlines, through an example, the value of information in reverse logistics. Section 16.4 presents the ICT systems that mainly address collection and update product data throughout the product's life cycle. Section 16.5 presents the ICT systems that support processes in reverse logistics and Section 16.6 presents the ICT systems that enable the formation of marketplaces in reverse logistics. In the following three sections, we examine systems that address a combination of two out of the three prime issues. In Section 16.7, we present ICT systems that support product data and reverse logistics processes, in Section 16.8, we present those that support reverse logistics processes and marketplaces, and in Section 16.9 we describe systems that enable both product data collection and marketplaces for reverse logistics. We conclude this chapter by identifying some issues for possible further development in Section 16.10.

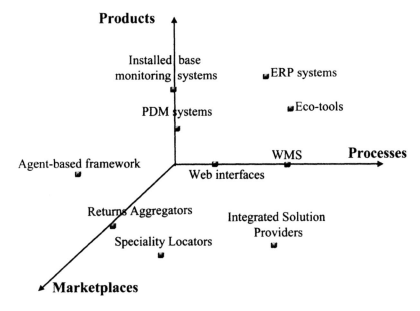

Fig. 16.1. A classification for ICT systems in reverse logistics

16.2 Uncertainty Issues

In this section, we discuss the notion of uncertainty and show how it arises in reverse logistics networks using the proposed framework in Section 16.1.

Planning of reverse logistics operations requires information on the (future) values of quantities such as processing times, arrival times, end-of-life product returns, customer demand, product quality, etc. In the scientific literature, the problem of 'uncertainty' attached to such quantities in reverse logistics processes has been addressed extensively (see, for example, Thierry et al. (1995); Fleischmann et al. (1997)). It is the aim of this section to clarify sources of uncertainty in reverse logistics networks.

Reverse logistics networks are merely part of demand and supply networks and are generally put in contrast to the classical forward supply chain (for an extensive discussion, see Chapter 2 on the Framework for reverse logistics). In the discussion below, we focus on reverse logistics flows, although it should be clear that not all sources of uncertainty are unique to such flows.

In many contributions to Operations Research, uncertainty has been described using concepts from probability theory. We shall adopt a description of uncertainty by Zimmerman (2000) as a situational property. A decision maker uses a model to perceive the information he actually receives from the object system. The following working definition of uncertainty is proposed.

Uncertainty arises from the fact that a decision-maker does not have information at his disposal which quantitatively or qualitatively is appropriate to describe, prescribe, or predict deterministically and numerically the object system, its behavior, or other characteristics.

Zimmermann elaborates on the fact that causes of uncertainty, types of uncertainty, and uncertainty models need be distinguished. Without going into too much detail, we mention the following causes of uncertainty from his paper. 1) lack of information, 2) abundance of information, 3) conflicting evidence, and 4) ambiguity.

Examples of each of these causes of uncertainty appear in reverse logistics networks. 1) A planner of a remanufacturing line may be confronted with lack of information on the quality status of an arriving batch of copiers. 2) A technician may not have the time to interpret a complete repair history of a computer server in order to decide not to dispose of a certain part. 3) An extensively used engine may show excellent test results. 4) The qualification 're-usable' for a computer screen may have several meanings referring to remanufacturing, cannibalization, or material recycling.

We will make the distinction between the case when required information is not present at all and the case when it is simply not available to the decision-maker. For example, the fraction of hazardous material that can be retrieved from a batch of refrigerators by means of a particular separation process may not be known in advance, since the process of extracting the material need not be deterministic. This is bound to be the case when the extraction process is destructive. On the other hand, the quality status of a part within a product may be fixed but only accessible after disassembly.

It is conceivable that uncertainty arises from a specific lack of information, namely a proper understanding of the mechanisms that determine certain outcomes. Forecasting methods that employ probabilistic methods can be used to relate product demand and return patterns; see Toktay (2001) or Chapter 3 in this book. An issue there is whether the forecasting models provide, on an aggregated level, a satisfying description of the outcome of immensely complicated processes or not.

Uncertainty, as described in this chapter, confronts the decision-maker with various ways of dealing with it. The decision-maker may invest in information systems, say, to get more or better information, he may adjust his model (e.g. by incorporating uncertainty in probabilistic terms) to achieve a better understanding of the information he receives from the object system, or he may try to manipulate the object system.

We give a concise overview of sources of uncertainty in reverse networks, discuss their impact qualitatively, and mention some possible measures a decision-maker may take. We use the three dimensions discussed in the introduction.

Product Data. Product design influences uncertainty in product quality. Here we define product quality from an instrumental point of view. The quality of a product, part, or material is high whenever it can be allocated to a reuse option with high added value. For example, a product that can be re-manufactured is assumed to be of better quality than a product that can only be used for material recycling. Canonical ways of determining product quality (such as technical functioning of the product) usually are consistent with this approach. The development of a multitude of product specifications has a negative impact on product quality. For example, certain electronic equipment may contain a large diversity of part types, so that allocation of required parts from the disassembly process becomes almost impossible. Modular design of products, where functional parts can be reused in several generations of the product if updates are not truly required, has a positive impact on the reusability of parts.

Processes facilitation. Logistics processes are influenced by uncertainty. We focus on the uncertainty aspects that are typical of reverse logistics. In the collection phase, the quantity and timing of returns through specific collection channels (retail shops, municipal waste companies, repair, etc.) may depend on non-deterministic factors such as customer behavior and product failure. Since the quality of collected products may determine the allocation to specific reuse options, it needs to be specified. This may involve assessment of product data (usage or repair history, product type specifications) or physical tests. Recovery processes such as disassembly, separation, extraction, and repair involving materials, parts, and products may be non-deterministic in released quantities and quality due to the nature of the physical processes. In most of these cases, product design and systematic collection of product data seem appropriate measures to deal with these uncertainty issues.

Marketplace consolidation. From a marketing perspective, several uncertainty issues come up. The release of end-of-life products does not only depend on the technical life cycle of the product. In fact, the economic life cycle depends on many factors such as termination dates of lease contracts and underlying sales strategies related to the release of new products. Furthermore, prices for reusable products, parts, and materials are notorious for their volatility (see, for example, http://euro.recycle.net). Customer behavior provides several instances of uncertainty in logistics systems. The description of uncertainty in timing, place, and volume of customer demand is well documented in the logistics literature, and can be found in textbooks such as Silver et al. (1998). In reverse logistics, the customer has become an integral part of the logistics network, resulting in additional instances of uncertainty, such as timing, place, channel, and quality of returns. The first point drawn from this is that customer behavior is never fully autonomous. Indeed, customer demand is influenced by marketing efforts and availability of products. Furthermore, the setup of the collection network has an impact on product and packaging returns. For example, old-for-new discounts at retailer shops make this return channel favorable to alternative channels. Besides manipulation of customer behavior, the aforementioned forecasting methods are also of use

Table 16.1. Average masses of components in computer monitors

components	m_1	m_2	m_4	m_5	m_6	m_7	m_8	m_9	m_{10}	m_{11}
masses (kg)	0.76	1.90	0.07	0.29	4.45	0.19	1.03	0.94	1.01	0.06

here. Information systems can play a role in monitoring planned or actual returns at an early stage in the collection process.

Although it is obvious that uncertainty is a complicating factor in the management of reverse logistics processes, it is not obvious how to assess the value of reducing it. This value needs to be assessed to support investment decisions in ICT. An example of how this can be done for a typical reverse logistics process, namely disassembly, is shown in the next section.

16.3 Value of Information

In this section, an example is given of a quantitative assessment of the value of information in disassembly. A planner of a disassembly line is faced with a heterogeneous supply of brown goods, computer monitors in this case. The objective of the disassembly activities is to enhance the recovered value from these products by offering components to specific recovery processes or markets. The disassembly process is part of a variety of recovery strategies, including cannibalization, remanufacturing, and material recycling. The disassembly process is steered by two types of decision questions. 1) Should a particular product or product component be (further) disassembled, and 2) to which recovery process should the product or product component be forwarded? The decisions are made taking into account

1) the objective of revenue maximization,
2) capacity of recovery installations, available resources, and markets, and
3) recovery targets set by corporate policy or governmental regulations.

Observe that, for example, separation of specific materials from the product prior to shredding may enhance the purity of the released fractions and henceforth the market value of these outbound flows. Of course, disassembly activities induce costs (manual or robotized labor), so an optimal degree of disassembly needs to be found.

The heterogeneity of the product flow comes from diversity in brands and product types. The characteristics of the products are modelled by means of a disassembly tree (see Figure 16.2). Each component of the product has a certain (average) mass measured in kilograms (see Table 16.1). Here the total average mass of the whole product $m_0 = 10.71$ kg and the total average mass of the chassis unit $m_3 = 3.05$ kg. The batch of products considered here consists of 48 computer screens. This data set originates from a pilot experiment and has been used in a previous study in Krikke (1998). In order to

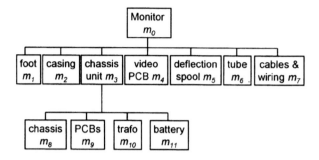

Fig. 16.2. Simplified disassembly tree for computer monitor

make the calculations below, these computer screens have been disassembled completely. Information on the mass of all components is available.

The disassembly decision question in this example comes down to 1) no disassembly, 2) disassembly of the first layer of the product (see Figure 16.2), or 3) complete disassembly. The estimated costs of disassembly are based on labor costs per minute and the average disassembly times for the whole product and the chassis unit (see Table 16.2).

The second decision question concerns the forwarding of product and components to recovery processes. We will assume that all components that are not further disassembled are forwarded to material recovery processes. Net prices are given in Table 16.3. Each component is put in a (possibly very specific) class of recyclable materials to which a price is attached. Observe that some prices are negative and that composite components are considered material mixes and henceforth provide less revenue per kg. (Note that a PCB is a Printed Circuit Board.)

We consider three scenarios that address the issue of the availability of information in the case of the batch of 48 computer monitors. In the first scenario, with full information, each monitor is disassembled in an optimal way, i.e. such that costs are minimized for that monitor. In the second scenario, with information on characteristics, two characteristics are known, namely whether a battery is present in the machine and whether a considerable amount of ferro material is present in the machine. Such information could be retrieved from the machines without prior disassembly, e.g. by X-ray scan or by using in-

Table 16.2. Labor costs and disassembly times

labor costs per minute (€)	0.30
average disassembly time, product (minutes)	5.25
average disassembly time, chassis unit (minutes)	3.50

Table 16.3. Material prices (in €)

components	product	foot	casing	chassis unit	video PCB	deflection spool
material class	mixed	plastics	plastics	mixed	PCBs	deflection spool
prices	-0.34	-0.10	-0.10	-0.34	0.11	0.22

components	tube	cables & wires	chassis	PCBs	trafo	battery
material class	tube	cables & wires	ferro	PCBs	trafo	battery
prices	-0.08	0.25	0.05	0.80	0.23	-0.90

formation available from the product itself (a chip containing an inventory of components inside). These two characteristics result in four classes (battery (y/n) and ferro (y/n)). For each of these four classes, an optimal disassembly strategy is given. All products in a class are treated in the same fashion. Finally, in the third scenario, all products are treated in the same fashion. Table 16.3 provides per scenario the average revenue per products and the average disassembly time per machine.

As mentioned before, establishing optimal disassembly strategies is often done under various constraints addressing resource capacities, regulations, etc. In order to model these constraints, linear programming techniques are used. We will not discuss the volatility of model parameters such as mass fractions and material prices. In Zuidwijk et al. (2001), stochastic programming techniques are discussed that deal with variations in material fractions for a similar example involving washing machines. A linear program that models the disassembly process with an additional labor constraint may also be found in Zuidwijk et al. (2001).

Current research addresses the fact that, from the pilot data, one should be able to come up with product design characteristics that are in line with optimal disassembly strategies for all individual products. The generation of a decision tree by means of data -mining techniques is an approach that is being pursued. Further, stochastic programming methods can also be used in this setting.

In this section, we demonstrate the value of information by studying disassembly costs for three scenarios with progressive amounts of information available. It has been indicated how, in this setting, capacity constraints can be modelled in a linear program. The analysis above indicates that information systems can be of value to the management of recovery processes. The

Table 16.4. Results from three scenarios

	full information	characteristics	no information
average revenue per monitor (€)	-2.10	-2.18	-2.25
average labor time (minutes)	5.72	7.51	8.75

remaining part of this chapter provides a systematic review of information systems that are used to support reverse logistics processes.

16.4 Information Management for Product Data

As shown in the previous section, it is important to have coherent, updated, and consistent information for returns management. The central question, however, is what kind of product-related information OEMs can make public without exposing their products to reverse engineering attempts, and what kind of information can be reconstructed upon the collection of a return. In this section, we examine the ICT approaches that collect and maintain product data.

16.4.1 PDM Systems

Product Data Management (PDM) systems focus on the engineering aspects of product development, dealing primarily with engineering data. The basic functionality of a PDM system, see Hameri (1998) and Peltonen (2000), is summarized as follows.

Secure storage of data objects in a database with controlled access. In many cases, the first motivation for the introduction of a PDM system comes when people cannot find the last updated document they are looking for.

Mechanism for associating objects with attributes. The attributes provide necessary information about an object, or they can be used in a query to locate objects.

Management of the temporal evolution of an object through sequential revisions. In a product development environment, users often modify existing designs as opposed to creating completely new ones. The evolution of drawings and other design objects is usually captured in the form of successive revisions in a PDM system.

Management of alternative variants of an object. Many products and documents have alternative variants. For example, a user's guide for a particular product can be available in different languages.

Management of the inspection and release procedures associated with a data object. Typically, engineering data must be checked and approved through elaborate procedures before they are released.

Management of the recursive division of an object into subassemblies in all possible ways. Almost all products have a hierarchical subassembly structure of components.

Management of changes that affect multiple related objects. One of the basic functions of a PDM system is to support change management, that is, management of the revisions and variants. However, it is also necessary to view a set of related objects as a single unit with respect to change management.

Management of multiple document representation. A PDM system must provide support for storing a data object in different formats. For example, a CAD tool can be available both in the original file format and in a stripped-down version for viewing and printing only.

Viewing tools. PDM systems can provide different views of the same data object to different user groups.

Tool integration. PDM systems provide integration capabilities with other software packages employed by the end-users.

Component and supplier management. The management of standard components provided by external suppliers is a rapidly growing field for PDM systems.

An indicative list of PDM software packages includes Agile PDM, Optegra, MatrixOne, and Metaphase.

PDM systems use a large number of concepts that are more or less related but for which there are no internationally accepted standards: for example, although PDM systems support change management there is no agreement on the meaning of the terms 'version,' 'revision,' 'variant,' etc. One standard-ization effort for the representation of data related to products is ISO 10303, also known as the STEP standard; see Ownen (1997). STEP provides a large representation of geometric data, but it is presently difficult to incorporate generic product structures in the general STEP framework.

Incompatibility of product data structures between actors in the supply chain that use different PDM software packages renders the exchange of such information problematic. PDM software packages support integration to ERP systems through Application Programming Interfaces (APIs). PDM systems are used in many industries, including airplanes (Boeing), vehicles (Ford), electronics and mobile phones (Nokia, Erickson, Siemens), and software de-velopment (Microsoft).

From the reverse logistics perspective, the main contribution of PDM sys-tems is their potential to provide an informational backbone to integrate eco-tools, MRP/ERP systems for returns, and installed base monitoring systems and waste-management systems for returns, and thus to support consistent and updated product data through its entire life cycle.

It is necessary to examine what extensions are required for the information systems that interact with PDM systems to update product data. In Table 16.5, we present the systems that interact with PDM systems and their main functional requirements.

16.4.2 Installed Base Monitoring Systems

Product data collection regarding the product's condition and configuration for the duration of its life cycle is generally referred to as condition monitoring of the installed base, where the installed base is defined as the total number of placed units of a particular product in the entire primary market. New

Table 16.5. Information systems for returns management

Information Systems	Functional requirements
Eco-tools	Product design for optimal end-of-life recovery
Installed Base Monitoring Systems	Support extended value-added services for returns
MRP/ERP Systems	Support different recovery options of returns
Waste-Management Systems	Environmentally friendly disposal

communications protocols can be combined with embedded control systems to monitor product data dynamically and transmit the data logged upon some event. For example, the remote monitoring system of OTIS elevators is based on sensors that constantly monitor the critical parameters of the system, and if a problem is detected a message is sent to the OTISLINE, a 24/7 communication center for further action. It is the same center that receives voice communication from passengers in a stalled elevator.

This comes as the latest development in this direction, where traditional installed base monitoring methods included end-user surveys, competitive interviews, trade associations, and maintenance records. Installed base monitoring is used for product improvements, product upgrades, introducing a peripheral device, and maintenance/repair services.

We use as a case example the Electronic Data Log (EDL) to demonstrate how condition monitoring is used for remanufacturing. The Bosch Group, a worldwide operation with 190 manufacturing locations and 250 subsidiary companies in 48 countries, is known for automotive supplies, power tools, household goods, thermo technology, and packaging technology supplies. In collaboration with Carnegie Mellon University, Bosch was looking for ways to capture data on the use of their range of power tools that would yield better decisions about their recovery options; see Klausner et al. (1998). As a result, the Electronic Data Log (EDL) was developed to measure, compute, and record usage-related parameters of motors in power tools aiming at the end-of-use of a tool to get the values of the recorded parameters for analysis and classification between reusable and non-reusable motors.

The following components are used in this approach.

- A low cost circuit (EDL) that measures, compiles, and stores parameters that indicate the degradation of the motor (including number of starts and stops), accumulated runtime of the motor, sensor information (such as temperature), and power consumption for an accumulated runtime of approximately 2300 hours of operation. Also, EDL computes and stores peak and average values for all recorded parameters.
- An easily accessible interface to retrieve data from the EDL circuits without disassembling the tool.

- An electronic device (reader) connected to the interface to retrieve the data values.
- Software data analysis and classification into two main categories: reusable motors and non-reusable motors.

The main benefits of EDL compared to conventional disassembly and testing can be summarized as follows.

1) Easily accessible interface to EDL, which does not require disassembly of the product, offers labor cost saving for non-reusable products.
2) Easy and speedy retrieval of information stored in EDL.
3) EDL data reconstruct the entire usage history of a product leading to the assessment of reuse for other components besides the motor.
4) EDL data can be used to improve design and support market research and quality management.

The economic efficiency of this method is the subject of Klausner et al. (1998). The trade-off between initially higher manufacturing costs and savings from reuse is analyzed, taking into account return and recovery rates as well as misclassification errors. The analysis shows that the use of EDL can result in large cost savings. Developments such as increased service obligations (product lease, extended warranties, service level agreements), environmental liabilities (producer responsibility/product stewardship, consumer demand for recycling), and product modularization will enable and necessitate the establishment of product life-cycle management and hence of closed-loop supply chains. The EDL case presented considered storing data on the product itself by inserting a chip. However, new sensor and transmission technology enables online and remote sensing. Moreover, the transaction capacity of (wireless) networks drastically increases and related transaction costs will decrease accordingly. This enables online monitoring not only of capital-intensive products but also of somewhat cheaper consumer goods such as cars or TVs.

16.5 ICT Systems that Support Reverse Logistics Processes

In forward logistics, Warehouse Management Systems (WMS) fulfill administrative tracking and handling support. To support the returns process, either special purpose reverse logistics ICT systems have been developed or WMS were extended with proprietary systems to control returns. We examine these issues in more detail in the following subsection. Also, we discuss some web applications, mainly interfaces, that aim to facilitate reverse logistics processes in Subsection 16.5.2.

16.5.1 Reverse Logistics/Warehouse Management Systems

Retailer returns, supplier refusals, and packaging returns are gathered to specialized warehouses called 'return centers,' or through conventional distribution centers. According to De Koster et al. (2000), returns cause additional labor-intensive processes in the warehouses, which constitute around 6% of total logistics costs.

Warehouse Management Systems (WMS) fulfill administrative tracking and handling support. WMS can provide decision-making for further recovery options and communicate this information to other actors involved. It also collects product information to optimize the processing of incoming returns. We present two examples of Warehouse Management Systems: the first one was specially developed to handle returns and the second refers to a proprietary system.

Genco has designed and developed R-log, a reverse logistics software program that controls returns and is often coupled with specialized warehouses called 'return centers,' which are also operated by Genco. Within R-log, each product delivered to a return center is labelled with a barcode. Within the return center, each product is tracked by its unique number and is routed to its container and/or storage area. R-log selects an optimal recovery or disposal option, based on secondary market channel opportunities (vendors, salvage, charity), cost, and constraints such as the condition of the returned product, hazardous contents, or specific customer instructions. R-log interfaces with other information systems to facilitate financial control of returns and their impact on planning and production through its powerful report-generating mechanism. R-log provides qualitative and quantitative information to management (compliance on recalls, reason for return, returns per vendor, etc.).

After determining optimal recovery/disposal channels, it bundles similar products to reduce handling and shipping costs, and provides management control in checking whether intended recovery/disposal operations are actually carried out. Because radio frequency computers and barcode scanners are used, the paperwork regarding the returns and the number of human errors are strongly reduced.

The proprietary system developed at Estèe Lauder for the control of the returns process required a 1.5-million-dollar investment in IT infrastructure (scanners, business intelligence tools, and an Oracle-based data warehouse); see Economist (1999). Based on these, the system leads to a yearly cost saving of 475,000 dollars through increased reuse and less handling. It also reduced the amount of products scrapped from 37% to 27%, improving the company's green image with the consumer. A summary of the functional requirements for WMS, as outlined in De Koster (2001), is shown in Table 16.6.

Table 16.6. WMS functional requirements for returns management

WMS must block lots with detected defects for forward delivery.
WMS must be able to delete lots of scrapped materials.
WMS must uniquely identify the lots of returns which will be forwarded to secondary markets.
WMS must be able to provide debit to suppliers with refused supplier lots.
WMS must be able to provide credit to customers with customer returns, if eligible by the return policy.
WMS must trace customer returns to supplier lots and keep track of return codes.
WMS must keep original product specifications for recovery operations.
WMS must uniquely identify returns kept at distinct locations.
Inventory levels must be adjusted according to process option (recover, sell, etc.).
WMS must support decision-making among recovery options (testing, inspection, and registration of decision taken).
Returns being recovered must be registered as (temporary) shipments and receivables, and recall functions must support control.
Repair/recovery instructions must be provided.
WMS must be able to analyze historical data of returns and generate comprehensive reports.

16.5.2 Web-enabled Applications for Returns Management Operations

Web-enabled applications for the support of reverse logistics operations are interfaces that interact with customers to address aspects of returns uncertainty.

Effective control on the volume of returns. Frequently, the concept of control of the volume of returns is interpreted as minimization of incoming returns. In general, however, we can not exclude the possibility for a firm to proactively search for returns (e.g. to use them as cores in new products). Online tracking and tracing of orders contributes to controlling forward logistics operations. Another way to minimize returns is to cross-examine each order for incompatibilities between ordered items and to notify the customer accordingly. For example, when someone orders a color printer and refill cartridges that do not fit that printer, then the user interface will point out the incompatibility and ask for customer confirmation. Finally, firms have set up gatekeepers for their returns on the web, that is, when customers declare their intention to return some products, they are directed to a web interface (www.e-rma.com, www.yantra.com) that minimizes customer returns due to some misunderstanding of its functionality. Dell's gatekeeping system contributes to keeping returns to around 5% for their computers sold online, versus about 10% for CompuUSA for computers sold in their stores.

Minimization of returns uncertainty factors. Uncertainty regarding returns makes planning and management a very difficult task. Web interfaces are used to minimize uncertainty associated with returns, i.e. when a customer declares a return, she is directed to a web interface that collects data on the product's condition, the intended collection method, and the time and the place of the return. These collected data support preliminary management for returns. Beyond a passive collection of data, however, the web interface can be designed in a way that gives financial incentives to customers to follow the optimal alternative for each return (www.return.com). In a more structured way, the experimental prototype we have developed, Kokkinaki et al. (2002c), demonstrates how processes for monitoring and benchmarking can lower uncertainties associated with the number, type, and quality status of returns in the sector of Information and Communication Technology (ICT).

Effective control on the processes of returns is the first step toward returns fulfillment. It is certain, however, that the topic of ICT systems for effective returns management remains still insufficiently addressed, and in the future we can expect to view a higher exploitation of ICT in this respect.

16.6 ICT Systems and Marketplaces for Reverse Logistics

In our research in Kokkinaki et al. (2002b), we have examined how e-commerce has enabled defragmentation of reverse logistics markets and in doing so has facilitated the redistribution of returns to the market.

In particular, the e-business model of returns aggregators brings together suppliers and customers, automates the procurement of returns, and creates value through high throughput and minimal transaction costs. Returns aggregators handle returns from many different OEMs, without owning products. There are returns aggregators for different returns flows, namely, production waste (www.metalsite.com), commercial returns (www.qxl.com), end-of-use products (www.ebay.com), or the combination of all return types (www.180commerce.com).

Typically, returns aggregators do not support logistic operations for returns, which are stored at their initial location until a transaction is completed. Third party logistics operators (3PLOs) support logistics services to returns aggregators. Most returns aggregators are open to all buyers, whereas sellers may have to register and pay fees. Overall, the level of control on trading parties remains low. A notable exemption is AUCNET (Lee (1997)) for the procurement of used cars in Japan. In AUCNET, the level of control is high, because it is restricted to partners who have established relations through conventional interaction or have been recommended by a member.

Web design and effective searching mechanisms are essential tools for increasing the visibility of returns aggregators and extending the participation of traders. Moreover, a dynamic price-determining mechanism (often an e-auction) enables high throughput of transactions and the de-fragmentation of a highly fragmented market. By employing auctions to track the demand and set the price, returns aggregators can overcome difficulties associated with conventional means for managing returns.

Returns aggregators offer added value in markets which have a large number of transactions with low individual transaction value. Their added value increases with the increase in the number of different SKUs they handle, because that increases the critical mass of their potential clientele. Despite their potentials to become truly global marketplaces, U.S.-based returns aggregators (like www.ebay.com) concentrate within the North American market due to their demographics or a managerial decision to simplify the logistics operations involved. Their global expansion is achieved via sister returns aggregators operating in different countries (i.e. Ebay sister companies in over 20 countries). In the EU, returns aggregators are member-state oriented (www.viavia.nl), as they follow cultural, logistic, linguistic, and financial diversities between different member states. For example, www.qxl.com offers different content in its sites in the UK, France, Netherlands, Germany, and Italy. Beyond the described functionality, some returns aggregators like Aucnet or Autodag (its U.S. equivalent) have promoted the redesign of business processes in their industry. Within these new e-business models, new processes have been devised to accomplish certified inspection and multimedia representation of the cars with registered flaws.

16.7 ICT Systems that Support Product Data and Reverse Logistics Processes

In this section, we start examining ICT systems that address a combination of two out of the three main aspects in reverse logistics. We start with those systems that address both product data and reverse logistics processes. In this category, we describe eco-tools and ERP systems for product recovery.

16.7.1 Eco-tools

Eco-tools are analytical tools that provide estimations of the environmental effects of production processes, recovery options, and the end-of-life disposal of products. Based on these estimations, users of eco-tools can examine alternative design scenarios and select a design strategy that is optimal with respect to environmental effects and production costs. Thus, eco-tools promote sustainability of products over their entire life cycle.

Eco-tools receive input information about a product, both its parts and subassemblies, including labor rates, disposal rates, cost benefits of reuse and recycling, item weights, and disassembly sequences. Details on the production, disassembly, and end-of-life processes for parts and subassemblies are also input to eco-tools. Eco-tools employ disassembly and environmental databases together with the provided input to calculate end-of-life recovery options and optimal product design, including disassembly optimization, by changing the input parameters. Eco-tools are distinguished into Life-Cycle Analysis (LCA) tools, Design for Disassembly (DfX) tools, and Waste-Management tools; De Caluwe (1997). LCA tools analyze products regarding the environmental behavior over the product life cycle. In general, LCA tools encompass the following phases.

1) Definition of goals and boundaries
2) Inventory analysis
3) Impact analysis
4) Improvement analysis

In phase 1, users specify the product to be analyzed. During the inventory analysis, quantifying inputs (processes) and outputs (e.g. emissions or toxicity) of the product's processes and activities are computed for the duration of its life cycle. In the phase of impact analysis, the qualitative or quantitative outputs of the previous phase are used for the computation of the environmental impact of the product, both in direct effects (e.g. ozone depletion) and long-term damage (e.g. increased risk of health hazards). The improvement step, which may not be included in each LCA tool, systematically evaluates opportunities to reduce the environmental impact over the entire life cycle, comprehending all product-related processes.

The massive data requirements of LCA tools raise major problems regarding their widespread applicability. A way to lower these data requirements is to develop mechanisms for the interactivity of PDM systems or CAD/CAM systems with LCA tools to enable them to extract product-related data. At another level, the use of simplified approximation methods, based on energy, waste, and toxicities parameters, could also lower the data requirements for related parameters of LCA tools.

Design for Disassembly (DfX) tools are similar to LCA tools. DfX tools focus on improvements of the product design regarding manufacturing methods, fastening methods, and materials used. A subcategory of DfX tools, namely the *Design for Disassembly/Recovery tools*, focus on the end-of-life/end-of-use stage and aim to enhance recovery options of components through reuse or recycling. If these options are not feasible, DfX tools can help reduce the environmental burden of incineration or disposal. *Waste management tools* receive input information on product design and production materials and processes, and present alternatives of EOL options at different levels, taking into account time- and context-dependent characteristics such as return quality, geographical data related to returns, legislation, and prices of returns in secondary markets. Most of the tools are related to regularity compliance and tracking of materials and emissions, and matching their values with defined checklists of materials and maximal emission values.

16.7.2 ERP Systems for Product Recovery

In this section, we discuss Enterprise Resource Planning (ERP) modules that support reverse logistics processes. ERP systems, if successfully integrated, link financial, manufacturing, human resources, distribution, and order-management systems into a tightly integrated single system with data shared horizontally across the business. Well-known ERP systems include SAP R/3, Oracle ERP system, and Siebel. Van Hillegersberg et al. (2001) reports on the use of ERP to support several process areas. Here we focus on the use of ERP to support manufacturing and recovery processes.

An important step in product recovery is disassembly. ERP systems are used to support disassembly scheduling to retrieve cores from returns. Cores are disassembled to a specified set of parts, which are then inspected and tested. Parts which meet the quality standards are stocked for reuse. Parts with minor defects need first to be scheduled for repair and are then stocked for reuse. Disapproved parts are scrapped. The disassembly process presents additional data requirements, including the yield of reusable/repairable components from the disassembly process, as well as the repair time of repairable components. These data are kept in the recovery BOMs. Recovery BOMs are not merely reversed BOMs, since in many cases disassembly is (partly) destructive.

Recovery BOMs cannot be uniquely defined. Recovery BOMs may have multiple versions for alternative recovery options which could address varying degrees of disassembly processes, quality standards, disassembly times, and yields of spare parts. ERP software packages do not support recovery BOMs, thus they need extensions to address the additional data requirements.

Furthermore, in order for ERP software packages to manage returns they need to be enhanced with new functional requirements. Based on Thierry et al. (1995). Krikke et al. (1999), and Van Hillegersberg et al. (2001) , the additional requirements to ERP systems are summarized below.

First of all. ERP software packages usually do not support disassembly scheduling, which needs to be performed periodically. A way to bypass these limitations is to insert the expected yield into the ERP software as negative requirements on a component level. In this way, unnecessary external orders for new components are prevented and hence reuse is maximized. Yield factors are monitored and, in the case of structural deviance, adapted. In the case that the actual yield after disassembly is more than expected (oversupply), the quantity of negative requirements is adjusted downward accordingly in the next planning cycle. When the actual yield is less than expected (undersupply), negative requirements are shifted backward in time, assuming that the average yield will be achieved over time (by oversupply in the next period). If not, a manual correction of the negative requirements is necessary, automatically leading to external orders.

Secondly, cores need to be registered under inventory control in the ERP system. Inventory control of returns has the additional complexity of multi-function, that is, the possibility to retrieve certain components from various types of cores or even from completely new ones. Identification of identical components retrievable from various cores can be solved by labelling these components. Current ERP software tools are not capable of distinguishing between different statuses of returns, i.e. 'to be dismantled,' 'for scrapping,' 'to be repaired,' or 'to be reused,' and between owners of returns inventory. This could be addressed by adding a status identifier to the article code and also by distinguishing different lead times for different states.

Thirdly, facilities for final assembly in manufacturing and remanufacturing are often shared. These introduce new requirements to ERP systems because the subassemblies from which final products are assembled can either be new or remanufactured. Due to dual sourcing, the integration of both input streams adds planning complexity to inventory control (as described above) and also to capacity planning, because in hybrid (re)manufacturing systems facilities are shared. Some of the aforementioned issues are discussed in more detail in Chapter 10 of this book.

16.8 ICT Systems that Support Reverse Logistics Operations and Marketplaces

Web-enabled applications that have been employed to facilitate reverse logistics processes have also introduced competitive advantage, which leads to the creation of new business processes in reverse logistics. As presented in Kokkinaki et al. (2002b), web-enabled applications have enabled structural modifications in reverse logistics activities and have resulted in new e-business models, which are described in this section.

16.8.1 Speciality Locators

Speciality locators are vertical portals which focus on highly specialized used parts or products. Usually they emerge from existing businesses that have been operating offline and have gathered expertise in their market; that is, they can provide some of the required services in reverse logistics. In that sense, internet exploitation is not essential for their core business processes, but it offers an omnipresent platform for conducting business.

Such electronic marketplaces can serve the need for authentic antiques, exact replicas parts or equipment in historic restoration projects, or the maintenance process for vehicles and industrial equipment. The main characteristics for this model are twofold. First, speciality locators are region-bounded and vertically structured, focusing on a limited range of used parts or equipment over a geographical region.

The major asset in this model is the ability to provide specialized (and thus highly priced) service. Provided services include training, frameworks for catalog search, selection and configuration, financing, and technical support.

Participation of suppliers in speciality locators is usually subject to financial contribution and this acts as a control mechanism on their trading partners. Besides participation fees, value is generated from reference fees, advertising fees, and from mining buyers profiles.

This business model has high entry barriers; for someone to enter this business model, he needs to design and impose new standards in a specialized topic, structure of information, and market liquidity. Identification of the part or product requested is a central issue to the success of this business process, and it implies the use of a common, unique, and unambiguous framework to describe requested products or parts. Independent schemes for parts identification offer high added value to customers, because contrary to suppliers coding systems, independent schemes enable customers to interchange spare parts from different suppliers. Standards and structure vary from one implementation (www.find-a-part.com) to another (www.bigmachines.com), whereas conventional catalogs of spare parts offer a unique coding system related to one supplier only. Identification of a part can be enhanced through special-purpose, web-accessible search engines that focus on some prominent features of the part (brand, description, code, etc.).

Some speciality locators address the issue of preventative or reactive maintenance for heavy industrial equipment which may operate in geographically remote places and under very stressful conditions. Remanufacturing industrial equipment is often a closed-loop process, in the sense that users send in a piece of their equipment and sometime later receive it back remanufactured. Severe time constraints and quality guarantees are important factors for remanufacturing. Addressing urgency is a determining factor for speciality locators' competitive advantage. In the future, we expect that speciality locators will address this point and their competitive advantage will likely shift to provide dynamically customized online expertise.

16.8.2 Integrated Solution Providers

The integrated solution providers capitalize on their distinctive expertise and integrate web technologies to offer unique services for handling returns. This positioning enables them to offer high value-added services that cannot easily be undertaken by their competitors. Furthermore, they actually become the owners of the returns instead of implementing a brokering mechanism as in the previous two models. For that, they also mark high in the degree of inclusion of reverse logistics activities.

This model aims to forge strong relationships with long-lasting customers in industries where the cost of a return itself may not be high, but its speedy handling is essential to its core business process. By definition, each integrated solution provider focuses on the reverse logistics network in an industry/sector, i.e. pharmaceuticals (www.returnlogistics.com, www.pharmacyreturns.com), machine tool manufacturers (www.milpro.com), or cellular phones (www.recellular.com). Returnlogistics provides an integrated solution to authorize, document, pack, and ship returns of pharmaceuticals by different manufacturers. Returnlogistics has included in its web-accessible databases around 80,000 product descriptions. Assisted by a search mechanism, users specify their returns and select a disposition method (return to OEM, destroy, sell, exchange, or donate). Returnlogistics enables returns processors to seamlessly track and document authorization for returns and enables credit managers to eliminate invoices of deduction. Furthermore, it assists users with the appropriate packaging and shipping documentation. Finally, it provides the full range of logistics services including destruction of controlled substances and unpacking and repackaging of products.

Along the same line of value-added services for returns, ReCelluLar acquires cellular phones through airtime providers or charity foundations (i.e. www.wirelessfoundation.org), grades and sorts them, remanufactures them if necessary, and repackages, distributes, and sells them globally. ReCellular exploits diffusion difference at a global level and creates value through cascades of reused or remanufactured cellular phones. Because the cellular communications industry is a very dynamic market, it is essential for ReCellular to develop fast responses. ReCellular B2B exchange shows the current stocks (model, price, grade, and quantity) and facilitates fast throughput of transactions.

The model for Integrated Solution Providers is still in its infancy, probably because it does not view e-commerce as a migration of existing practices and services over a new infrastructure but rather, as a new tool to restructure a business activity and offer new services. This e-business model creates value through escrow and processing fees and through locking in the customer for add-on services or products.

16.9 ICT Systems for Product Data Collection and Marketplace Formation

So far, we have examined how product data have facilitated operations in reverse logistics. In this section, we examine how e-marketplaces enriched with product data can facilitate both the collection of returns and their redistribution into the market. The added value of an e-marketplace in this dual role would be

1) to act as a consolidation channel for the collection of returns from a highly fragmented market;
2) to receive information from the users about the incoming returns;
3) to motivate users to follow optimal recovery policies for their returns;
4) to facilitate better planning and control of returns management;
5) to facilitate reintroduction of recovered products or parts back into the original or secondary markets; and

6) to collect historical data that would enable cost-efficient multi-echelons of the reverse logistics networks (i.e. to collect and disassemble products with known reusable components locally and forward only the reusable components to a centralized facility).

To outline these points, we present this example of an e-marketplace for end-of-use PCs.

16.9.1 An E-marketplace for End-of-use PCs

We have developed an electronic marketplace and an agent-based framework in Kokkinaki et al. (2002c,2002d) which uses web technologies to lower the uncertainty related to ICT returns. This electronic marketplace enables interested users to perform configuration detection and benchmarking of their PCs remotely, just by connecting them to the internet and accessing the electronic marketplace. The main concept is to provide users with

1) an accurate assessment of the condition and configuration of their end-of-use PC,
2) to inform them of their recovery options and to give them an indication of the value they can retrieve from each option, and
3) to bring them in contact with other users interested in their end-of-use PC.

Similar functionalities have appeared recently to Amazon, which has extended its services to reselling used books from their clients or partners. More specifically, Amazon customers, as soon as they access Amazon, get a message that informs them that they can resell their books and provides them with an estimated price. If they agree, Amazon lists their books together with the brand new copies. Amazon acts as a broker and brings interested parties together; for this service Amazon gets a transaction fee for each completed transaction.

Our system is more complicated because end-of-use ICT equipment has more recovery options than used books. The system builds on a set of communicating agents that act on behalf of the related actors (i.e. interested sellers and buyers) and perform a series of tasks for them (see Figure 16.3). Actors are logged onto the system, 'hire' their personal agents (by paying subscription fees that depend on the time they want them to 'live'), create a profile for them, and launch them into the e-market. A potential seller may initiate a configuration request; this triggers a trusted third-party agent to perform configuration detection and benchmarking of the PC currently used to access the e-marketplace. Upon the seller's agreement, the information on the screened PC is registered in the system's repository and is marked as a new offer. Potential buyers (i.e. OEMs or recyclers) can register their collection intentions as requests. Requests can be issued either for entire PC units or single PC modules. The system integrates mechanisms to match the above offers and

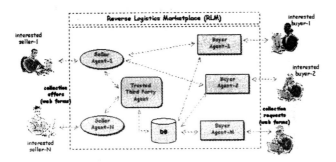

Fig. 16.3. The overall e-marketplace framework

requests. In addition, potential buyers may be instantly notified about incoming offers. Whenever a match is identified, the corresponding seller is notified through the mail services of the system. Then, the seller may launch an agent to comparatively evaluate all requests retrieved.

Transactions can be initiated either by human actors or software agents. An actor who has been notified that some requests match her offer may perform a comparative evaluation of all relevant requests before reaching an optimal decision. Agents are proactive and semi-autonomous, so they can also initiate transactions. For example, a buyer agent whose profile matches a seller's offers may take the initiative to contact its actor and ask his/her opinion on proceeding if an interesting offer has been inserted into the system's repository.

From a business perspective, the proposed system is very appealing, because it lowers transaction costs and automates several, previously manual, operations. Furthermore, it offers a trustworthy mechanism for configuration detection and quality assessment. Even more importantly, it provides a cohesive platform that consolidates the returns market and releases trapped value. From an ICT perspective, the proposed system maintains profiles for all potential buyers (through the personalization of the agents involved) to give recommendations in alignment with their interests and preferences. Also, agents may take the initiative to contact their actors when they identify a seemingly interesting transaction. Third, seller agents in this system may perform a progressive synthesis and comparative evaluation (across a set of attributes) of the existing requests. This requires a highly interactive tool, based on multiple-criteria decision theory, that enables customers to examine alternative scenarios (by selecting which of the attributes of the matched requests to be taken into account) and recommends the best recovery options, concerning reuse, remanufacturing, or recycling, according to the information at hand (an example of which is shown in Figure 16.4).

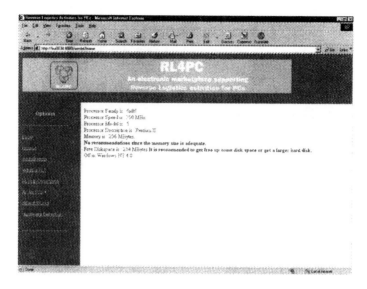

Fig. 16.4. Recommendations for a specific PC configuration

16.10 Conclusions

In this chapter, we have examined ICT systems related to returns management, covering the whole spectrum of applications for product data management, reverse logistics operations support, and formation of e-marketplaces. Special-purpose software tools have been developed, and their use is widespread in business practice.

As mentioned in Section 16.2, the development of techniques in order to handle uncertainty requires further research. In addition, information systems as described in the remaining sections require further systematic research in order to cater to the specific requirements of reverse logistics. Following the framework for the research in ICT systems for reverse logistics, we consider in this section challenges for further research regarding products, processes, and markets in reverse logistics.

As mentioned in Section 16.2, the role of information systems in reducing uncertainty requires further systematic investigation. For example, monitoring the customer base of a product provides both opportunities (proactive collection) and challenges (data management of field measurements). If one considers the extended product (including after-sales services, besides the physical product), then customer-base management and the management of product returns both belong to the package of services around a product.

The reverse logistics network is even more complex than the classical forward supply chain, since more organizations are involved and the underlying processes are more complex from an information processing point of view.

For those reasons, the use of inter-organizational information systems can be both rewarding and challenging. Standardization of information exchange on reverse logistics processes, such as proposed by the Reverse Logistics Executive Council (see www.rlec.org, which reports on EDI disposition and return codes that enable information exchange on product returns), show that solutions are under development. Beyond standardization of information exchange, it is also necessary to provide dynamic and efficient communication infrastructure for enterprise information inter-operability; that is, to enable sharing of data (product design, product development, product use, and conditions of disposal) among value-chain enterprises in a way that is more efficient than the point-to-point communication embedded in EDI.

The implementation of 1) environmental performance indicators, 2) planning and control of recovery processes, and 3) cost accounting in reverse logistics in standard ERP modules certainly is an area that requires further exploration. These issues have already received some attention in the literature (see e.g. Van Hillegersberg et al. (2001) and Lambert et al. (2000)), but require further research.

The development of business models that support the match of supply and demand of returned products is still an open research issue.

Table 16.7. A comprehensive list of all e-businesses mentioned in this chapter

E-Business Name	E-Business URL
180Commerce	www.180commerce.com
Aucnet	accessible to trade only
Autodag	www.autodag.com
Bigmachines	www.bigmachines.com
Dell	ww.dell.com
E-rma	wwww.e-rma.com
Ebay	www.ebay.com
Fairmarket	www.fairmarket.com
Find-a-part	www.find-a-part.com
Genco	www.e-genco.com
IBM	www.ibm.com
Metalsite	www.metalsite.com
Pharmacy Returns	www.pharmacyreturns.com
QXL	www.qxl.com
ReCellular	accessible to trade only
Returnlogistics	www.returnlogistics.com
TheReturnExchange	www.thereturnexchange.com
Yantra	www.yantra.com
Viavia	www.viavia.nl

Notation

The following notation is used throughout the book. Additional specific notation is introduced where needed.

Parameters

α	discount factor
d	demand rate
$(.)_d$	subscript for demand–related quantities (general)
D	cumulative demand
$\delta(.)$	geographical demand density
γ	recovery yield
$\rho(.)$	geographical return density
u	return rate
u_r	return rate of recoverable products
u_w	return rate of defective products
$(.)_u$	superscript for return–related quantities (general)
U	cumulative returns
τ	time/period index
$(.)(\tau)$	time dependence (general)
T	time horizon

Processes

I	inventory (general)
I_n	net stock serviceables
I_o	on-hand stock serviceables
I_s	inventory position serviceables
I_u	on-hand stock recoverables
L	lead time (general)
L_p	production lead time
L_r	recovery lead time

p	production rate
$(.)_p$	subscript for production–related quantities (general)
$\overline{p}$	production capacity
$\overline{(.)}$	capacity (general)
Q_p	production lot size
Q_r	recovery lot size
Q_w	disposal lot size
r	recovery (remanufacturing, recycling, ...) rate
$(.)_r$	subscript for recovery–related quantities (general)
w	disposal rate
$(.)_w$	subscript for disposal–related quantities (general)
X	material flow (general)
X_{ik}	material flow from location $i \in \mathcal{I}$ to $k \in \mathcal{K}$
Y	location / assignment variable (general)

Costs

c_b	unit backorder cost rate
c_p	unit variable production cost
c_r	unit variable recovery cost
c_t	unit transportation cost (general)
c_{ik}^t	unit transportation cost from location $i \in \mathcal{I}$ to $k \in \mathcal{K}$
c_u	unit buy–back price
c_w	unit variable disposal cost
f_i	fixed installation cost for location $i \in \mathcal{I}$ (general)
h_s	holding cost rate serviceables (if undistinguished)
h_p	holding cost rate produced items
h_r	holding cost rate recovered items
h_u	holding cost rate recoverable items
K_p	setup cost production
K_r	setup cost recovery
K_w	setup cost disposal

Index sets

$\mathcal{K}$	set of customer locations
$\mathcal{I}$	general index set;
	set of facility locations (general)
Ω	scenario space

References

Adler, P.S. and Clark, K.B. (1991). Behind the learning curve: A sketch of the learning process. *Management Science*, 37(3):267–281.

Almemark, M., Bjuggren, C., Granath, J., Olsson, J., Röttorp, J., and Lindfors, L.-G. (2000). Analysis and development of the interpretation process in LCA. Technical report, IVL Swedish Environmental Research Institute Ltd.

Amazon.com (2003). www.amazon,com.

Ammons, J.C., Realff, M.J., and Newton, D. (1997). Reverse production system design and operation for carpet recycling. Working paper, Georgia Institute of Technology.

Anderson, S., Browne, M., and Allen, J. (1999). Logistics implications of the UK packaging waste regulations. *International Journal of Logistics: Research and Applications*, 2(2):129–145.

Andrews, C.J. and Swain, M. (2000). Institutional factors affecting life-cycle impacts of microcomputers. *Resources, Conservation and Recycling*, 31:171–188.

Anily, S. (1996). The vehicle routing problem with delivery and backhaul options. *Naval Research Logistics*, 43:415–434.

Anily, S. and Bramel, J. (1999). Vehicle routing and the supply chain. In Tayur, S.R., Ganeskan, R., and Magazine, M.J., editors, *Quantitative Models for Supply Chain Management*, pages 147–196. Kluwer Academic Publishers, Boston, MA.

Ashayeri, R., Heuts, R., Jansen, A., and Szczerba, B. (1996). Inventory management of repairable service parts for personal computers: A case study. *International Journal of Operations & Production Management*, 16(12):74–96.

Assad, A.A. and Golden, B.L. (1995). Routing methods and applications. In Ball, M.O., Magnanti, T.L., Monma, C.L., and Nemhauser, G.L., editors, *Network Routing*, pages 375–483. Elsevier Science Publishers, Amsterdam, The Netherlands.

Azzone, G., Noci, G., Manzini, R., and Weelford, R. (1996). Defining environmental performance indicators: An integrated framework. *Business Strategy and the Environment*, 5:69–80.

Ball, M.O. (1988). Allocation/routing: Models and algorithms. In Golden, B.L. and Assad, A.A., editors, *Vehicle Routing: Methods and Studies*, chapter 10. Elsevier Science Publishers, Amsterdam, The Netherlands.

Baptista, S., Oliveira, R.C., and Zuquete, E. (2002). A period vehicle routing case study. *European Journal of Operational Research*, 139:220–229.

Barad, M. and Braha, D. (1996). Control limits for multi-stage manufacturing processes with binomial yield (single and multiple production runs). *Journal of the Operational Research Society*, 47:98–112.

Barlas, Y. (1996). Formal aspects of model validity and validation in system dynamics. *System Dynamics Review*, 12(3):183–210.

Barros, A.I., Dekker, R., and Scholten, V. (1998). A two-level network for recycling sand: A case study. *European Journal of Operational Research*, 110:199–214.

Bartel, T. (1995). Recycling program for toner cartridges and optical photoconductors. In *Proceedings IEEE Symposium on Electronics and the Environment*, pages 225–228, Orlando, FL.

Bartholdi, J.J., Bunimovich, L.A., and Eisenstein, D.D. (1999). Dynamics of two- and three-worker 'bucket brigade' production lines. *Operations Research*, 47(3):488–491.

Bartholdi, J.J. and Eisenstein, D.D. (1996). A production line that balances itself. *Operations Research*, 44(1):21–34.

Bartholdi, J.J. and Gue, K.R. (2000). Reducing labor costs in an LTL crossdocking terminal. *Operations Research*, 48(6):823–832.

Batta, R. and Chiu, S. (1988). Optimal obnoxious paths on a network: Transportation of hazardous materials. *Operations Research*, 36:84–93.

Bayindir, Z.P., Erkip, N., and Güllü, R. (2003). A model to evaluate inventory costs in a remanufacturing environment. *International Journal of Production Economics*, 81–82:597–607.

Beardwood, J., Halton, H., and Hammersley, J.M. (1959). The shortest path through many points. In *Proceedings of the Cambridge Philosophical Society 55*, pages 299–327.

Belenguer, J.M. and Benavent, E. (2000). A cutting plane algorithm for the capacitated arc routing problem. Working paper, Departamento de Estadística e Investigación Operativa, Universitat de València, Spain.

Bell, W.J., Dalberto, L.M., Fisher, M.L., Greenfield, A.J., Jaikumar, R., Kedia, P., Mack, R.G., and Prutzman, P.J. (1983). Improving the distribution of industrial gases with an on-line computerized routing and scheduling optimizer. *Interfaces*, 13(6):4–23.

Bellman, R.E. (1957). *Dynamic Programming*. Princeton University Press, Princeton, NJ.

Beltran, J.L. and Krass, D. (2002). Dynamic lot sizing with returning items and disposals. *IIE Transactions*, 34:437–448.

Benavent, E., Corberan, A., and Sanchis, J.M. (2000). Linear programming based methods for solving arc routing problems. In Dror, M., editor, *Arc Routing: Theory, Solutions and Applications*, pages 231–275. Kluwer Academic Publishers, Boston, MA.

Bensoussan, A., Crouhy, M., and Proth, J.M. (1983). *Mathematical Theory of Production Planning*. North-Holland, Amsterdam, The Netherlands.

Bensoussan, A., Hurst, E.G., and Näslund, B. (1974). *Management Applications of Modern Control Theory*. North-Holland, Amsterdam, The Netherlands.

Berger, C., Chauny, F., Langevin, A., Loulou, R., Riopel, C., Savard, G., and Waaub, P. (1998). EUGENE: An optimisation based DSS for long term integrated regional solid waste management planning. In *Proceedings of the International*

Workshop Systems Engineering Models for Waste Management, Gothenburg, Sweden.

Berger, T. and Debaillie, B. (1997). Location of disassembly centres for re-use to extend an existing distribution network. Master's thesis, University of Leuven, Belgium. (In Dutch).

Berkhaout, F. (1995). Life cycle assessment and industrial innovation. In *The 1995 Eco-Management and Auditing Conference*, pages 50–64, Leeds, U.K.

Bettac, E., Mayers, K., Beullens, P., and Bopp, R. (1999). Current practice and management of take-back logistics and electronics recycling in Europe. The RELOOP Consortium. Submitted to the European Commission as Deliverable D3, 122 pages.

Beullens, P. (2001). *Location, Process Selection, and Vehicle Routing Models for Reverse Logistics*. PhD thesis, Katholieke Universiteit Leuven, Belgium.

Beullens, P., Muyldermans, L., Cattrysse, D., and Van Oudheusden, D. (2003). A guided local search heuristic for the capacitated arc routing problem. *European Journal of Operational Research*, 147(3):629–643.

Beullens, P., Van Oudheusden, D., and Cattrysse, D. (1999). Bi-destination waste collection: Impact of vehicle type and operations on transportation costs. In Flapper, S.D.P. and de Ron, A.J., editors, *Proceedings of the Second International Working Seminar on Reuse*, pages 5-14. Eindhoven University of Technology, The Netherlands.

Birge, J.R. and Louveaux, F. (1997). *Introduction to Stochastic Programming*. Springer-Verlag, New York.

Bloemhof-Ruwaard, J.M. and Salomon, M. (1997). Reverse Logistics. In Ploos van Amstel, M.J., Duijker, J.P., and de Koster, M.B.M., editors, *Praktijkboek Magazijnen en Distributiecentra*. Kluwer Bedrijfswetenschappen, Deventer, The Netherlands. (In Dutch).

Bloemhof-Ruwaard, J.M., Salomon, M., and Van Wassenhove, L.N. (1996a). The capacitated distribution and waste disposal problem. *European Journal of Operational Research*, 88:490–503.

Bloemhof-Ruwaard, J., Van Wassenhove, L.N., Gabel, H.L., and Weaver, P.M. (1996b). An environmental life cycle optimization model for the European pulp and paper industry. *Omega*, 24(6):615–629.

Boon, J.E., Isaacs, J.A., and Gupta, S. (2001). Economic impact of aluminum-intensive vehicles on the U.S. automotive recycling infrastructure. *Journal of Industrial Ecology*, 4(2):117–134.

Bourjault, A. (1984). *Contribution à une approche méthodologique de l'assemblage automatisé: élaboration automatique des séquences opératoires*. Phd-Thesis, Université de Besancon, France. (In French).

Box, G.E.P. and Jenkins, G.M. (1976). *Time Series Analysis, Forecasting and Control*. Holden-Day, San Francisco, CA, revised edition.

Bozer, Y.A., Meller, R.D., and Erlebacher, S.J. (1994). An improvement-type layout algorithm for single and multiple floor facilities. *Management Science*, 40(7):918–932.

Bozer, Y.A. and White, J. (1984). Travel time models for automated storage/retrieval systems. *IIE Transactions*, 16(4):329–338.

Bozer, Y.A. and White, J. (1990). Design and performance models for end-of-aisle order picking systems. *Management Science*, 36(7):852–866.

Bras, B. and McIntosh, M.W. (1999). Product, process, and organizational design for remanufacture: An overview of research. *Robotics and Computer Integrated Manufacturing*, 15:167–178.

Brouwers, W.C.J. and Stevels, A.L.N. (1995). Cost model for the end-of-life stage of electronic goods for consumers. In *Proceedings of the IEEE Symposium on Electronics and the Environment*, pages 279–284, Orlando, FL.

Brundland, G.H. (1998). European Union and the Environment. Report, European Union. Luxemburg.

Burns, L.D., Hall, R.H., Blumenfield, D.E., and Daganzo, C. (1985). Distribution strategies that minimize transportation and inventory costs. *Operations Research*, 33(3):469–490.

Buschmann, F., Meunier, R., Rohnert, H., Sommerland, P., and Stal, M. (2001). *Pattern-Oriented Software Architecture: A System of Patterns*. John Wiley and Sons Ltd, New York.

Cachon, G.P. (2003). Supply Chain Coordination with Contracts. In Graves, S. and de Kok, A.G., editors, *Handbooks of Operations Research and Management Science: Supply Chain Management*. North-Holland. (Forthcoming).

Cachon, G.P. and Lariviere, M.A. (1999). Capacity choice and allocation: Strategic behavior and supply chain performance. *Management Science*, 45:1091–1108.

Cachon, G.P. and Zipkin, P.H. (1999). Competitive and cooperative inventory policies in a two stage supply chain. *Management Science*, 45:936–953.

Caldwell, B. (1999). Reverse logistics. *InformationWeek*, 729:48–56.

Camm, F. (2001). Environmental management in proactive commercial firms. Report MR-1038-OSD, RAND.

Campbell, A.M., Clarke, L.W., Kleywegt, A., and Savelsbergh, M.W.P. (1998). The inventory routing problem. In Crainic, T.G. and Laporte, G., editors, *Fleet Management and Logistics*, chapter 4. Kluwer Academic Publishers, Boston, MA.

Campbell, A.M., Clarke, L.W., and Savelsbergh, M.W.P. (2002). Inventory routing in practice. In Toth, P. and Vigo, D., editors, *The Vehicle Routing Problem*, chapter 12. Society for Industrial and Applied Mathematics, Philadelphia, PA.

Carrol, A.B. (1979). A three-dimensional conceptual model of corporate performance. *Academy of Management Review*, 4(4):497–505.

Carter, C.R. and Ellram, L.M. (1998). Reverse logistics: A review of the literature and framework for future investigation. *International Journal of Business Logistics*, 19(1):85–102.

Caruso, C., Colorni, A., and Paruccini, M. (1993). The regional urban solid waste management system: A modelling approach. *European Journal of Operational Research*, 70:16–30.

Casco, D.O., Golden, B.L., and Wasil, E.A. (1988). Vehicle routing with backhauls: Models, algorithms, and case studies. In Golden, B.L. and Assad, A.A., editors, *Vehicle Routing: Methods and Studies*, chapter 7. Elsevier Science Publishers. Amsterdam, The Netherlands.

Cattani, K. and Souza, G. (2003). Inventory rationing and shipment flexibility alternatives for direct market firms. *Production and Operations Management*. (Forthcoming).

ChainStoreAge (2002). Returns are products too. *Chain Store Age*, Feb., 2B-3B.

Chen, F. (1999). Decentralized supply chains subject to information delays. *Management Science*, 45:1076–90.

Chen, F., Drezner, Z., Jennifer, R.K., and Simchi-Levi, D. (2000). Quantifying the bullwhip effect in a simple supply chain: The impact of forecasting, lead times, and information. *Management Science*, 3:436–443.

Chen, F. and Zheng, Y.-S. (1994). Evaluating echelon stock (r, nQ) policies for serial production/inventory systems with stochastic demand in multi-echelon stochastic inventory systems. *Management Science*, 40(10):1262–1275.

Chen, R.W., Navin-Chandra, D., and Prinz, F.B. (1993). Product design for recyclability: A cost benefit analysis model and its application. In *Proceedings IEEE Symposium on Electronics and the Environment*, pages 178–183, Arlington, VA.

Chern, C.C. and Yang, P. (1999). Determining a threshold control policy for an imperfect production system with rework jobs. *Naval Research Logistics*, 46:273–301.

Chopra, S. and Meindl, P. (2001). *Supply Chain Management*. Prentice Hall, Upper Saddle River, NJ.

Clark, A.J. and Scarf, H. (1960). Optimal policies for a multi-echelon inventory problem. *Management Science*, 6:475–490.

Clegg, A.J., Williams, D.J., and Uzsoy, R. (1995). Production planning for companies with remanufacturing capability. In *Proceedings of the IEEE Symposium on Electronics and the Environment*, pages 186–191, Orlando, FL.

Coffman, E.G. Jr., Courcoubetis, C., Garey, M.R., Johnson, D.S., Shor, P.W., Weber, R.R., and Yannakakis, M. (2000). Bin packing with discrete item sizes, part 1: Perfect packing theorems and the average case behavior of optimal packings. *SIAM Journal on Discrete Mathematics*, 13(3):384–402.

Corberan, A., Letchford, A.N., and Sanchis, J.M. (2001). A cutting plane algorithm for the general routing problem. *Mathematical Programming, Series A*, 90:291–316.

Corbett, C.J. and DeCroix, G.A. (2001). Shared savings contracts for indirect materials in supply chains: Channel profits and environmental impacts. *Management Science*, 47:881–893.

Corbett, C.J. and Savaskan, C. (2003). Contracting and coordination in closed-loop supply chains. In Guide, Jr.. V.D.R. and Van Wassenhove, L.N., editors, *Business Aspects of Closed-Loop Supply Chains*, pages 93–113. Carnegie Mellon University Press, Pittsburgh, PA.

Corbett, C.J. and Tang, C.S. (1999). Designing supply contracts: Contract type and information asymmetry. In Tayur, S.R., Ganesham, R., and Magazine, M.J., editors, *Quantitative Models for Supply Chain Management*. Kluwer Academic Publishers, Boston, MA.

Corbey, M., Inderfurth, K., van der Laan, E.A., and Minner, S. (1999). The use of accounting information in production and inventory control for reverse logistics. Working paper, Erasmus University Rotterdam, The Netherlands.

Cordeau, J.F., Gendreau, M., and Laporte, G. (1997). A tabu search heuristic for periodic and multi-depot vehicle routing problems. *Networks*, 30:105–119.

Coyle, R.G. (1978). *Management System Dynamics*. J. Wiley and Sons, New York.

Daganzo, C.F. (1984). The distance travelled to visit n points with a maximum of c stops per vehicle: An analytical model and application. *Transportation Science*, 18:331–350.

Daganzo, C.F. (1999). *Logistics Systems Analysis*. Springer-Verlag, Berlin, Germany, 3rd edition.

Daganzo, C.F. and Hall, R.W. (1993). A routing model for pickups and deliveries: No capacity restrictions on the secondary items. *Transportation Science*, 27(4):315–329.

Daniel, S.E. and Pappis, C.P. (2003). Applying life cycle impact assessment to reverse supply chains — a case study. Working Paper.

Daniel, S.E., Pappis, C.P., and Voutsinas, T.G. (2003a). Applying life cycle inventory to reverse supply chains: A case study of lead recovery from batteries. *Resources, Conservation and Recycling*, 37(4):251–340.

Daniel, S.E., Tsoulfas, G.T., Pappis, C.P., and Rahaniotis, N.P. (2003b). Aggregating and evaluating the results of different environmental impact assessment methods. *Ecological Indicators*. (Forthcoming).

Dantzig, G.B. and Ramser, J.H. (1959). The truck dispatching problem. *Management Science*, 6:80–91.

Daskin, M.S. (1995). *Network and Discrete Location*. Wiley, New York.

Davis, S., Gerstner, E., and Hagerty, M. (1995). Money back guarantees in retailing: Matching products to consumer tastes. *Journal of Retailing*, 71(1):7–22.

Davis, S., Hagerty, M., and Gerstner, E. (1998). Return policies and the optimal level of 'hassle'. *Journal of Economics and Business*, 50:445–460.

de Brito, M.P. and Dekker, R. (2002). A framework for reverse logistics. Econometric Institute Report EI2002-38, Erasmus University Rotterdam, The Netherlands.

de Brito, M.P. and Dekker, R. (2003). Modelling product returns in inventory control — exploring the validity of general assumptions. *International Journal of Production Economics*, 81–82:225–241.

de Brito, M.P., Flapper, S.D.P., and Dekker, R. (2003). Reverse logistics: A review of case studies. Report Series Research in Management ERS-2003-012-LIS, Erasmus University Rotterdam, The Netherlands.

de Brito, M.P. and van der Laan, E.A. (2002). Inventory management with product returns: The impact of (mis)information. Econometric Institute Report EI 2002-29, Faculty of Economics, Erasmus University Rotterdam, The Netherlands.

de Caluwe, N. (1997). Ecotools manual - a comprehensive review of DfE tools. Metropolitan University of Manchester
http://sun1.mpce.stu.mmu.ac.uk/pages/projects/dfe/pubs/dfe33 ecotools.htm.

de Kool, R. (2002). ERP and reverse logistics. Master's thesis, Rotterdam School of Management, Erasmus University Rotterdam, The Netherlands. (In Dutch).

de Koster, M.B.M. (1994). Performance estimation of pick-to-belt orderpicking systems. *European Journal of Operational Research*, 72:558–573.

de Koster, M.B.M., de Brito, M.P., and van Vendel, M.A. (2002a). How to organise return handling: An exploratory study with nine retailer warehouses. *International Journal of Retail and Distribution Management*, 30:407–421.

de Koster, M.B.M., Jacobs, F.H.W.M., and van Dort, H.E. (2000). Informatie binnen magazijnen. In Ploos van Amstel, M.J., Duijker, J.P., and de Koster, M.B.M., editors, *Praktijkboek Magazijnen en Distributiecentra*. Kluwer Bedrijfswetenschappen, Deventer, The Netherlands. (In Dutch).

de Koster, M.B.M. and Neuteboom, J. (2001). *The Logistics of Supermarket Chains*. Elsevier, Doetinchem, The Netherlands.

de Koster, M.B.M., van der Poort, E.S., and Wolters, M. (1999). Efficient order batching methods in warehouses. *International Journal of Production Research*, 37(7):1479–1504.

de Koster, R. (2001). Erasmus University Rotterdam. Personal communication.

de Koster, R.B.M., Flapper, S.D.P, Krikke, H.R., and Vermeulen, W.S. (2002b). Networks for the collection and processing of end-of-life large white goods. Working paper, Erasmus University Rotterdam, The Netherlands.

Debo, L.G. (2002). *Topics in Product Remanufacturing, Repair and Disposal*. PhD thesis, INSEAD, Fontainebleau, France.

Debo, L.G., Toktay, L.B., and Van Wassenhove, L.N. (2001). Market segmentation and production technology selection for remanufacturable products. Working Paper 47/TM/CIMSO 18, INSEAD, Fontainbleau, France.

DeCroix, G., Song, J., and Zipkin, P. (2002). A series system with returns: Stationary analysis. Working paper, Fuqua School of Business, Durham, NC.

DeCroix, G. and Zipkin, P. (2002). Inventory management for an assembly system with product or component returns. Working paper, Fuqua School of Business, Durham, NC.

Deif, I. and Bodin, L.D. (1984). Extension of the Clarke and Wright algorithm for solving the vehicle routing problem with backhauling. In Kidder, A., editor, *Proceedings of the Babson Conference on Software Uses in Transportation and Logistic Management*, pages 75–96, Babson Park, FL.

Dekker, R., Bloemhof-Ruwaard, J.M., Fleischmann, M., van Nunen, J.A.E.E., van der Laan, E.A., and Van Wassenhove, L.N. (1998). Operational research in reverse logistics: Some recent contributions. *International Journal of Logistics: Research and Applications*, 1:141–155.

Dekker, R., de Koster, M. B. M., van Kalleveen, H., and Roodbergen, K. J. (2002). Quick response practices at the warehouse of Ankor. Working paper, Erasmus University Rotterdam, The Netherlands.

Dempster, A.P., Laird, N.M., and Rubin, D.B. (1977). Maximum likelihood from incomplete data via the EM algorithm (with discussion). *J. Royal Stat. Soc. Series B*, 339:1–22.

Desrochers, M., Lenstra, J.K., Savelsbergh, M.W.P., and Soumis, F. (1988). Vehicle routing with time windows: Optimization and approximation. In Golden, B.L. and Assad, A.A., editors, *Vehicle routing: Methods and Studies. Studies in Management Science and Systems*, pages 65–84. Elsevier.

Dethloff, J. (2001). Vehicle routing and reverse logistics: The vehicle routing problem with simultaneous delivery and pick-up. *OR Spektrum*, 23:79–96.

Dethloff, J. (2002). Relation between vehicle routing problems: An insertion heuristic for the vehicle routing problem with simultaneous delivery and pick-up applied to the vehicle routing problem with backhauls. *Journal of the Operational Research Society*, 53:115–118.

Dijkhuizen, H.P. (1997). Reverse Logistics bij IBM. In Van Goor, A.R., Flapper, S.D.P., and Clement, C., editors, *Reverse Logistics*, chapter E1310. Kluwer, Deventer, The Netherlands. (In Dutch).

Diks, E.B., de Kok, A.G., and Lagodimos, A.G. (1996). Multi-echelon systems: A service measure perspective. *European Journal of Operational Research*, 95:241–263.

Dockner, E., Joergensen, S., Van Long, N., and Sorger, G. (2000). *Differential Games in Economics and Management Science*. Cambridge University Press, Cambridge, U.K.

Dowlatshahi, S. (2000). Developing a theory of reverse logistics. *Interfaces*, 30(3):143–155.

Dowling, M. (1999). Getting less in return. *Catalog Age*, 16(4):1–18.

Driesch, H.M., van Oyen, J.E., and Flapper, S.D.P. (1997). Control of the Daimler-Benz MTR product recovery operation. In *Proceedings 3. Magdeburger Logistik-Tagung, Logistik auf Umweltkurs: Chancen und Herausforderungen*, pages 157–165, Magdeburg, Germany.

Dror, M. (2000). *Arc Routing: Theory, Solutions and Applications*. Kluwer Academic Publishers, Boston, MA.

Dror, M. and Ball, M. (1987). Inventory/routing: Reduction from an annual to a short-period problem. *Naval Research Logistics*, 34:891–905.

Dror, M. and Trudeau, P. (1990). Split delivery routing. *Naval Research Logistics*, 37:383–402.

Duhaime, R., Riopel, D., and Langevin, A. (2001). Value analysis and optimization of reusable containers at Canada Post. *Interfaces*, 31(3):3–15.

Dumas, Y., Desrosiers, J., and Soumis, F. (1990). The pickup and delivery problem with time windows. Les cahiers du GERAD G-89-17. Ecole des Hautes Etudes Commerciales, Montreal, Canada.

Economist (1999). Cash from trash. The Economist, February 6th.

Edmonds, J. and Johnson, E.J. (1973). Matching, Euler tours and the Chinese postman problem. *Mathematical Programming*, 5:88–124.

Eggleston, H.S. et al., editors (1991). *CORINAIR Working group on Emission Factors for Calculating 1990 Emissions from Road Traffic. Vol.1: Methodology and Emission Factors*. CEC DG XI, Contract no. B4-3045, 10PH.

Eiselt, H.A., Gendreau, M., and Laporte, G. (1995). Arc routing problems. Part ii: The rural postman problem. *Operations Research*, 43:399–414.

Eisenstein, D.D. and Iyer, A.V. (1997). Garbage collection in Chicago: A dynamic scheduling model. *Management Science*, 43(7):922–933.

ELA (1999). Insight to impact. results of the fourth quinquennial European logistics study. European Logistics Association (ELA). Brussels, Belgium.

Emmons, H. and Gilbert, S.M. (1998). Note. The role of return policies pricing and inventory decisions for catalogue goods. *Management Science*, 44(2):276–283.

Erkut, E. (1996). The road not taken. *OR/MS Today*, Dec.:22–28.

Europa (2003). Europa, the European Union online. http://www.europa.eu.int/.

Farrow, P.H. and Jonhson, R.R. (2000). Entrepreneurship, innovation and sustainability strategies at Walden Paddlers, Inc. *Interfaces*, 30(3):215–225.

Faucheux, S. and Nicolai, I. (1998). Environmental technological change and governance in sustainable development policy. *Ecological Economics*, 27:243–256.

Fearon, H.E. and Leenders, M.R. (1993). *Purchasing and Materials Management*. Irwin, Homewood, IL.

Federgruen, A. and Simchi-Levi, D. (1995). Analysis of vehicle routing and inventory-routing problems. In Ball, M.O., Magnanti, T.L., Monma, C.L., and Nemhauser, G.L., editors, *Network Routing*, pages 297–373. Elsevier Science Publishers, Amsterdam, The Netherlands.

Feichtinger, G. and Hartl, R.F. (1986). *Optimale Kontrolle ökonomischer Prozesse: Anwendungen des Maximumprinzips in den Wirtschaftswissenschaften*. de Gruyter, Berlin, Germany. (In German).

Ferguson, N. and Browne, J. (2001). Issues in end-of-life product recovery and reverse logistics. *Production Planning & Control*, 12(5):534–547.

Fernandéz (2003). The concept of Reverse Logistics. A review of litertaure. In *Proceedings of the NOFOMA 2003*, pages 464–478. Oulu, Finland. NOFOMA.

Fernie, J. and Hart, C. (2001). UK packaging waste legislation. Implications for food retailers. *British Food Journal*, 103(3):187–197.

Ferrer, G. (1996). Market segmentation and product line design in remanufacturing. Working Paper 96/66/TM, INSEAD, Fontainbleau, France.

Ferrer, G. (1997). The economics of PC manufacturing. *Ressources, Conservation and Recycling*, 21:79-108.

Fisher, M.L. (1997). What's the right supply chain for your product? *Harvard Business Review*, pages 105–116.

Flapper, S.D.P. (1994). On the logistics aspects of integrating procurement, production and recycling by lean and agile-wise manufacturing companies in the automotive industries. In *Proceedings of the 27th ISATA International Dedicated Conference on Lean/Agile Manufacturing*, pages 749–756, Aachen, Germany.

Flapper, S.D.P., Fransoo, J.C., Broekmeulen, R.A.C.M., and Inderfurth, K. (2002). Planning and control of rework in the process industries: A review. *Production Planning & Control*, 13:26–34.

Flapper, S.D.P. and Jensen, T. (2003). Logistic planning and control of rework. *International Journal of Production Research*, (forthcoming).

Flapper, S.D.P. and Pels, H.J. (1997). Information aspects of reuse: An overview. *Proceedings of ASL '97 conference Life Cycle Approaches to Production Systems, Budapest, Hungary*, pages 535–541.

Flapper, S.D.P. and Teunter, R.H. (2001). Logistics planning and control of in-line rework in deterministic one-stage, one-product production situations with deteriorating work-in-process. Working Paper, Technical University of Eindhoven, The Netherlands.

Flapper, S.D.P., van Nunen, J.A.E.E., and Van Wassenhove, L.N. (2003). *Managing Closed-Loop Supply Chains*. Springer-Verlag, Heidelberg, Germany. (Forthcoming).

Fleischmann, M. (2001). *Quantitative Models for Reverse Logistics*. Lecture Notes in Economics and Mathematical Systems 501. Springer Verlag, Berlin, Germany.

Fleischmann, M. (2003). Reverse logistics network structures and design. In Guide, Jr., V.D.R. and Van Wassenhove, L.N., editors, *Business Aspects of Closed-Loop Supply Chains*, pages 117–148. Carnegie Mellon University Press, Pittsburgh, PA.

Fleischmann, M., Beullens, P., Bloemhof-Ruwaard, J.M., and Van Wassenhove, L.N. (2001). The impact of product recovery on logistics network design. *Production and Operations Management*, 10(2):156–173.

Fleischmann, M., Bloemhof-Ruwaard, J.M., Dekker, R., van der Laan, E.A., van Nunen, J.A.E.E., and Van Wassenhove, L.N. (1997). Quantitative models for reverse logistics: A review. *European Journal of Operational Research*, 103:1–17.

Fleischmann, M., Krikke, H.R., Dekker, R., and Flapper, S.D.P. (2000). A characterisation of logistics networks for product recovery. *Omega*, 28(6):653–666.

Fleischmann, M. and Kuik, R. (2003). On optimal inventory control with stochastic item returns. *European Journal of Operational Research*, 151(1):25–37.

Fleischmann, M., Kuik, R., and Dekker, R. (2002). Controlling inventories with stochastic item returns: A basic model. *European Journal of Operational Research*, 138:63–75.

Fleischmann, M. and Minner, S. (2003). Inventory management in closed-loop supply chains. In Dyckhoff, H., Lackes, R., and Reese, J., editors, *Supply Chain Management and Reverse Logistics*. Springer-Verlag, Heidelberg, Germany.

Fleischmann, M., van Nunen, J.A.E.E., and Gräve, B. (2003). Integrating closed-loop supply chains and spare parts management at IBM. *Interfaces*. (Forthcoming).

Forrester, J.W. (1961). *Industrial Dynamics*. MIT Press, Cambridge, MA.

Fox, B.R. and Kempf, K.G. (1985). Opportunistic scheduling for robotic assembly. In *Proceedings of IEE International Conference on Robotics an Automation*, pages 880–889.

Fuller, D.A. and Allen, J. (1995). A typology of reverse channel systems for post-consumer recyclables. In Polonsky, J. and Mintu-Winsatt, A.T., editors, *Environmental Marketing: Strategies, Practice, Theory, and Research*, pages 241–266. Haword Press, Binghamton, NY.

Gabel, H.L., Weaver, P.M., Bloemhof-Ruwaard, J.M., and Van Wassenhove, L.N. (1996). Life cycle analysis and policy options: The case of the European pulp and paper industry. *Business Strategy and the Environment*, 5:156–167.

Gademann, A.J.R.M., van den Bergand, J.P., and van der Hoff, H.H. (2001). An order batching algorithm for wave picking, in a parallel-aisle warehouse. *IIE Transactions*, 33(5):385–398.

Gaimon, C. (1988). Simultaneous and dynamic price, production, inventory and capacity decisions. *European Journal of Operational Research*, 35:426–441.

Gandolfo, G. (1993). *Continuous Time Econometrics*. Chapman & Hall, London.

Ganeshan, R., Jack, E., Magazine, M.J., and Stephens, P. (1999). A taxonomic review of supply chain management research. In Tayur, S.R., Ganesham, R., and Magazine, M.J., editors, *Quantitative Models for Supply Chain Management*. Kluwer Academic Publishers, Boston, MA.

Garey, M.R. and Johnson, D.S. (1979). *Computers and Intractability, A Guide to the Theory of NP-Completeness*. Bell Telephone Laboratories, Inc., Murray Hill, NJ.

Garner, A. and Keoleian, G.A. (1995). *Industrial Ecology: An Introduction*. National Pollution Prevention Center, Ann Arbor, MI.

Gaudioso, M. and Paletta, G. (1992). A heuristic for the periodic vehicle routing problem. *Transportation Science*, 26(2):86–92.

Gelders, L.F. and Cattrysse, D. (1991). Public waste collection: A case study. *Belgian Journal of Operations Research, Statistics and Computer Science*, 31(1-2):3–15.

Georgakellos, D.A. (1997). *Waste Packaging Managemant. Life Cycle Analysis of Different Packages in Greece and the Consequences in the Environmental Quality*. PhD thesis, University of Piraeus, Dept. of Business Administration, Piraeus, Greece. (In Greek).

Georgiadis, P. and Vlachos, D. (2003). The effect of environmental parameters on product recovery networks. *European Journal of Operational Research*. (Forthcoming).

Georgiadis, P., Vlachos, D., and Karatsis, I. (2002a). Reverse logistics modeling: A new approach, using system dynamics. In *6th Balkan Conference on Operational Research Proceedings*.

Georgiadis, P., Vlachos, D., and Tagaras, G. (2002b). Long-term analysis of closed-loop supply chains: A system dynamics approach. Working paper, Aristotle University of Thessaloniki, Greece.

German Federal Environment Ministry (1996). German recycling and waste management act. http://www.bmu.de/files/promoting.pdf.

Geyer, R. and Van Wassenhove, L.N. (2000). Product take-back and component reuse. Working Paper 34/TM/CIMSO12, INSEAD, Fontainbleau, France.

Ghiani, G. and Laporte, G. (2000). A branch and cut algorithm for the undirected rural postman problem. *Mathematical Programming, Series A*, 87:467–481.

Giangrande, A. (1994). How to assess the weights of the criteria in the AHP. In Paruccini, M., editor, *Applying Multiple Criteria Aid for Decision to Environmental Management*. Kluwer Academic Publishers.

Ginter, P.M. and Starling, J.M. (1978). Reverse distribution channels for recycling. *California Management Review*, 20(3):72–81.

Goetschalckx, M. (1992). An interactive layouot heuristic based on hexagonal adjacency graphs. *European Journal of Operational Research*, 63:304–321.

Goetz, W.G. and Egbelu, P.J. (1990). Guide path design and location of load pick-up/drop-off points for an automated guided vehicle system. *International Journal of Production Research*, 28(5):927–941.

Goggin, K. and Browne, J. (2000). Towards a taxonomy of resource recovery from end-of-life products. *Computers in Industry*, 42:177–191.

Goh, T.N. and Varaprasad, N. (1986). A statistical methodology for the analysis of the life-cycle of reusable containers. *IIE Transactions*, 18(1):42–47.

Golany, B., Yang, J., and Yu, G. (2001). Economic lot-sizing with remanufacturing options. *IIE Transactions*, 33(11):995–1003.

Golden, B., Baker, E., Alfaro, J., and Schaffer, J. (1985). The vehicle routing problem with backhauling: Two approaches. In Hammerfahr, R.D., editor, *Proceedings of the Twenty-First Annual Meeting of S.E. TIMS*, pages 90–92, Myrtle Beach, SC.

Golden, B.L. and Wong, R.T. (1981). Capacitated arc routing problems. *Networks*, 11:305–315.

Goldstein, L. (1994). The strategic management of environmental issues: A case study of Kodak's single-use cameras. Master's thesis, Sloan School of Management, MIT, Cambridge, MA.

Gooley, T.G. (2001). Diminishing returns. *Logistics Management and Distribution Report*, 40(6):43.

Gotzel, C. and Inderfurth, K. (2002). Performance of MRP in product recovery systems with demand, return and leadtime uncertainties. In Klose, A., Speranza, M.G., and Van Wassenhove, L.N., editors, *Quantitative Approaches to Distribution Logistics and Supply Chain Management*, pages 99–114. Springer, Heidelberg.

Graham, B.P.E. (2001). Collection equipment and vehicles. In Lund, H.F., editor, *The McGraw-Hill Recycling Handbook*, chapter 27. McGraw-Hill, New York.

Groenevelt, H. and Majumder, P. (2001). Competition in remanufacturing. *Production and Operations Management*, 2:125–141.

Grosfeld-Nir, A. and Gerchak, Y. (2002). Multistage production to order with rework capability. *Management Science*, 48(5):652–664.

Guide, Jr., V.D.R. (1996). Scheduling using drum-buffer-rope in a remanufacturing environment. *International Journal of Production Research*, 34:1081–1091.

Guide, Jr., V.D.R. (2000). Production planning and control for remanufacturing: Industry practice and research needs. *Journal of Operations Management*, 18(4):467–483.

Guide, Jr., V.D.R. and Srivastava, R. (1997a). Buffering from material recovery uncertainty in a recoverable manufacturing environment. *Journal of the Operational Research Society*, 48:519–529.

Guide, Jr., V.D.R. and Srivastava, R. (1997b). Repairable inventory theory: Models and applications. *European Journal of Operational Research*, 102:1–20.

Guide, Jr., V.D.R., Srivastava, R., and Kraus, M.E. (1997a). Scheduling policies for remanufacturing. *International Journal of Production Economics*, 48:187–204.

Guide, Jr., V.D.R., Srivastava, R., and Kraus, M.E. (1997b). Product structure complexity and scheduling of operations in recoverable manufacturing. *International Journal of Production Research*, 35:3179–3199.

Guide, Jr., V.D.R. and Van Wassenhove, L.N. (2001). Managing product returns for remanufacturing. *Production and Operations Management*, 10(2):142–155.

Guide, Jr., V.D.R. and Van Wassenhove, L.N. (2003). *Business Aspects of Closed-Loop Supply Chains*. volume 2 of *Carnegie Bosch Institute International Management Series*. Carnegie Mellon University Press, Pittsburgh, PA.

Guiltinan, J. and Nwokoye, N. (1974). Reverse channels for recycling: An analysis for alternatives and public policy implications. In Curhan, R.G., editor, *New Marketing for Social and Economic Progress, Combined Proceedings*. American Marketing Association.

Guiltinan, J.P. and Nwokoye, N.G. (1975). Developing distribution channels and systems in the emerging recycling industries. *International Journal of Physical Distribution*, 6(1):28–38.

Gungor, A. and Gupta, S.M. (1999). Issues in environmentally conscious manufacturing and product recovery: A survey. *Computers & Industrial Engineering*, 36:811–853.

Hackman, S.T. and Platzman, L.K. (1990). Near-optimal solution of generalized resource allocation problems with large capacities. *Operations Research*, 38(5):902–910.

Hafeez, K., Griffiths, M., Griffiths, J., and Naim, J.J. (1996). System design of a two-echelon steel industry supply chain. *International Journal of Production Economics*, 45:121–130.

Hameri, A-P. and Nihtila, J. (1998). Product data management exploratory study on state of the art in one of a kind industry. *Computers in Industry*, 35(3):195–206.

Hanafi, S., Arnaud, F., and Vaca, P. (1998). Municipal solid waste collection: A local search approach for solving the sectorisation problem. Université de Valenciennes and Escuela Politécnica Nacional.

Hanssen, O.J., Forde, J.S., and Thoresen, J. (1994). Environmental indicator and index systems. An overview and test of different approaches. A pilot study for Staoil. Technical Report Research Report OR 17.94, Ostfold Research Foundation.

Hanssen, O.J., Forde, J.S., and Thoresen, J. (1998). Environmental impacts of product systems in a life cycle perspective: A survey of five product types based on life cycle assessments studies. *Journal of Cleaner Production*, 6:299–311.

Harker, P.T. (1987). Alternative modes of questioning in the analytic hierarchy process. *Mathematical Modelling*, 9:335–360.

Hartl, R.F., Feichtinger, G., and Kirakossian, G.T. (1992). Optimal recycling of tailings for the production of buyilding materials. *Czechoslovak Journal of Operations Research*, 3:181–192.

Hartl, R.F. and Sethi, S.P. (1984). Optimal control problems with differential inclusions: Sufficiency conditions and an application to a product inventory model. *Optimal Control Appl. Meth.*, 5:289–307.

Hausman W.H., L.B. Schwarz, S.C. Graves (1976). Optimal storage assignment in automatic warehousing systems. *Management Science*, 22(6):629–638.

Hausschild, M. and Wenzel, H. (1998). *Environmental Assessment of Products. Volume 2: Scientific Background.* Kluwer Academic Publishers, Boston, MA.

Heisig, G. (2002). *Planning Stability in Material Requirements Planning Systems.* Lecture Notes in Economics and Mathematical Systems 515. Springer, Berlin, Germany.

Heisig, G. and Fleischmann, M. (2001). Planning stability in a product recovery system. *OR Spektrum*, 23:25–50.

Heskett, J.L. (1963). Cube-per-order index - a key to warehouse stock location. *Transportation and Distribution Management*, 3:27–31.

Hess, J.D., Chu, W., and Gerstner, E. (1996). Controlling product returns in direct marketing. *Transportation and Distribution Management*, 7(4):307–317.

Hess, J.D. and Mayhew, G.E. (1997). Modeling merchandise returns in direct marketing. *Journal of Direct Marketing*, 11(2):20–35.

Heyman, D.P. (1977). Optimal disposal policies for single-item inventory system with returns. *Naval Research Logistics Quarterly*, 24:385–405.

Hong, J.D. and Hayya, J.C. (1992). Just-in-time purchasing: Single or multiple sourcing? *International Journal of Production Economics*, 27:175–181.

Hoshino, T., Yura, K., and Hitomi, K. (1999). Optimization analysis for recycle-oriented manufacturing systems. *International Journal of Production Research*, 33:2069–2078.

IBM (2000). Press release. Nov. 14.

IBM (2001). Annual corporate environmental report. http://www.ibm.com.

IGD.com (2002). Information, research and education, fact sheets. http://www.igd.com/.

IISD (2000). Annual report 1999-2000. www.iisd.org.

Inderfurth, K. (1997). Simple optimal replenishment and disposal policies for a product recovery system with leadtimes. *OR Spektrum*, 19:111–122.

Inderfurth, K. (1998). The performance of simple MRP driven policies for a product recovery system with leadtimes. Preprint 32, Faculty of Economics and Management, Otto-von-Guericke University of Magdeburg, Germany.

Inderfurth, K. (2002). Optimal policies in hybrid manufacturing/remanufacturing systems with product substitution. Working Paper 1/2001, Faculty of Economics and Management, Otto-von-Guericke University Magdeburg, Germany.

Inderfurth, K., de Kok, A.G., and Flapper, S.D.P. (2001). Product recovery policies in stochastic remanufacturing systems with multiple reuse options. *European Journal of Operational Research*, 102:130–152.

Inderfurth, K. and Jensen, T. (1999). Analysis of MRP policies with recovery options. In Leopold-Wildburger, U., Feichtinger, G., and Kistner, K.P., editors, *Modelling and Decicions in Economics*, pages 189–228. Springer, Heidelberg.

Inderfurth, K., Lindner, G., and Rahaniotis, N.P. (2003). Lotsizing in a production system with rework and product deterioration. Working Paper 1, Otto-von-Guericke-University Magdeburg, Germany.

Inderfurth, K. and Teunter, R.H. (2003). Production planning and control of closed-loop supply chains. In Guide, Jr., V.D.R. and Van Wassenhove, L.N., editors,

Business Aspects of Closed-Loop Supply Chains, pages 149–174. Carnegie Mellon University Press, Pittsburgh, PA.

Inderfurth, K. and van der Laan, E.A. (2001). Leadtimes effects and policy improvement for stochastic inventory control with remanufacturing. *International Journal of Production Economics*, 71(1-3):381–390.

ISO 14042 (2000). Environmental Management - Life Cycle Assessment - Life Cycle Impact Assessment.

Jacobs-Blecha, C. and Goetschalckx, M. (1992). The vehicle routing problem with backhauls: Properties and solution algorithms. Working paper 30332, Georgia Tech Research Corporation, Altanta, Georgia.

James, R. (2001). Cleanaway, U.K. Personal communication.

Jayaraman, V., Guide, Jr., V.D.R., and Srivastava, R. (1999). A closed–loop logistics model for remanufacturing. *Journal of the Operational Research Society*, 50(5):497–508.

Jenkins, L. (1982). Parametric mixed integer programming: An application to solid waste management. *Management Science*, 28(11):1270–1284.

Johnson, M.R. and Wang, M.H. (1995). Planning product disassembly for material recovery opportunities. *International Journal of Production Research*, 33:3119–3142.

Kambil, A. and van Heck, E. (2002). *Making Markets: How Firms Can Design and Profit from Online Auctions and Exchanges*. Harvard Business School Press.

Kamien, M.I. and Schwartz, N.L. (1991). *Dynamic Optimization*. North-Holland, Amsterdam, The Netherlands, 2nd edition.

Kantor, P.B. and Zangwill, W.I. (1991). Theoretical foundation for a learning rate budget. *Management Science*, 37(3):315–330.

Kaplan, S. and Sawhney, M. (2000). E-Hubs: The new B2B marketplaces. *Harvard Business Review*, 78(May-June):97–103.

Karacapilidis, N.I. and Papadias, D. (2001). Computer supported argumentation and collaborative decision making: The HERMES system. *Information Systems*, 26(4):259–277.

Kelle, P. and Silver, E.A. (1989a). Purchasing policy of new containers considering the random returns of previously issued containers. *IIE Transactions*, 21(4):349–354.

Kelle, P. and Silver, E.A. (1989b). Forecasting the returns of reusable containers. *Journal of Operations Management*, 8:17–35.

Kiesmüller, G.P. (2003a). A new approach for controlling a hybrid stochastic manufacturing remanufacturing system with inventories and different leadtimes. *European Journal of Operational Research*, 147(1):62–71.

Kiesmüller, G.P. (2003b). Optimal control of a one product recovery system with leadtimes. *International Journal of Production Economics*, 81-82:333–340.

Kiesmüller, G.P. and Minner, S. (2002). Simple expressions for finding recovery system inventory control parameters. *Journal of the Operational Research Society*. (Forthcoming).

Kiesmüller, G.P., Minner, S., and Kleber, R. (2000). Optimal control of a one product recovery system with backlogging. *IMA Journal of Mathematics Applied in Business and Industry*, 11:189–207.

Kiesmüller, G.P. and Scherer, C.W. (2002). Computational issues in a stochastic finite horizon one product recovery inventory model. *European Journal of Operational Research*. (Forthcoming).

Kiesmüller, G.P. and van der Laan, E.A. (2001). An inventory model with dependent product demands and returns. *International Journal of Production Economics*, 72(1):73–87.

Kimrey, E. (1996). Rethinking the refuse/recycling ratio. *BioCycle*, 37(7):44–47.

Kistner, K.-P. and Dobos, I. (2000). Optimal production-inventory strategies for a reverse logistics system. In Dockner, E.J. and R.F.Hartl, M. Luptacik, G. Sorger, editors, *Optimization, Dynamics, and Economic Analysis*, pages 246–258. Physica, Heidelberg, Germany.

Klausner, M., Grimm, W.M., and Henderson, C. (1998). Reuse of electric motors in consumer products. design and analysis of an electronic data log. *Journal of Industrial Ecology*, 2(2):89 –102.

Klausner, M. and Hendrickson, C. (2000). Reverse-logistics strategy for product take-back. *Interfaces*, 30(3):156–165.

Kleber, R. (2002). Balancing of returns with demands in a capacitated dynamic product recovery system. Working Ppaper, Otto-von-Guericke University of Magdeburg, Germany.

Kleber, R., Minner, S., and Kiesmüller, G.P. (2002). A continuous time inventory model for a product recovery system with multiple reuse options. *International Journal of Production Economics*, 79:121–141.

Kleijn, M. and Dekker, R. (1999). An overview of inventory systems with several demand classes. In Grazia Speranza, M. and Stähly, P., editors, *New Trends in Distribution Logistics*, volume 480 of *Lecture Notes in Economics and Mathematical Systems*. Springer Verlag, Berlin, Germany.

Klein Haneveld, W.K. and Teunter, R.H. (1998). Effects of discounting and demand rate variability on the EOQ. *International Journal of Production Economics*, 54:173–192.

Koehorst, H., de Vries., H., and Wubben, E. (1999). Standardisation of crates: Lessons learned from the versfust (freshcrate) project. *Supply Chain Management*, 49(2):95–101.

Koh, S.-G., Hwang, H., and Sohn, K.-I. (2002). An optimal ordering and recovery policy for reusable items. *Computers & Industrial Engineering*, 43:59–73.

Kokkinaki, A.I., Dekker, R., de Koster, M.B.M., Pappis, C., and Verbeke, W. (2002a). E-business models for reverse logistics: Contributions and challenges. *Proceedings of IEEE Computer Society (ITCC) 2002 International Conference on Information Technology, Las Vegas, Nevada, USA*, pages 470–476.

Kokkinaki, A.I., Dekker, R., Lee, R., and Pappis, C. (2002b). Design issues on e-marketplaces for returns. *Proceedings of the 5th International Conference on Electronic Commerce Research*. (Forthcoming).

Kokkinaki, A.I., Karakapilides, N., Dekker, R., and Pappis, C. (2002c). A web-based recommender system for end-of-use ICT products. *Towards the Knowledge Society. E-commerce, E-business, E-government*, pages 601–614.

Kopczak, L.R. and Johnson, M.E. (2003). The supply-chain management effect. *Sloan Management Review*, Spring 2003:27–34.

Kopicky, R.J., Berg, M.J., Legg, L., Dasappa, V., and Maggioni, C. (1993). *Reuse and Recycling: Reverse Logistics Opportunities*. Council of Logistics Management, Oak Brook, IL.

Kostecki, M. (1998). *The Durable Use of Consumer Products: New Options for Business and Consumption*. Kluwer Academic Publishers, Boston, MA.

Kouvelis, P. and Yu, G. (1997). *Robust Discrete Optimization and its Applications.* Kluwer Academic Publishers, Boston, MA.

Krikke, H.R. (1998). *Recovery Strategies and Reverse Logistic Network Design.* PhD thesis, University of Twente, BETA Institute for Business Engineering and Technology Application, Enschede, The Netherlands.

Krikke, H.R., Bloemhof-Ruwaard, J., and Van Wassenhove, L.N. (2001). Dataset of the refrigerator case (Design of closed loop supply chains: A production and return network for refrigerators). Working paper, Erasmus Research Institute of Management, Erasmus University Rotterdam.

Krikke, H.R., Bloemhof-Ruwaard, J., and Van Wassenhove, L.N. (2002). Design of closed loop supply chains: A production and return network for refrigerators. Working Paper.

Krikke, H.R., Kokkinaki, A.I., and van Nunen, J.A.E.E. (2003). Information technology in closed-loop supply chains. In Guide, Jr., V.D.R. and Van Wassenhove, L.N., editors, *Business Aspects of Closed-Loop Supply Chains*, pages 255–290. Carnegie Mellon University Press, Pittsburgh, PA.

Krikke, H.R., van Harten, A., and Schuur, P.C. (1998). On a medium term product recovery and disposal strategy for durable assembly products. *International Journal of Production Research*, 36(1):111–139.

Krikke, H.R., van Harten, A., and Schuur, P.C. (1999a). Business case Océ: Reverse logistic network re-design for copiers. *OR Spektrum*, 21(3):381–409.

Krikke, H.R., van Harten, A., and Schuur, P.C. (1999b). Business case Roteb: Recovery strategies for monitors. *Computers and Industrial Engineering*, 36:739–757.

Kroon, L. and Vrijens, G. (1995). Returnable containers: An example of reverse logistics. *International Journal of Physical Distribution & Logistics Management*, 25(2):56–68.

Krumwiede, D.W. and Sheu, C. (2002). A model for reverse logistics entry by third-party providers. *Omega*, 30(5):325–333.

Kuik, R., Salomon, M., and Van Wassenhove, L.N. (1994). Batching decisions: Stucture and models. *European Journal of Operational Research*, 75:243–263.

Lambert, A.J.D. (1997). Optimal dissassembly of complex products. *International Journal of Production Research*, 35:2509–2523.

Lambert, A.J.D. (1999). Linear programming in disassembly/clustering sequence generation. *Computers and Industrial Engineering*, 36:723–738.

Lambert, A.J.D. (2001a). Life cycle analysis including recycling. In Sarkis, J., editor, *Greener Manufacturing and Operations: From Design to Delivery and Back*, pages 36–55. Greenleaf Publishers, Sheffield, UK.

Lambert, A.J.D. (2001b). Automatic determination of transition matrices in optimal disassembly sequence generation. In *Proceedings of IEEE International Symposium on Assembly and Task Planning*, pages 220–225.

Lambert, A.J.D. (2002a). Determining optimum disassembly sequences in electronic equipment. *Computers and Industrial Engineering*, 43:553–575.

Lambert, A.J.D. (2002b). Disassembly sequencing: A review. *International Journal of Production Research*, Special Issue on Product Recovery, forthcoming.

Lambert, A.J.D. and Gupta, S.M. (2002). Demand-driven disassembly optimization for electronic products. *Journal of Electronics Manufacturing*, 2:121–135.

Lambert, A.J.D., Jansen, M.H., and Splinter, M.A.M. (2000). Environmental information systems based on enterprise resource planning. *Integrated Manufacturing Systems*, 11(2):105–111.

Lambert, A.J.D. and Stoop, M.L.M. (2001). Processing of discarded household refrigerators: Lessons from the Dutch example. *Journal of Cleaner Production*, 9:243–252.

Lambert, D.M. and Stock, J.R. (1981). *Strategic Physical Distribution Mangement*. Irwin, Homewood, IL.

Langevin, A., Mbaraga, P., and Campbell, J.F. (1996). Continuous approximation models in freight distribution: An overview. *Transportation Research B*, 30(3):163–188.

Laporte, G. (1992). The vehicle routing problem: An overview of exact and approximate algorithms. *European Journal of Operational Research*, 20:58–67.

Lau, H.S. and Zhao, L.G. (1993). Optimal ordering with two suppliers when lead times and demand are all stochastic. *European Journal of Operational Research*, 68:120–133.

Lee, C.-H. (1998). Formulation of resource depletion index. *Resources, Conservation and Recycling*, 24:285–298.

Lee, H.L. (1992). Lot sizing to reduce capacity utilization in a production process with defective items, process corrections, and rework. *Management Science*, 38(9):1314–1328.

Lee, H.G. (1997). Aucnet: Electronic intermediary for used-car transactions. *Electronic Markets*, 7(4):24–28.

Lee, H.L., Billington, C.B., and Carter, B. (1993). Hewlett-Packard gains control of inventory and service through design for localization. *Interfaces*, 23:1–11.

Lee, H.L., Padmanabhan, V., and Whang, S. (1997). Information distortion in a supply chain: The bullwhip effect. *Management Science*, 43(4):546–558.

Lee, H.L. and Rosenblatt, M.J. (1987). Simultaneous determination of production cycle and inspection schedules in a production system. *Management Science*, 33(9):1125–1136.

Lee, H.L. and Whang, S. (1999). Decentralized multi-echelon inventory control systems: Incentives and information. *Management Science*, 45:633–640.

Lee, J.J., O'Callaghan, P., and Allen, D. (1995). Critical review of life cycle analysis and assessment techniques and their application to commercial activities. *Resources, Conservation and Recycling*, 13:37–56.

Lenstra, J.K. and Rinnooy Kan, A.H.G. (1976). On general routing problems. *Networks*, 6:273–280.

Lifset, R. and Lombardi, D.R. (1997). Who should pay and why? Some thoughts on the conceptual foundations for the assignment of extended producer responsibility. In *OECD Internatiuonal Workshop on EPR*, Ottawa, Canada.

Lindfors, L.-G. (1995). Nordic guidelines on product life-cycle assessments. Nord 1995:20, Nordic Council of Ministers, Århus, Denmark.

Lindfors, L.-G., Christiansen, L., and Hoffmann, L. (1995a). LCA-Nordic Technical Reports No 1-9. TemaNord 1995:502, Nordic Council of Ministers, Copenhagen, Denmark.

Lindfors, L.-G., Christiansen, L., and Hoffmann, L. (1995b). LCA-Nordic Technical Reports No 10 and Special Reports No 1-2. TemaNord 1995:503, Nordic Council of Ministers, Copenhagen, Denmark.

Lindner, G. (2002). Optimizing single stage production in the presence of rework. Working paper, Otto-von-Guericke-University Magdeburg, Germany.

Lindner, G. and Buscher, U. (2002). An optimal lot and batch size policy for a single item produced and remanufactured on one machine in the presence of limitations on the manufacturing and handling capacity. Working Paper, Otto-von-Guericke-University Magdeburg, Germany.

Listes, O. and Dekker, R. (2001). Stochastic approaches for product recovery network design: A case study. Econometric Institute Series 2001-08, Erasmus University Rotterdam, The Netherlands.

Liu, J.J. and Yang, P. (1996). Optimal lot-sizing in an imperfect production system with homogeneous reworkable jobs. European Journal of Operational Research, 91:517-527.

Louveaux, F.V. (1986). Discrete stochastic location models. Annals of Operations Research, 6(1-4):23-34.

Louwers, D., Kip, B.J., Peters, E., Souren, F., and Flapper, S.D.P. (1999). A facility location allocation model for reusing carpet materials. Computers & Industrial Engineering, 36:855-869.

Lund, R. (1998). Remanufacturing: An American resource. In Proceedings of the Fifth International Congress for Environmentally Conscious Design and Manufacturing, Rochester Institute of Technology, Rochester, NY.

Luo, Z.-Q., Pong, J.-S., and Ralph, D. (1996). Mathematical Programming with Equilibrium Constraints. Cambridge University Press, Cambridge, U.K.

Mabini, M.C., Pintelon, L.M., and Gelders, L.F. (1992). EOQ type formulations for controlling repairable inventories. International Journal of Production Economics, 28:21-33.

Mahadevan, B., Pyke, D.F., and Fleischmann, M. (2002). Periodic review, push inventory policies for remanufacturing. ERIM Report Series ERS-2002-35-LIS, Erasmus University Rotterdam, The Netherlands.

Mansini, R. and Speranza, M.G. (1998). A linear programming model for the separate refuse collection service. Computers and Operations Research, 25(7/8):659-673.

Marín, A. and Pelegrín, B. (1998). The return plant location problem: Modelling and resolution. European Journal of Operational Research, 104:375-392.

Marx-Gómez, J., Rautenstrauch, C., Nurnberger, A., and Kruse, R. (2002). Neuro-fuzzy approach to forecast returns of scrapped products to recycling and remanufacturing. Knowledge-Based Systems, 15:119-129.

Matson, J.O. and White, J.A. (1982). Operational research and material handling. Omega, 11:309-318.

McConaghy, R. and Mills, L. (1997). Calculating curbside recycling's bottom line. World Wastes, 40(4):30-34.

McMillen, A.P. and Skumatz, L.A. (2001). Separation, collection, and monitoring systems. In Lund, H.F., editor, The McGraw-Hill Recycling Handbook, chapter 5. McGraw-Hill, New York, second edition.

Meijer, H.W. (1998). Green logistics. In Van Goor, A.R., Flapper, S.D.P., and Clement, C., editors, Handboek of Reverse Logistics. Kluwer B.V., Deventer, The Netherlands. (In Dutch).

Melissen, F.W. and de Ron, A.J. (1999). Defining recovery practices - definitions and terminology. International Journal on Environmentally Conscious Manufacturing and Design, 8(2):1-18.

Meller, R.D. (1997). Optimal order-to-lane assignment in an order accumulation/sortation system. *IIE Transactions*, 29:2930–301.

Meyer, H. (1999). Many happy returns. *The Journal of Business Strategy*, 20(4):27–31.

MHIA.org (2002). Material handling industry of America. http://www.mhia.org.

Michalewicz, Z. and Fogel, D. (2000). *How to Solve It: Modern Heuristics*. Springer, New York.

Miettinen, P. and Hamalainen, R.P. (1997). How to benefit from decision analysis in environmental life cycle assessment (LCA). *European Journal Operational Research*, 102:274–285.

Min, H. (1989). The multiple vehicle routing problem with simultaneous delivery and pickup points. *Transportation Research*, 23A:377–386.

Minegishi, S. and Thiel, D. (2000). System dynamics modeling and simulation of a particular food supply chain. *Simulation - Practice and Theory*, 8:321–339.

Minner, S. (2001). Economic production and remanufacturing lot-sizing under constant demands and returns. In Fleischmann, B., Lasch, R., Derigs, U., Domschke, W., and Rieder, U., editors, *Operations Research Proceedings 2000*, pages 328–332. Springer-Verlag, Berlin, Germany.

Minner, S. (2002a). On the implementation of economic ordering quantities for recoverable item inventory systems. Working paper, Otto-von-Guericke-University Magdeburg, Germany.

Minner, S. (2002b). Optimal production and remanufacturing lot-sizing under constant demands and returns. Working paper, Otto-von-Guericke-University of Magdeburg, Germany.

Minner, S. and Kiesmüller, G.P. (2002). Dynamic product acquisition in closed loop supply chains. Working Paper 9/2002, Faculty of Economics and Management, Otto-von-Guericke University of Magdeburg, Germany.

Minner, S. and Kleber, R. (2001). Optimal control of production and remanufacturing in a simple recovery model with linear cost function. *OR Spektrum*, 23:3–24.

Mirchandani, P.B. and Francis, R.L. (1989). *Discrete Location Theory*. Wiley Publications, New York.

Moorthy, S. and Srinivasan, K. (1995). Signaling quality with a money-back guarantee: The role of transaction costs. *Marketing Science*, 14:442–446.

Morphy, E. (2001). Newgistics gears up to deliver many happy returns. *CRM-Daily.com*. http://www.crmdaily.com.

Morrell, A.L. (2001). The forgotten child of the supply chain. *Modern Materials Handling*, 56(6):33–36.

Mosheiov, G. (1994). The travelling salesman problem with pickup and delivery. *European Journal of Operational Research*, 79:299–310.

Mostard, J. and Teunter, R. (2002). The newsboy problem with resalable returns. ERIM Report Series ERS-2002-89-LIS, Erasmus University Rotterdam, The Netherlands.

Muckstadt, J.A. and Isaac, M.H. (1981). An analysis of single item inventory system with returns. *Naval Research Logistics Quarterly*, 28:237–254.

Murphy, P.R. (1986). A preliminary study of transportation and warehousing aspects of reverse distribution. *Transportation Journal*, 35(4):12–21.

Murphy, P.R. and Poist, R.P. (1989). Managing of logistics retromovements: An empirical analysis of literature suggestions. *Transportation Research Forum*, 29(1):177–84.

Muther, R. (1973). *Systematic Layout Planning*. Cahners Books, Boston, MA, 2nd edition.

Muyldermans, L., Beullens, P., Cattrysse, D., and Van Oudheusden, D. (2001). The k-opt approach for the general routing problem. Working paper 01/18. Center for Industrial Management, Katholieke Universiteit Leuven, Belgium.

Muyldermans, L., Beullens, P., Cattrysse, D., and Van Oudheusden, D. (2002). Exploring variants of 2- and 3-opt for the general routing problem. Working paper 02/20. Center for Industrial Management, Katholieke Universiteit Leuven. Belgium.

Naddor, E. (1966). *Inventory Systems*. John Wiley and Sons, New York.

Nagel, C. and Meyer, P. (1999). Caught between ecology and economy: End-of-life aspects of environmentally conscious manufacturing. *Computers & Industrial Engineering*. 36(4):781–792.

Nahmias, S. (1997). *Production and Operations Analysis*. Irwin, Chicago, IL, 3rd edition.

Nahmias, S. and Rivera, H. (1979). A deterministic model for a repairable inventory system with a finite repair rate. *International Journal of Production Research*, 17(3):215–221.

Nakashima, K., Arimitsu, H., Nose, T., and Kuriyama, S. (2002). Analysis of a product recovery system. *International Journal of Production Research*, 40(15):3849–3856.

Narisetti, R. (1998). Printer wars: Toner discount incites rivals. Wall Street Journal, April 10. New York, NY.

Neck. R. (1984). Stochastic control theory and operational research. *European Journal of Operational Research*. 17:283–301.

Nederland ICT (2002). The ICT take back system http://www.nederlandict.nl/ictinzamelgb.htm. (Feb.25, 2002).

Newell, G.F. (1971). Dispatching policies for a transportation route. *Transportation Science*. 5:91–105.

Office for fair trading (2003). http://www.oft.gov.uk/.

Orloff, C.S. (1974). A fundamental problem in vehicle routing. *Networks*, 4:35–64.

Ossenbruggen, P.J. and Ossenbruggen, P.M. (1992). Swap, a computer package for solid waste management. *Computers. Environment and Urban Systems*. 16:83–99.

Owen, J. (1997). *STEP - An Introduction*. Information Geometers, Winchester, UK, 2nd edition.

Pasternack, B.A. (1985). Optimal pricing and return policies for perishable commodities. *Marketing Science*, 4:166–176.

Peltonen, H. (2000). *Concepts and Implementation for Product Data Management*. PhD thesis, Helsinki University of Technology.

Pohlen, T.L. and Farris II, M. (1992). Reverse logistics in plastic recycling. *International Journal of Physical Distribution & Logistics Management*, 22(7):35–47.

Pontryagin, L.S., Boltyanskii, V.G., Gamkrelidze. R.V., and Mishchenko, E.F. (1962). *The Mathematical Theory of Optimal Processes*. Wiley, New York.

Porteus, E.L. (1986). Optimal lot sizing, process quality improvement and setup cost reduction. *Operations Research*, 34(1):137–144.

Porteus, E.L. (1990). The impact of inspection delay on process and inspection lot sizing. *Management Science*, 36(8):999–1007.

Porteus, E.L. and Whang, S. (1993). On manufacturing/marketing incentives. *Management Science*, 9:1166–1181.

Prejean, M. (1989). Half-soles, kettles and cures. milestones in the tire retreading and repairing industry. *American Retreaders' Association*.

Rabinovich, E., Windle, R., Dresner, M., and Corsi, T. (1999). Outsourcing of integrated logistics functions. *International Journal of Physical Distribution and Logistics Management*, 29:353–273.

Ratliff, H.D. and Rosenthal, A.S. (1983). Order picking in a rectangular warehouse: A solvable case of the traveling salesman problem. *Operations Research*, 31(3):507–521.

Realff, M.J., Ammons, J.C., and Newton, D. (1999). Carpet recycling: Determining the reverse production system design. *The Journal of Polymer–Plastics Technology and Engineering*, 38:547–567.

Realff, M.J., Ammons, J.C., and Newton, D.J. (2002). Robust reverse production system design for carpet recycling. *IIE Transactions*. (Forthcoming).

REVLOG (1998). The European working group on reverse logistics. http://www.fbk.eur.nl/OZ/REVLOG/.

Richter, K. (1996a). The EOQ repair and waste disposal model with variable setup numbers. *European Journal of Operational Research*, 96:313–324.

Richter, K. (1996b). The extended EOQ repair and waste disposal model. *International Journal of Production Economics*, 45:443–448.

Richter, K. and Dobos, I. (1999). Analysis of the EOQ repair and waste disposal problem with integer setup numbers. *International Journal of Production Economics*, 59:463–467.

Richter, K. and Sombrutzki, M. (2000). Remanufacturing planning for the reverse Wagner/Whitin models. *European Journal of Operational Research*, 121(2):304–315.

Richter, K. and Weber, J. (2001). The reverse Wagner/Whitin model with variable manufacturing and remanufacturing cost. *International Journal of Production Economics*, 71(1-3):447–456.

Robinson, L.W., McLain, J.O., and Thomas, L.J. (1990). The good, the bad and the ugly: Quality on an assembly line. *International Journal of Production Research*, 28:963–980.

Rodrigue, J.-P., Slack, B., and Comtois, C. (2001). Green logistics. In Brewer, A.M., Button, K.J., and Henshe, D.A., editors, *The Handbook of Logistics and Supply-Chain Management*. Handbooks in Transport N. 2. Pergamon/Elsevier, London, UK.

Rogers, D.S. and Tibben-Lembke, R.S. (1999). *Going Backwards: Reverse Logistics Trends and Practices*. Reverse Logistics Executive Council, Pittsburgh, PA.

Rogers, D.S. and Tibben-Lembke, R.S. (2001). An examination of reverse logistics practices. *Journal of Business Logistics*, 22(2):129–248.

Romijn, P.M. (1999). Return flows of electrical and electronic equipment in the Netherlands. Master's thesis, Erasmus University Rotterdam, The Netherlands. (In Dutch).

Rondinelli, D.A. and Vastag, G. (1996). International environmental standards and corporate policies: An integrative framework. *Sloan Management Review*, 39(1):106–122.

Roodbergen, K.J. (2001). *Layout and Routing Methods for Warehouses*. PhD thesis, Erasmus University Rotterdam, The Netherlands.

Rouwenhorst, B., Reuter, B., Stockrahm, V., van Houtum, G.J., Mantel, R.J., and Zijm, W.H.M. (2000). Warehouse design and control: Framework and literature review. *European Journal of Operational Research*, 122:515–533.

Rouwenhorst, B., van den Berg, J.P., van Houtum, G.J., and Zijm, W.H.M. (1996). Performance analysis of a carousel system. In Graves, R.J. et al., editors, *Progress in Material Handling Research*, pages 495–511. Material Handling Institute, Charlotte, NC.

Saaty, T.L. (1986). Axiomatic foundations of the analytic hierarchy process. *Management Science*, 32:841–855.

Saaty, T.L. (1988a). *Decision Making for Leaders*. RWS Publications, Pittsburgh, PA.

Saaty, T.L. (1988b). *Multicriteria Decision Making: The Analytic Hierarchy Process*. RWS Publications, Pittsburgh, PA.

Salhi, S. and Nagy, G. (1999). A cluster insertion heuristic for single and multi-depot vehicle routing problems with backhauling. Working paper, University of Birmingham, University of Greenwich, U.K.

Salhi, S. and Sari, M. (1997). A multi-level composite heuristic for the multi-depot vehicle fleet mix problem. *European Journal of Operational Research*, 103:95–112.

Salhi, S. and Wade, A.C. (2002). The restricted and the fully mixed vehicle routing problem with backhauls: Some results with ant systems. In *IFORS 2002 Conference*, Edinburgh, U.K.

Sasse, H., Karl, U., and Renz, O. (1999). Cost efficient and ecological design of cross-company recycling systems applied to sewage sludge re-integration. In Despotis, D.K. and Zouponidis, C., editors, *Proceedings of the DSI Conference*, pages 1418–1420, Athens, Greece.

Savaskan, R.C. (2000). *Management of Closed-Loop Supply Chains for Recoverable Products*. PhD thesis, INSEAD, Fontainebleau, France.

Savaskan, R.C., Bhattacharya, S., and Van Wassenhove, L.N. (1999). Channel choice and coordination in a remanufacturing environment. Working paper 99/14/TM, INSEAD, Fontainebleau, France.

Savaskan, R.C., Bhattacharya, S., and Van Wassenhove, L.N. (2001). Closed-loop supply chain models with remanufacturing. Math Center Working Paper Series, Kellogg School of Management, Northwestern University, Evanston, IL.

Savaskan, R.C. and Van Wassenhove, L.N. (2000). Strategic decentralization of reverse cannels and price discrimination through buyback payments. Math Center Working Paper Series, Kellogg School of Management, Northwestern University, Evanston, IL.

Savelsbergh, M.W.P. and Sol, M. (1995). The general pickup and delivery problem. *Transportation Science*, 29(1):474–490.

Schiffeleers, S. (2001). Reduction of distribution costs for EMI Music in Europe. Master's thesis, Eindhoven University of Technology, The Netherlands.

Schmidt, G. and Wilhelm, W.E. (2000). Strategic, tactical and operational decisions in multinational logistics networks: A review and discussion of modelling issues. *International Journal of Production Research*, 38(7):1501–1523.

Schmidt, R.A., Sturrock, F., Ward, P., and Lea-Greenwood, G. (1999). Deshopping - the art of illicit consumption. *International Journal of Retail & Distribution Management*, 27(8):290–301.

Schrady, D.A. (1967). A deterministic inventory model for reparable items. *Naval Research Logistics Quarterly*, 14:391–398.

Schultz, G. (2002). The recurring side of supply chain management. *APICS - The Performance Advantage*, 10:35–38.

Sculli, D. and Wu, S.Y. (1981). Stock control with two suppliers and normal lead times. *Journal of Operational Research Society*, 32:1003–1009.

Seierstadt, A. and Sydsaeter, K.K. (1987). *Optimal Control Theory with Economic Applications*. North-Holland, Amsterdam, The Netherlands.

SETAC (1991). *A technical Framework for Life-Cycle Assessments*. Society of Environmental Toxicology and Chemistry, Pensacola, FL.

SETAC (1993). *Guidelines for Life-cycle Assessmenmt: Code of Practice*. Society of Environmental Toxicology and Chemistry, Pensacola, FL.

Sethi, S.P. and Thompson, G.L. (2000). *Optimal Control Theory: Applications to Management Science and Economics*. Kluwer Academic Publishers, Boston MA, 2nd edition.

Shih, L. (2001). Reverse logistics system planning for recycling electric appliances and computers in Taiwan. *Resources, Conservation and Recycling*, 32:55–72.

Silver, E.A., Pyke, D.F., and Peterson, R. (1998). *Inventory Management and Production Planning and Scheduling*. John Wiley & Sons, New York, 3rd edition.

Simchi-Levi, D., Kaminsky, P., and Simchi-Levi, E. (2002). *Designing and Managing the Supply Chain*. McGraw-Hill, 2nd edition.

Simon, H. (1989). *Price Management*. Elsevier Science, Amsterdam, The Netherlands.

Simpson, V.P. (1970). An ordering model for recoverable stock items. *AIIE Transactions*, 2(4):315–320.

Simpson, V.P. (1978). Optimum solution structure for a repairable inventory problem. *Operations Research*, 26(2):270–281.

Slovin, J. (1994). Communities make the switch to bi-weekly collection. *World-Wastes*, 37(9):14–16.

Sniezek, J. (2002). *The Capacitated Arc Routing Problem with Vehicle/Site Dependencies: An Application of Arc Routing and Partitioning*. PhD thesis, University of Maryland, College Park, MD.

So, K.C. and Tang, C.S. (1995a). Optimal batch sizing and repair strategies for operations with repairable jobs. *Management Science*, 41(5):894–908.

So, K.C. and Tang, C.S. (1995b). Optimal operating policy for a bottleneck with random rework. *Management Science*, 41(4):620–636.

Sodhi, M.S. (1999). University of Rhode Island, Kingston, RI. Personal communication.

Sodhi, M.S. and Reimer, B. (1999). Truck sizing models for recyclables pickup. In *25th Computer and Industrial Engineering Conference*, pages 95–99, New Orleans, LA.

Solomon, M.M. and Desrosiers, J. (1988). Time window constrained routing and scheduling problems. *Transportation Science*, 22(1):1–13.

Spengler, T., Püchert, H., Penkuhn, T., and Rentz, O. (1997). Environmental integrated production and recycling management. *European Journal of Operational Research*, 97:308–326.

Stedge, G.D., Halstead, J.M., and Morris, D.E. (1993). Alternative approaches to collecting recyclables. *BioCycle*, 43(5):68–70.

Sterman, J. (2000). *Business Dynamics - System Thinking and Modeling for a Complex World*. Irwin Mcgraw-Hill.

Stock, J.R. (1992). *Reverse Logistics*. Council of Logistics Management, Oak Brook, IL.

Stock, J.R. (1998). *Development and Implemmentation of Reverse Logistics Programs*. Council of Logistics Management, Oak Brook, IL.

Szendrovits, A.Z. (1976). On the optimality of sub-batch sizes for a multi-stage EPQ model - A rejoinder. *Management Science*, 23(3):334–338.

Szendrovits, A.Z. and Truscott, W.G. (1989). Fundamentals of scheduling: The manufacturing cycle time. In Wild, R., editor, *International Handbook of Production and Operations Management*, pages 324–347. London.

Tagaras, G. and Vlachos, D. (2001). A periodic review inventory system with emergency replenishments. *Management Science*, 47(3):415–429.

Tapscott, D., Ticoll, D., and Lowy, A. (2000). *Digital Capital Harnessing the Power of Business Webs*. Harvard Business School Press, Boston, MA.

Taylor, H. (1999). *Modeling Paper Material Flows and Recycling in the US Macroeconomy*. PhD thesis, MIT, Cambridge, MA.

Tayur, S.R., Ganeshan, R., and Magazine, M.J., editors (1998). *Quantitative Models for Supply Chain Management*. Kluwer Academic Publishers, Boston, MA.

Tersine, R.J. (1988). *Principles of Inventory and Materials Management*. Elsevier Science Publishing Company, New York.

Teunter, R.H. (2001a). Economic ordering quantities for recoverable item inventory systems. *Naval Research Logistics*, 48(6):484–495.

Teunter, R.H. (2001b). A reverse logistics valuation method for inventory control. *International Journal of Production Research*, 39(9):2023–2035.

Teunter, R.H. (2002). Economic ordering quantities for stochastic discounted cost inventory systems with remanufacturing. *International Journal of Logistics*. (Forthcoming).

Teunter, R.H. and Flapper, S.D.P. (2003). Lot-sizing for a single-stage single-product production system with rework of perishable production defectives. *OR Spectrum*, 25:85–96.

Teunter, R.H., Inderfurth, K., Minner, S., and Kleber, R. (2000a). Reverse logistics in a pharmaceutical company: A case study. Working Paper 15, Otto-von-Guericke-University Magdeburg, Germany.

Teunter, R.H. and van der Laan, E.A. (2000). Average costs versus net present value: A comparison for multi-source inventory systems. ERIM report series 2000-47-LIS, Rotterdam School of Management, Erasmus University Rotterdam, The Netherlands.

Teunter, R.H. and van der Laan, E.A. (2002). On the non-optimality of the average cost approach for inventory models with remanufacturing. *International Journal of Production Economics*. (Forthcoming).

Teunter, R.H., van der Laan, E.A., and Inderfurth, K. (2000b). How to set the holding cost rates in average cost inventory models with reverse logistics? *OMEGA The International Journal of Management Science*, 28:409–415.

Teunter, R.H., van der Laan, E.A., and Vlachos, D. (2002). Inventory strategies for hybrid manufacturing/remanufacturing systems with unequal leadtimes. ERIM Report Series ERS-2002-77-LIS, Erasmus University Rotterdam, The Netherlands.

Teunter, R.H. and Vlachos, D. (2002). On the necessity of a disposal option for returned items that can be remanufactured. *International Journal of Production Economics*, 75(3):257–266.

Thierry, M.C. (1997). *An Analysis of the Impact of Product Recovery Management on Manufacturing Companies*. PhD thesis, Erasmus University Rotterdam, The Netherlands.

Thierry, M., Salomon, M., van Nunen, J.A.E.E., and Van Wassenhove, L.N. (1995). Strategic issues in product recovery management. *California Management Review*, 37(2):114–135.

Tirole, J. (1988). *The Theory of Industrial Organizations*. MIT Press, Cambridge, MA.

Toktay, L.B. (2003). Forecasting product returns. In Guide, Jr., V.D.R. and Van Wassenhove, L.N., editors, *Business Aspects of Closed-Loop Supply Chains*, pages 203–219. Carnegie Mellon University Press, Pittsburgh, PA.

Toktay, L.B., Wein, L.M., and Zenios, S.A. (2000). Inventory management of remanufacturable products. *Management Science*, 46(11):1412–1426.

Tompkins, J.A., White, J.A., Bozer, Y.A., Frazelle, E.H., Tanchoco, J.M.A., and Trevino, J. (1996). *Facilities Planning*. Wiley, New York.

Toth, P. and Vigo, D. (2002). *The Vehicle Routing Problem*. SIAM Society for Industrial and Applied Mathematics, Philadelphia, PA.

Towill, D. (1995). Industrial dynamics modeling of supply chains. *International Journal of Physical Distribution and Logistics Management*, 26(2):23–42.

Tsoulfas, G.T. and Pappis, C.P. (2001). Application of environmental principles to reverse supply chains. In *Proceedings of the 3rd Aegean Conference*, Tinos, Greece.

Tsoulfas, G.T., Pappis, C.P., and Daniel, S.E. (2002a). The use of decision-making tools in the interpretation phase of life cycle analysis of extended supply chains. IFORS 2002 Conference, Edinburgh, Scotland.

Tsoulfas, G.T., Pappis, C.P., and Minner, S. (2002b). An environmental analysis of the reverse supply chain of the SLI batteries. *Resources, Conservation and Recycling*, 36(2):135–154.

Tushman, M.L. and O'Reilly III, C.A. (1996). Ambidextrous organizations: Managing evolutionary and revolutionary change. *California Management Review*, 38(4):8–30.

Umeda, Y., Nonomura, A., and Tomiyama, T. (2000). A study on life cycle design for the post mass production paradigm. *AIEDAM (Artificial Intelligence for Engineering Design, Analysis and Manufacturing)*, 14:149–161.

Uzsoy, R. and Venkatachalam, G. (1998). Supply chain management for companies with product recovery and remanufacturing capability. *International Journal of Environmentally Conscious Design and Manufacturing*, 7:59–72.

van den Berg, J.P. (1999). A literature survey on planning an control of warehousing systems. *IIE Transactions*, 31:751–762.

van den Berg, J.P. and Zijm, W.H.M. (1999). Models for warehouse management: Classification and examples. *International Journal of Production Economics*, 59:519–528.

van der Laan, E.A. (1997). *The Effects of Remanufacturing on Inventory Control*. PhD thesis, Rotterdam School of Management, Erasmus University Rotterdam, The Netherlands.

van der Laan, E.A. (2000). An NPV and AC analysis of a stochastic inventory system with joint manufacturing and remanufacturing. *ERIM report series 2000-38-LIS.*

van der Laan, E.A., Dekker, R., Ridder, A., and Salomon, M. (1996a). An (s,Q) inventory model with remanufacturing and disposal. *International Journal of Production Economics.* 46-47:339-350.

van der Laan, E.A., Dekker, R., and Salomon, M. (1996b). Production planning and inventory control with remanufacturing and disposal: A numerical comparison between alternative control strategies. *International Journal of Production Economics.* 45:489-498.

van der Laan, E.A. and Salomon, M. (1997). Production planning and inventory control with remanufacturing and disposal. *European Journal of Operational Research,* 102(2):264-278.

van der Laan, E.A., Salomon, M., and Dekker, R. (1999a). Lead-time effects in push and pull controlled manufacturing/remanufacturing systems. *European Journal of Operational Research,* 115:195-214.

van der Laan, E.A., Salomon, M., Dekker, R., and Van Wassenhove, L.N. (1999b). Inventory control in hybrid systems with remanufacturing. *Management Science.* 45(5):733-747.

van der Laan, E.A. and Teunter, R.H. (2002). Average costs versus net present value: A comparison for multi-source inventory models. In Klose, A., Speranza, M.G., and Van Wassenhove, L.N., editors, *Quantitative Approaches to Distribution Logistics and Supply Chain Management,* pages 359-378. Springer-Verlag, Heidelberg, Germany.

van der Meer, J.R. (2000). *Operational Control of Internal Transport.* PhD thesis, Erasmus University Rotterdam, The Netherlands.

van Hillegersberg, J., Zuidwijk, R., van Nunen, J.A.E.E., and van Eijk, D. (2001). Supporting return flows in the supply chains. *Communications of the ACM.* 44(6):74-79.

van Houtum, G.J., Inderfurth, K., and Zijm, W.H.M. (1996). Materials coordination in stochastic multi-echelon systems. *European Journal of Operational Research,* 95:1-23.

van Laarhoven, P., Berglund, M., and Peters, M. (2000). Third-party logistics in Europe - five years later. *International Journal of Physical Distribution and Logistics Management.* 30(5):425-442.

Van Wassenhove, L.N., Guide, Jr., V.D.R., and Neeraj, K. (2002). Managing product returns at HP. Case, INSEAD, Fontainebleau, France.

Veerakamolmal, P. and Gupta, S.M. (1998). Optimal analysis of lot-size balancing for multiproducts selective disassembly. *International Journal of Flexible Automation and Integrated Manufacturing,* 6:245-269.

Verter, V., Aras, N., and Boyaci, T. (2003). Designing distribution systems with reverse flows. Research working paper, Faculty of Management, McGill University, Montreal, Canada.

Vis, I.F.A (2002). *Planning and Control Concepts for Material Handling Systems.* PhD thesis, Erasmus University Rotterdam, The Netherlands.

Vlachos, D. and Dekker, R. (2000). Return handling options and order quantities for single period products. Report Econometric Institute EI2000-29/A, Erasmus University Rotterdam, The Netherlands.

Vlachos, D. and Tagaras, G. (2001). An inventory system with two supply modes and capacity constraints. *International Journal of Production Economics*, 72(1):41–58.

Voutsinas, T.G. and Pappis, C.P. (2000). A branch and bound method for scheduling with exponentially deteriorating parameters. Internal Report, University of Piraeus, Greece.

Voutsinas, T.G. and Pappis, C.P. (2002). Scheduling jobs with values exponentially deteriorating over time. *International Journal of Production Economics*, 79:163–169.

VVAV (2003). Dutch waste processing association. www.vvav.nl.

Wagner, H.M. and Whitin, T.M. (1958). Dynamic version of the economic lot size model. *Management Science*, 5(1):89–96.

Wäscher, G. (2002). Order picking: A survey of planning problems and methods. Working paper 13/2002, Faculty of Economics and Management, Otto-von-Guericke University Magdeburg, Germany.

Wein, A.S. (1992). Random yield, rework and scrap in a multistage batch manufacturing environment. *Operations Research*, 40:551–563.

Wenzel, H., Hauschild, M., and Alting, L. (1997). *Environmental Assessment of Products. Volume 1: Methodology, Tools, and Case Studies*. Kluwer Academic Publishers, Boston, MA.

Wise, R. and Morrison, D. (2000). Beyond the exchange: The future of B2B. *Harvard Business Review*, Nov–Dec:86–96.

Wojanowski, R. (1999). Zur Güte von dynamischen Losgrößenheuristiken in einem deterministischen Produktions-Lagerhaltungssystem mit Recycling. Master's thesis, Faculty of Economics and Management, Otto-von-Guericke-University Magdeburg, Germany. (In German).

Wood, S.L. (2001). Remote purchase environments: The influence of return policy leniency on two-stage decision processes. *Journal of Marketing Research*, 38(2):157–169.

Yang, J., Golany, B., and Yu, G. (2001). A concave-cost production planning problem with remanufacturing options. Working Paper, University of Texas at Austin.

Yano, C.A. and Lee, H.L. (1995). Lot sizing with random yields: A review. *Operations Research*, 43(2):311–334.

Yong, J. and Zhou, X.Y. (1999). *Stochastic Controls. Hamiltonian Systems and HJB Equations*. Springer, New York.

Yuan, X.-M. and Cheung, K.L. (1998). Modeling returns of merchandise in an inventory system. *OR Spektrum*, 20(3):147–154.

Zamudio-Ramirez, P. (1996). The economics of automobile recycling. Master's thesis, MIT, Cambridge, MA.

Zellner, A. (1987). *An Introduction to Bayesian Inference in Econometrics*. Robert E. Krieger Publishing Company, Malabar, Florida.

Zhiquiang, L.U. (2003). *Hierarchical Planning and Optimization of Logistics with Reverse Flows*. PhD thesis, Science et Technologies de l'Information et des Matériaux, Université de Nantes, France.

Zimmermann, H. J. (2000). An application-oriented view of modeling uncertainty. *European Journal of Operational Research*, 122:190–198.

Zoller, K. and Robrade, A. (1988a). Efficient heuristics for dynamic lot sizing. *International Journal of Production Research*, 26:249–265.

Zoller, K. and Robrade, A. (1988b). Dynamic Lot Sizing Techniques: Survey and Comparison. *Journal of Operations Management*, 7:125–148.

Zuidwijk, R.A. and Krikke, H.R. (2001). Disassembly for recovery under uncertainty. In Gupta, S.M., editor, *SPIE Proceedings 4569: Environmentally Conscious Manufacturing II*, pages 44–43.